Mechanisms of Gene Expression

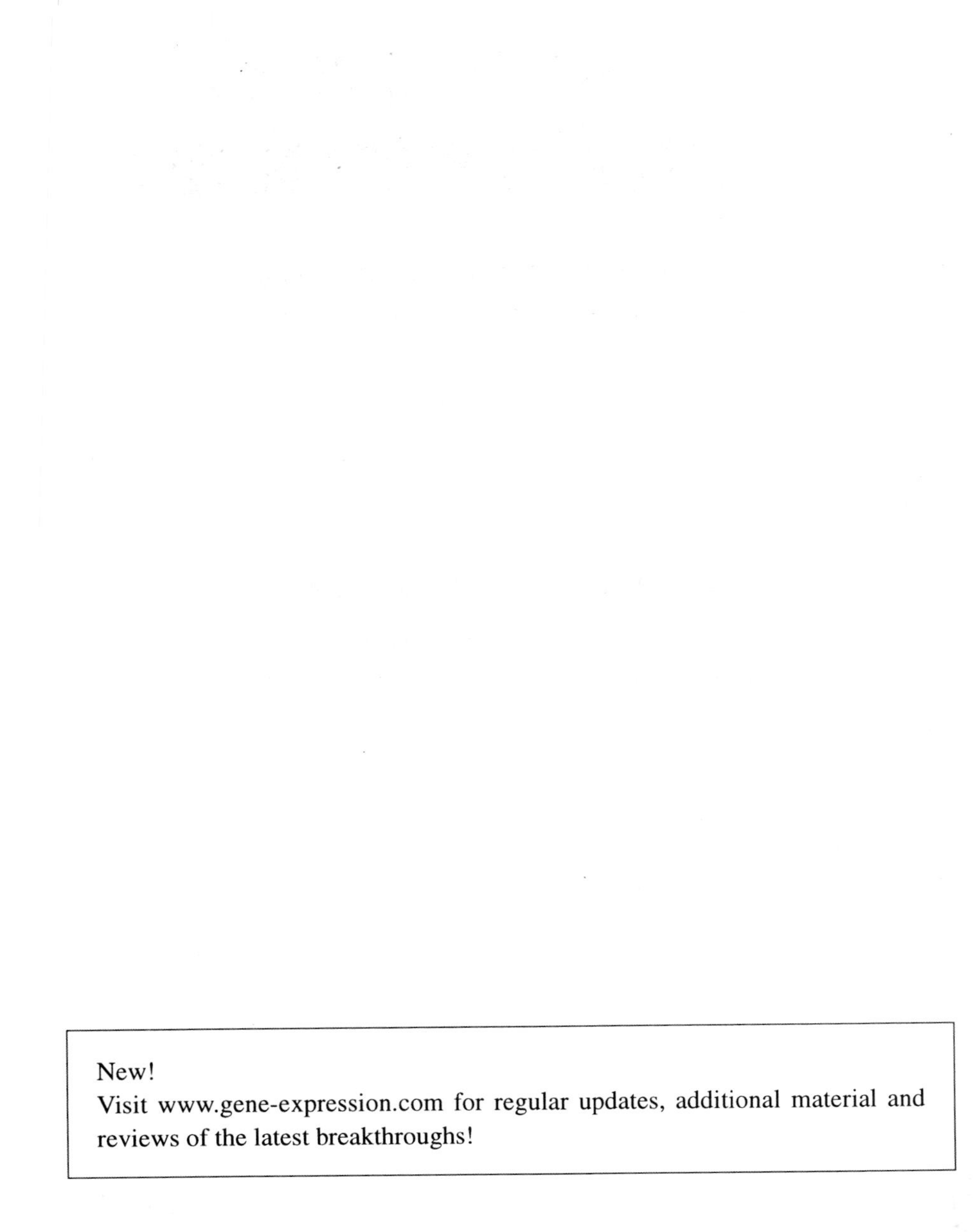

Mechanisms of Gene Expression

Structure, Function and Evolution of the Basal Transcriptional Machinery

Robert O. J. Weinzierl

Department of Biochemistry
Imperial College of Science, Technology and Medicine

ICP

Imperial College Press

Published by

Imperial College Press
57 Shelton Street
Covent Garden
London WC2H 9HE

Distributed by

World Scientific Publishing Co. Pte. Ltd.
P O Box 128, Farrer Road, Singapore 912805
USA office: Suite 1B, 1060 Main Street, River Edge, NJ 07661
UK office: 57 Shelton Street, Covent Garden, London WC2H 9HE

British Library Cataloguing-in-Publication Data
A catalogue record for this book is available from the British Library.

MECHANISMS OF GENE EXPRESSION
Structure, Function and Evolution of the Basal Transcriptional Machinery

ISBN 1-86094-126-5

Printed in Singapore.

Preface

The mechanisms underlying the transcriptional control of gene expression are of fundamental importance to many areas of contemporary biomedical research, ranging from understanding fundamental issues (such as regulation of embryonic development) to practical applications in industry and medicine. As a research scientist and university lecturer I was always particularly aware of the sparseness of suitable textbooks, that would allow both undergraduate and postgraduate students to obtain some knowledge about molecular transcription mechanisms in various pro- and eukaryotic systems. True, the fundamental concepts of gene expression are covered in all basic Molecular Biology textbooks, but the depth of coverage is usually rather limited and the material frequently out-of-date. Some of the textbooks still focus extensively on bacterial transcriptional control of gene expression and mostly ignore the wealth of new information about the archael and eukaryotic transcriptional machineries that has recently become available. Other (usually edited) books do contain a lot of specialized and up-to-date information about particular model systems but are often too fragmented, insufficiently comprehensive or assume too much background knowledge to be suitable as undergraduate textbooks.

I therefore set myself the task to write a book that would present much of the current thinking about the molecular mechanisms of gene expression in a form that would be easily accessible for undergraduates with an understanding of basic Molecular Biology concepts. In the end, I probably ended up with more than I bargained for, because I realized very quickly that there were some areas of which I was only rather dimly aware, and of which other transcription experts frequently also knew very little about. On quite a few occasions it became necessary to go beyond the well-trodden reviews and dig up some rather unusual books and magazines from dusty library corners.

I therefore hope that this book will not only serve a function as an undergraduate textbook, but will also give some of my colleagues the feeling after reading it that they have learned something new and interesting.

Another ambitious hope is, that the existence of this book will stimulate the creation of more advanced university courses dealing with the subject area. Especially here in Europe it is very easy to get the impression that many students hear very little about this exciting research area, simply because it is considered as a 'difficult' and 'controversial' subject. The absence of suitable advanced courses on gene expression affects the number of good quality students who are willing to enter the field as PhD and postdoctoral students. Although there are certainly plenty of controversies about details, I believe that during the last few years some of the major concepts have successfully started to emerge and are likely to withstand the test of time. If this book can stimulate an increased level of interest and awareness of transcriptional control mechanisms it will have fulfilled an important task.

I wish to acknowledge the many people who have made direct or indirect contributions to this book. Although it is a single-author publication, it would not have been possible without the participation of many others. Some helped by patiently explaining their ideas to me, others by providing me with preprints of research papers and reviews, photographs and digital computer images. I am especially gratefull to Dr. Jeremy Baum for helping me with the design of the high-resolution Silicon Graphics images of transcription factor structures. I owe much of my understanding of the eukaryotic transcriptional machinery to Robert Tjian and numerous ex-members of his laboratory, especially Brian Dynlacht, Tim Hoey, Grace Gill, Lucio Comai, Peter Verrijzer and Joost Zomerdjik. I am also grateful for the numerous discussions and collaborations involving current and past members of my own laboratory, the Department and from other institutions.

Finally, in an age of computers and e-mail it is quite easy for readers to provide feedback, that may help to correct any mistakes and provide fresh ideas for the next edition. I am sure that other people would have chosen different examples, would have put a different emphasis on the various topics or may even disagree with some of the my interpretations. Comments are welcome! (e-mail: r.weinzierl.ic.ac.uk).

Robert O.J. Weinzierl
April 1999

Contents

Chapter 1

RNA Polymerases

RNA polymerases (RNAPs) are the key enzymes around which all transcription processes revolve. Before we start investigating the intricate mechanisms that control the activities of RNAPs in various transcription systems, we will briefly focus on some of the most essential aspects of the structure and function of RNAPs in different organisms. In the subsequent chapters we will then explore the diverse group of accessory proteins ('basal' factors, gene-specific transcription factors, chromatin etc.) that assist or prevent RNAPs from carrying out their task.

1.1. The Evolutionary History of RNA Polymerases

All known life forms on Earth belong exclusively to one of the three major evolutionary domains: the bacteria, archaea and eukaryotes (Figure 1.1; Woese *et al.*, 1990; Klenk and Doolittle, 1994; Brown and Doolittle, 1997). Biochemical and genome sequencing studies have established that these biological domain boundaries are also distinctly reflected in the generic subunit composition of cellular RNAPs found in the various organisms representing each of these domains (Figure 1.2). Bacteria, *such as E. coli*, contain the simplest forms of cellular RNA polymerases consisting of two large subunits (β and β') that associate with two copies of a smaller subunit, α, to form 'core' RNAP (e.g. Heyduk *et al.*, 1996). This core RNAP contains all the activities required for the catalysis of RNA synthesis, but an additional subunit (σ) is needed to initiate transcription at specific start sites on template DNAs. The four different subunits (α, β, β′ and σ) form a complete transcriptional machinery capable of starting transcription at a specific start site (promoter), elongating through several kilobases of DNA and finally stopping transcription at specific sequences on (ρ-independent) terminators. Bacterial RNAPs therefore provide a relatively simple solution to the problem of transcribing a complex genome in a controlled manner. In contrast, the RNAPs found in archaea and eukaryotes contain at least ten different subunits and need additional proteins, the 'basal'

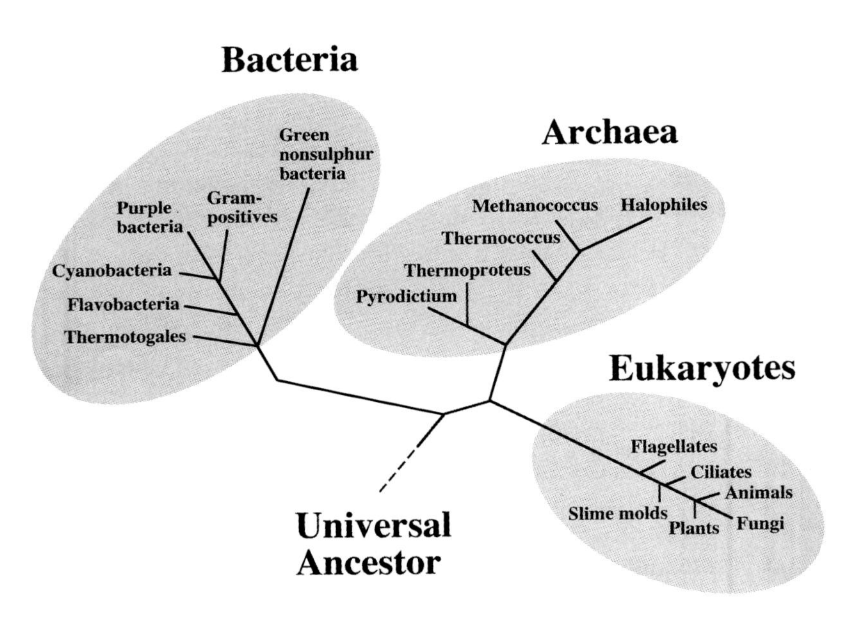

Figure 1.1. Overview of the Evolutionary History of Life on Earth.
All living organisms can be subdivided into three distinct evolutionary 'domains' that originated from a single ancestral organism. The archaea (also sometimes referred to as archaebacteria) are mostly found in hot springs and other hostile environments. The domain 'bacteria' contains many well-studied strains, such as *E.coli*. Finally, the eukaryotic domain contains all the organisms with internal compartmentalization and separate organelles, such as mitochondria and golgi stacks. Note that most of the organisms familiar from everday life (plants and animals) are on the peripheral outbranches of the eukaryotic tree and have only diverged rather 'recently' from each other on the evolutionary timescale of life on earth.

factors, to specifically initiate transcription on promoters (Chapter 2). In eukaryotes the task of transcribing the nuclear genome is shared between three different types of nuclear RNAPs that carry out distinct and nonoverlapping transcription programs (Figure 1.3).

1.2. Structure and Function of Cellular RNA Polymerase Subunits

The Large RNAP Subunits are Universally Conserved and Contain the Catalytic Center

The two 'large' subunits (β and β′ in bacteria, 'A' and 'B' in archaea, RPA1/RPB1/RPC1 and RPA2/RPB2/RPC2 in eukaryotes; see Figure 1.2) are highly conserved across all three evolutionary domains and are therefore the only polypeptides present in all known type of cellular RNAPs (e.g. Allison *et al.*, 1985; Sweetser *et al.*, 1987; Memet *et al.*, 1988; Zillig *et al.*, 1979 and 1993; Puhler *et al.*, 1989; Lanzendorfer *et al.*, 1994). None of the other RNAP subunits or other transcription factors are universally conserved across the whole evolutionary spectrum in a similar manner. Although we will in the subsequent chapters encounter several other highly conserved proteins, such as the TATA-binding protein (TBP; Chapter 2) and the histones (Chapter 7), the complete absence of these proteins from the bacterial transcriptional machineries prevent them from occupying a similar central position as the large RNAP subunits in terms of absolute evolutionary conservation.

In some archaea the large subunits ('A' and 'B') are for unknown reasons split into two separately produced proteins that, when taken together, contain the same protein domains that are found organized within the uninterrupted polypeptide chains in the bacterial and eukaryotic RNA polymerases (Figure 1.4). Interestingly, a similar split has also been found in in the large subunits of chloroplast RNAPs, that are otherwise very similar to the enzymes found in modern bacteria (reflecting their symbiotic origin; see below). Severinov *et al.* (1996) showed that a functionally active *E.coli* RNAP can be generated from β and β′ subunits that are artificially split in the same positions as the archaeal/chloroplast large subunits. At present it is not clear whether the splitting of the large archaeal/chloroplast RNAP subunits into two separate

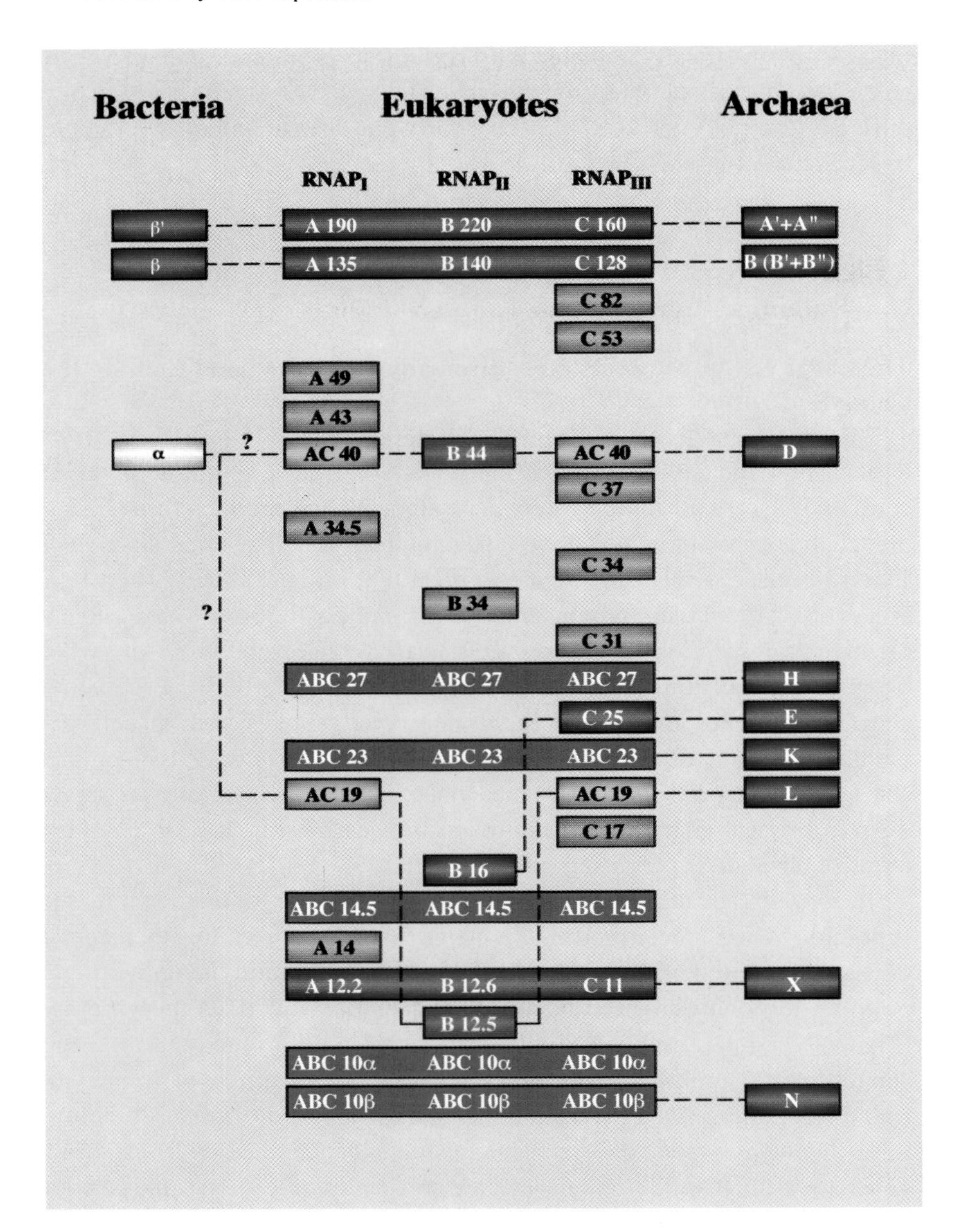
Bacteria
Eukaryotes
Archaea
RNAP$_I$
RNAP$_{II}$
RNAP$_{III}$
β'
A 190
B 220
C 160
A'+A"
β
A 135
B 140
C 128
B (B'+B")
C 82
C 53
A 49
A 43
α
?
AC 40
B 44
AC 40
D
C 37
A 34.5
C 34
?
B 34
C 31
ABC 27
ABC 27
ABC 27
H
C 25
E
ABC 23
ABC 23
ABC 23
K
AC 19
AC 19
L
C 17
B 16
ABC 14.5
ABC 14.5
ABC 14.5
A 14
A 12.2
B 12.6
C 11
X
B 12.5
ABC 10α
ABC 10α
ABC 10α
ABC 10β
ABC 10β
ABC 10β
N

Figure 1.2. Subunit Composition of RNA Polymerases From Organisms Representing All Three Evolutionary Domains.

The RNA polymerase subunits from *Escherichia coli* (representing bacteria), *Saccharomyces cerevisiae* (yeast, representing eukaryotes) and *Methanococcus jannaschii* (representing Archaea) are displayed to show the sequence similarities between them. Subunits that are homologous to each other, either within a single domain or between different domains, are shown in blue adjacently to each other or are linked with dashed lines. The subunits are arranged, wherever possible, according to their approximate size in the eukaryotic enzymes (the absolute sizes of homologous subunits fluctuate between different species within a given domain). Unique eukaryotic subunits, that are specific for a single type of RNAP, are shown in purple. The AC40/AC19 subunits shared between $RNAP_{I}$ and $RNAP_{III}$ are shown in yellow, and their $RNAP_{II}$ homologs (B44/RPB3 and B12.5/RPB11) in green, to highlight their unusual type-specific properties (see text for more explanantions). The five subunits that are present in all three eukaryotic RNAPs are shown in red. The bacterial α subunit is shown in brown to indicate the lack of extensive sequence homology to other RNAP subunits. Data compiled from multiple sources, Thuriaux & Sentenac, (1992) and Thuriaux, personal communication.

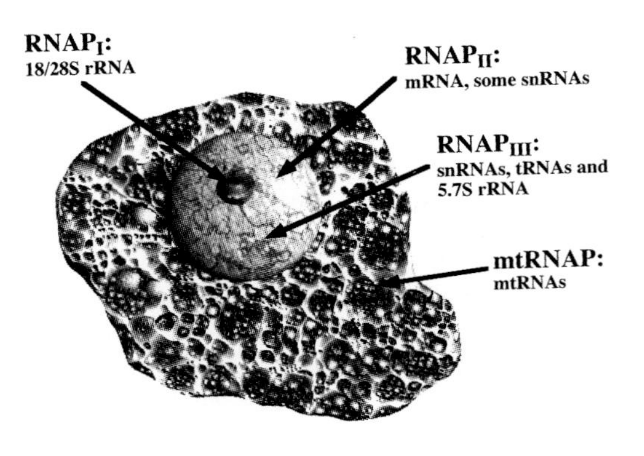

Figure 1.3. RNAPs Present in Eukaryotic Cells.
The three distinct types of nuclear RNAPs ($RNAP_I$, $RNAP_{II}$ and $RNAP_{III}$) transcribe different, non-overlapping sets of genes as indicated. $RNAP_{II}$ and $RNAP_{III}$ carry out their functions in the nucleoplasm, wheras $RNAP_I$ is specificially compartmentalized into the nucleolus. Some cytoplasmic organelles, such as mitochondria (and chloroplasts, not shown), contain independent transcription systems (see text for more information).

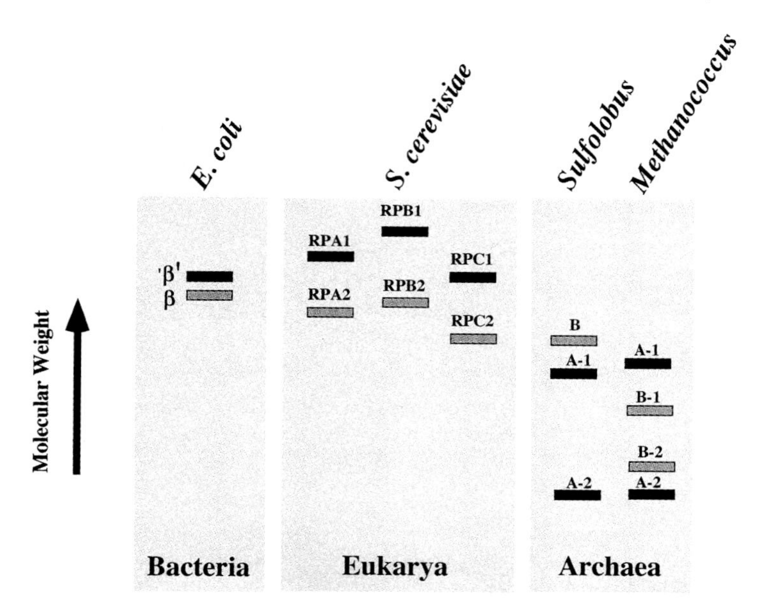

Figure 1.4. Molecular Organization of the Large RNAP Subunits.
The majority of large RNAP subunits found in bacteria and eukaryotes are organized as two large (>100 kDa), single polypeptide chains. In archaeal RNAPs (and some chloroplast-RNAPs) one or both of the large subunit polypeptides are encoded as two distinct polypeptides that, when taken together, contain the same sequence information as the uninterupted subunits present in other species. The data suggests that the large RNAP subunits are organized into distinct domains that do not have to be encoded as a single, continuous polypeptide chain (see also Severinov *et al.*, 1996).

portions is an evolutionary remnant reflecting a 'primitive' feature of a common ancestral RNAP, or occured only in an isolated instance in a common precursor cell giving rise to these two lineages.

The high degree of primary amino acid sequence conservation of the large RNAP subunits already strongly hints that they play a crucial role in RNAP function. This view has been confirmed by numerous biochemical studies designed to map the enzymatically active sites of RNAPs. The large subunits contain the main catalytic centre responsible for DNA-directed RNA biosynthesis, as well as the nucleotide- and DNA-binding sites, and thus constitute the 'business end' of RNAPs (e.g. Figure 1.5; Cho and Kimball, 1982; Carroll and Stollar, 1983; Riva *et al.*, 1987; Treich *et al.*, 1992b; Zaychikov *et al.*, 1996). Although some of the other small RNAP subunits

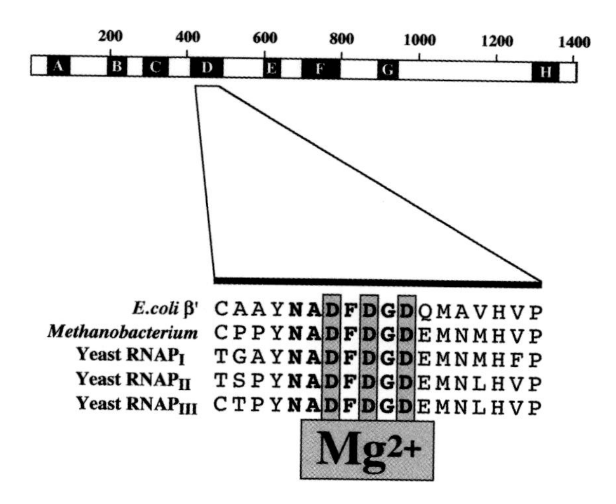

Figure 1.5. The Catalytic Center of Cellular/Nuclear RNAPs is Highly Conserved and Contains Specifically Coordinated Metal Ions.

The schematic diagram in the top left half of the figure shows the position of highly conserved domains in the largest (β′) RNAP subunit of *E. coli*. One of these domains contains several aspartate residues (D) that are invariably present in all bacterial (*E.coli*), archaeal (*Methanobacterium*) and all three types of eukaryotic(*S. cerevisiae*, yeast) large subunits. The conserved aspartate residues are almost certainly used to coordinate the binding of one of the three Mg2+ ions participating in the catalytic reaction of RNA synthesis. After Zaychikov *et al.*, 1996.

contain additional nucleic acid binding sites that are used to for the specific recognition of promoter motifs (e.g. the α and σ subunits of *E. coli* RNAP; Ross *et al.*, 1993; see also Chapter 2), it is likely (though not yet proven) that the large RNAP subunits contain all the structural domains sufficient for the non-specific synthesis of an RNA copy from a DNA template.

The Functions of Conserved Small RNAP Subunits in Archeal/ Eukaryotic RNAPs Remain Largely Unknown

Despite of the central role of RNAPs in the transcription process, the functions of many of the small subunits in archaeal and eukaryotic RNAPs are currently not well-understood. It is likely that they participate in the recruitment of RNAPs to the promoter, control the interaction of RNAPs with other transcription factors and possibly also participate in various catalytic activities. The only functional clue has come from genetic studies in yeast, where it was

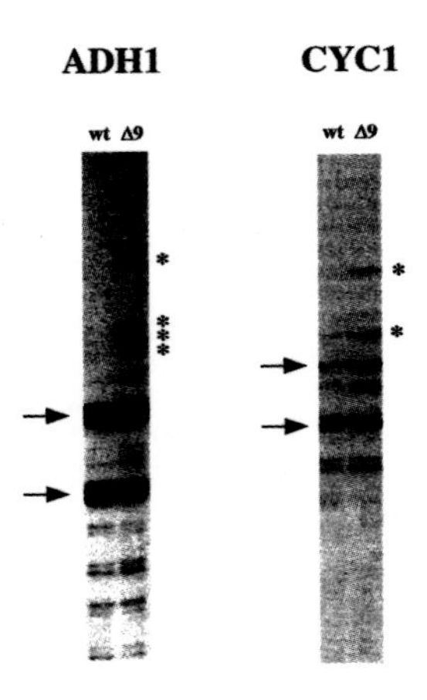

Figure 1.6. Mutations in the Small Eukaryotic RNAP Subunit B12.6/RPB9 Affect the Accuracy of Start Site Selection.
Transcription extracts prepared from wildtype yeast cells direct the accurate transcription initiation from two major distinct sites of the *ADH1* and *CYC1* promoters *in vitro* (indicated by arrows). Extracts from yeast cells lacking a functional B12.6/RPB9 gene behave similarly, but many transcripts are less accurately initiated (additional minor sites marked with asterisks). This implies that the B12.6/RPB9 subunit of $RNAP_{II}$ is directly or indirectly involved in start site selection and accuracy of transcript initiation. Data from Hull *et al.*, 1995.

shown that deletions or other severe mutations of the subunit B12.6 (RPB9) reduce the efficiency of start-site selection, possibly through an indirect effect on the activity of the large RNAP subunit (Figure 1.6; Furter-Graves *et al.*, 1994; Hull *et al.*, 1995).

Although many of these subunits are highly conserved between archaea and eukaryotes (thus proving their ancient evolutionary origin several billion years ago; Figure 1.7), they are completely absent from bacterial RNAPs (see

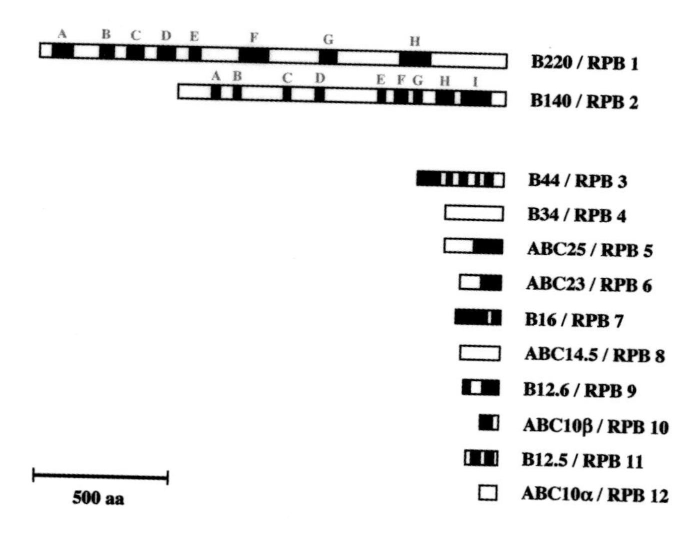

Figure 1.7. Homologies between Eukaryotic $RNAP_{II}$ and Archaeal RNAP Subunits. This diagram displays a comprehensive summary of all regions of yeast $RNAP_{II}$ subunits that are conserved in their homologs encoded by the genome of the archaeon *Methanococcus jannaschii* (Bult *et al.*, 1996). Regions containing a high degree of sequence homology to their archaeal counterparts are shown in black. Note that six (B220/RPB1; B140/RPB2; B44/RPB3; B16/RPB7; B12.6/RPB9 and B12.5/RPB11) of the 12 yeast $RNAP_{II}$ subunits are very similar in their overall primary structural organization to their archaeal homologs. These proteins contain multiple extensive blocks of sequence similarity dispersed evenly over the entire coding region, indicating that the overall tertiary fold of the encoded proteins will be highly similar. Two other archaeal RNAP subunits are highly homologous to the C-terminal domains of the yeast subunits ABC25/RPB5 and ABC23/RPB6, indicating that the N-terminal domains of the yeast proteins may carry out a eukaryotic-specific function. Archaeal homologs of the remaining three yeast $RNAP_{II}$ subunits (B34/RPB4, ABC14.5/RPB8 and ABC10α/RPB12) are either absent or highly diverged in their primary amino acid sequence.

Figure 1.2). There are small intriguing patches of sequence similarity between the bacterial α subunits and the B44 (RPB3)/RPAC40 subunits of eukaryotic RNAPs (Figure 1.8), but it is far from clear at this stage what degree of functional and structural similarities such homologies reflect. Overall, it is likely that the bacterial α and σ subunits are quite dissimilar in structure and function in comparison to the archaeal/eukaryotic small RNAP subunits, and are probably specifically adapted to bacterial gene expression mechanisms that are very different from the archaeal/eukaryotic systems in many fundamental aspects (see Chapters 2, 4 and 7).

yRPB 3	DLAMANSLRRVMIAEIPT
yAC 40	DTSIANAFRRIMISEVPS
yRPB 11	DHTLGNLIRAELLNDRKV
yAC 19	DHTLGNALRYVIMKNPDV
***E. coli* α**	**GHTLGNALRRILLSSMPG**

Figure 1.8. Localized Sequence Homology Between Bacterial a and the Eukaryotic RPB3/ RPAC40 and RBP11/RPAC19 Subunits.
A portion of the *E. coli* α-subunit sequence is shown in alignment with sequences from the yeast $RNAP_{II}$ subunits B44/RPB3, B12.5/RPB11, and the $RNAP_{I/III}$ subunits AC40 and AC19. Identical amino acid residues are black, nonconserved ones are shown in grey. It is not clear at this stage how significant these sequence similarities are.

All three types of eukaryotic nuclear RNAPs characteristically range in size between 500 and 600 kDa and contain between 12 and 17 distinct polypeptide subunits each (Figure 1.2, reviewed in Young, 1991; Thuriaux and Sentenac, 1992). The molecular mapping and characterization of the nuclear RNAP subunits has progressed most extensively in yeast and is now essentially completed (Figure 1.2; Thuriaux and Sentenac, 1992). All the available data suggests that the yeast enzymes are typical representatives of the all eukaryotic RNAPs and the high degree of structural and functional conservation across the eukaryotic evolutionary range is well illustrated by the experimental observations that several human RNAP subunits can substitute for their mutant homologs in yeast (Khazak *et al.*, 1995; McKune *et al.*, 1995; Shpakovski *et al.*, 1995). It is therefore likely that many of the results obtained in the yeast model system are directly applicable to understanding the RNAPs found in higher eukaryotes. Similarly, biochemical studies and genome sequencing efforts have resulted in the complete molecular characterization of all subunits present in archaeal RNAPs (e.g. Bult *et al.*, 1996). We are therefore in the fortunate position of having access to complete sets of primary sequence data for the RNAP subunits of species representing all three evolutionary domains.

Despite their large number, most of the small archaeal/eukaryotic RNAP subunits vary only between 7–30 kDa in size. Apart from the presence of metal-binding motifs in some of them, as indicated by regularly-spaced cysteine residues and an ability to bind zinc *in vitro* (Treich *et al.*, 1991; Werner *et al.*, 1992; Thuriaux and Sentenac, 1992), they do not contain any recognizable domains homologous to other protein families. The precise stoichiometry of these subunits in RNAPs has still not been sufficiently investigated in a large enough number of eukaryotic and archaeal species, but the available data (especially from *S. cerevisiae*, Kolodziej *et al.*, 1990) suggests that the small subunits make up less than half of the protein mass in the complete RNAP (Figure 1.9).

Two of the subunits, B220/RPB1 and ABC23/RPB6 are phosphorylated (Figure 1.9; Kolodziej *et al.*, 1990). We will see later (Chapter 6) that extensive phosphorylation of the C-terminal domain of B220/RPB1 is responsible for converting $RNAP_{II}$ into an transcript elongation-competent form, but the functional consequences of the post-translational modification of ABC23/RPB6 are currently not understood.

Several of these subunits are type-specific (e.g. they only occur in a single type of RNAP) and two of them are shared between $RNAP_{I}$ and $RNAP_{III}$ (Figure 1.2; AC19 and AC40; Sentenac, 1985; Mann *et al.*, 1987; Dequard-Chablat *et al.*, 1991), but are not present in $RNAP_{II}$ (which contains the homologous subunits B44/RPB3 and B12.5/RPB11 instead; Figure 1.2). Remarkably, five of the remaining subunits are shared between all three eukaryotic RNAPs (Figure 1.2; ABC27/RPB5, ABC23/RPB6, ABC14.5/RPB8, ABC10β/RPB10 and ABCα/RPB12; Woychik *et al.*, 1990; Carles *et al.*, 1991; Treich *et al.*, 1992a). Inspection of their primary sequences and elucidation of

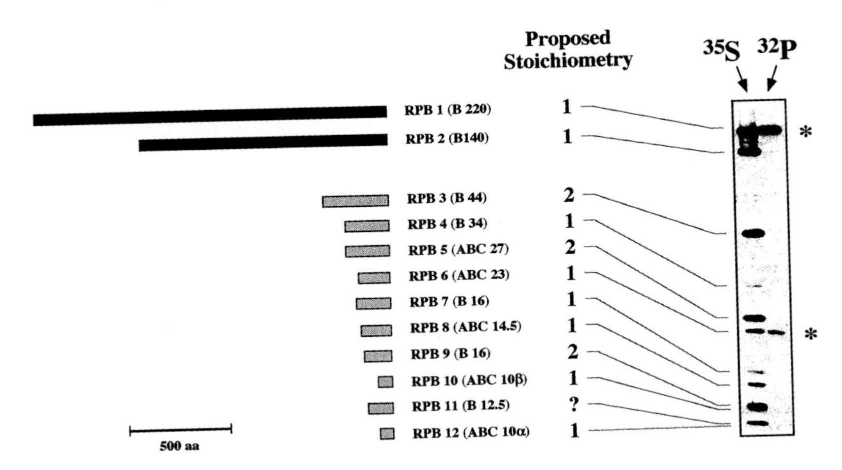

Figure 1.9. Protein Mass Distribution, Subunit Stoichiometry and Phosphorylation in Eukaryotic RNAP$_{II}$.

The left hand side shows a schematic diagram of the relative sizes of the various RNAP$_{II}$ subunits and their stoichiometry in *S. cerevisiae*. The two large subunits (B220/RPB1 and B140/RPB2) are shown in black, and the ten small subunits in grey (B44/RPB3-ABC10α/RPB12). Note that the two large subunits make up approximately 57% of total RNAP$_{II}$ mass and the combined protein mass of the ten 'small' subunits contribute the remaining 43%. The right hands side shows an SDS protein gel of purified yeast RNAP$_{II}$ labelled in presence of ^{35}S and ^{32}P, respectively. Only two subunits (RPB1/B220 and RPB6/ABC23; labelled with asterisks) are labelled in the presence of ^{32}P, indicating that they are post-translationally modified by phosphorylation. Gel data from Kolodziej *et al.*, 1990.

structural motifs present in a selection of small RNAP subunits (Figure 1.10; Krapp *et al.*, 1998; Wang *et al.*, 1998) has not revealed much about their functions yet, but it is reasonable to assume that the five shared subunits are almost certainly involved in common functions carried out by all three eukaryotic polymerases. Three of them (ABC27/RPB5, ABC23/RPB6 and ABC10β/RPB10) are also at least partially conserved in archaea, hinting at their ancient evolutionary origin (Figure 1.7).

Subunit Interactions and Type-Specific Assembly of Eukaryotic RNAPs

Presumably the type-specific subunits have evolved for at least two distinct purposes. Firstly, eukaryotic RNAPs need to interact specifically with basal

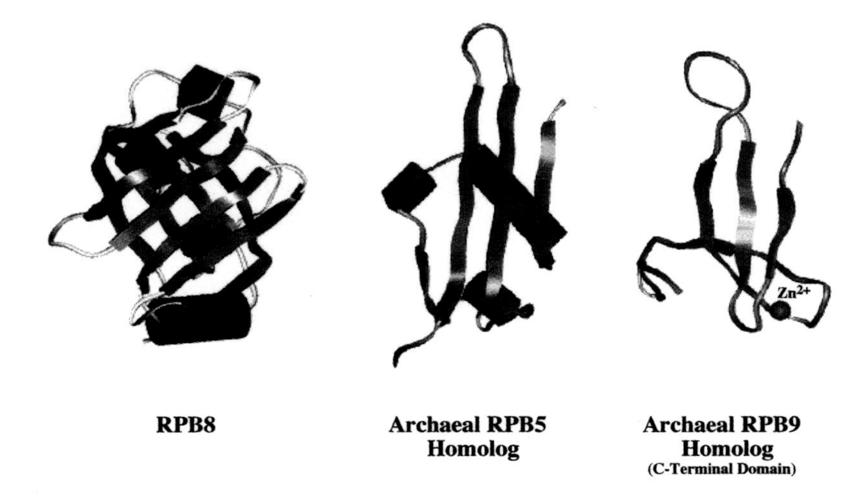

Figure 1.10. Solution Motifs Found in Various Archaeal and Eukaryotic RNAP Subunits. *Left:* Solution structure of the yeast ABC14.5/RPB8 subunit (PDB Access No. 1A1D). Note the striking degree of C_2 symmetry that arises from the barrel-like arrangement of the antiparallel β-sheets. The overall topology is similar to that displayed by a family of proteins containing the 'OB' ('oligonucleotide/oligosaccharide binding')-fold, suggesting that ABC14.5/RPB8 may participate in nucleic acid interaction events. *Middle:* Solution structure of the RNAP subunit H, an archaeal homolog of the yeast RNAP subunit ABC27/RPB5, featuring a novel β-ribbon fold (PDB Access No. 1HMJ). *Right:* Solution structure of the C-terminal domain of subunit 'X', a homolog of the yeast RNAP subunit B12.6/RPB9 (PDB Access No. 1QYP). Note the Zn^{2+}-binding site which is characteristic for several different archaeal and eukaryotic RNAP subunits.

factors (see Chapters 2 and 7) that are instrumental in ensuring that e.g. $RNAP_{II}$ only transcribes mRNA encoding genes, but not rRNA genes. Secondly, the type-specific factors play an important role in determining the identity of the three distinct RNAPs while they are assembled from individual polypeptides in the cell. They must ensure that the various RNAP precursor structures are guided into type-specific assembly pathways resulting in the three different nuclear RNAPs and avoiding the assembly of 'hybrid' RNAPs that may lack the ability and specificity to transcribe distinct sets of genes. Although we currently know very little about the RNAP assembly *in vivo* and the control mechanisms regulating the production of the various RNAP subunits, the relatively large number of type-specific subunits (especially in $RNAP_I$ and $RNAP_{III}$, see Figure 1.2) supports the view that the divergence of $RNAP_I$, $RNAP_{II}$ and $RNAP_{III}$ from a single precursor RNAP may have occurred during a rather early stage of eukaryotic evolution. Selection pressures favouring the diversification of a single nuclear transcription system into three distinct ones probably included a need to 'compartmentalize' transcription of the genome into distinct sets of genes with independent transcriptional control mechanisms (see also Chapter 6). The common RNAP subunits shared by all three types of eukaryotic RNAP almost certainly represent a major regulatory link integrating the overall expression levels of total RNA biosynthesis on a genome-wide basis. Increased production of these subunits would result in increased levels of all three RNAPs, which is an important prerequisites for supporting rapid growth rates, especially during embryonic development and tumor growth (see Chapter 6 for more details).

The existence of two subunits (AC40 and AC19), that are shared between $RNAP_I$ and $RNAP_{III}$ (Figure 1.2), suggests that these subunits are probably used to control a coordinated joint production of $RNAP_I$ and $RNAP_{III}$. The $RNAP_I$ and $RNAP_{III}$ transcription systems deliver many of the RNA components (rRNAs and all tRNAs) that are necessary and rate-limiting for ribosome assembly and function. The $RNAP_{II}$ subunit homologs (B44 [RPB3] and B12.5 [RPB11]) of AC40 and AC19 could be used in eukaryotic cells for decoupling the assembly pathway of $RNAP_{II}$ from $RNAP_I$ and $RNAP_{III}$ production. The detection of specific protein-protein interactions between the AC40/AC19 subunits (Lalo *et al.*, 1993) and the B44/RBP3 and B12.5/RPB11 homologs in various species (Ulmasov *et al.*, 1996; Korobko *et al.*, 1997) supports the view that these RNAP subunit pairs may indeed be able to nucleate separate assembly pathways for $RNAP_{I/III}$ and $RNAP_{II}$ (Figure 1.11).

Archaeal RNAPs contain a highly conserved homolog (subunit 'D'; Figure 1.2) of the B44/RPB3 and AC40 subunits that is capable of specifically interacting with either the eukaryotic B12.5/RPB11 or the AC40 subunits (Eloranta *et al.*, 1998). This observation suggests that some of the essential elements of the binding surface between these proteins are exceptionally highly conserved across the archaeal/eukaryotic domain boundary. The type-specific

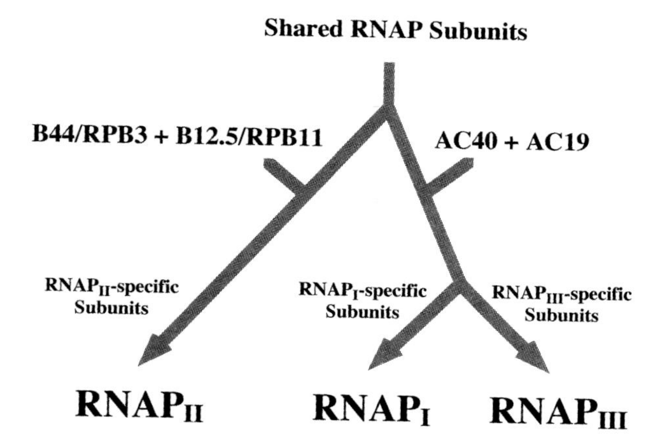

Figure 1.11. A Possible Assembly Pathway for the Three Distinct Eukaryotic RNA Polymerases From Pools of Type-specific and Shared Subunits.
The unusual structural properties of the RNAP$_{I/III}$ AC40 and AC19 subunits, and their RNAP$_{II}$-specific homologs B44/RPB3 and B12.5/RPB11, suggest that these subunit pairs may nucleate the formation of subcomplexes that channel various other common and type-specific RNAP subunits into distinct assembly pathways.

interaction properties of the eukaryotic subunits may have therefore evolved specifically during the early stages of eukaryotic cell evolution in order to support the diversification of the eukaryotic RNAPs into three distinct enzyme types.

1.3. Other RNA Polymerases

We have seen in the previous section that many structural and functional aspects of the RNAPs involved in the transcription of bacterial, archaeal and eukaryotic genomes are still shrouded in mystery. The high degree of complexity of the cellular RNA polymerases could easily create the impression that the biosynthesis of RNA is complicated and requires large enzymes to catalyze the necessary reactions. But, as we will see below, such an assumption is not entirely true.

Reinventing the Wheel Twice? Single Polypeptide RNAPs in Bacteriophages and Mitochondria

All bacteriophages encode various macromolecular components that facilitate a hostile take-over of bacterial transcriptional machineries. The T7 bacteriophage produces a RNAP that allows the high level transcription from specific bacteriophage promoters and provided the first example of structurally 'simple' RNAP (Chamberlain *et al.*, 1970). Although 'merely' 99 kDa in size, T7 RNAP is perfectly capable of, just like any of the cellular RNAPs discussed earlier, recognizing promoters in a sequence-specific manner and carrying out all the subsequent initiation and elongation and termination steps. The unwinding of template DNA and 'abortive' phase of transcription is functionally comparable to the mechanisms known to be employed by the bacterial and eukaryotic enzymes (e.g. Krummel and Chamberlain, 1989). The existence of such 'simple' RNAPs thus proves that large size is not an absolute physical necessity for RNA synthesis.

Despite the functional similarities, it has not been possible to find evidence for any significant primary sequence homologies between these simple RNAPs and their more complex, multisubunit cellular counterparts. An X-ray crystallographic study revealed, however, an extensive structural homology of T7 RNAP to various DNA polymerases and reverse transcriptase (Figure 1.12; Delarue *et al.*, 1990; Beese *et al.*, 1993; Sousa *et al.*, 1993; Davies *et al.*, 1994). Bacteriophage RNAPs are thus closely related to a large family of proteins that encompasses all known nucleic acid polymerases with the sole exception of the multisubunit RNAPs. This result serves to highlight the rather exclusive status of the multisubunit RNAPs within the broad family of nucleic acid polymerases and deepens the mystery of their evolutionary origin further. The multisubunit enzymes seem to have either evolved completely separately from the single-chain polymerases, or diverged radically from them at a very early stage during the evolution of life on earth. Interestingly, despite the differences in primary structure between single- and multisubunit RNAPs, they all display a distinct channel, which is wide enough to hold double-helical DNA (e.g. Darst *et al.*, 1989 and 1991; Sousa *et al.*, 1993; Schultz *et al.*, 1993; Polyakov *et al.*, 1995). Future structural research, especially of the active center of cellular RNAPs, and a detailed comparison of the catalytic mechanisms

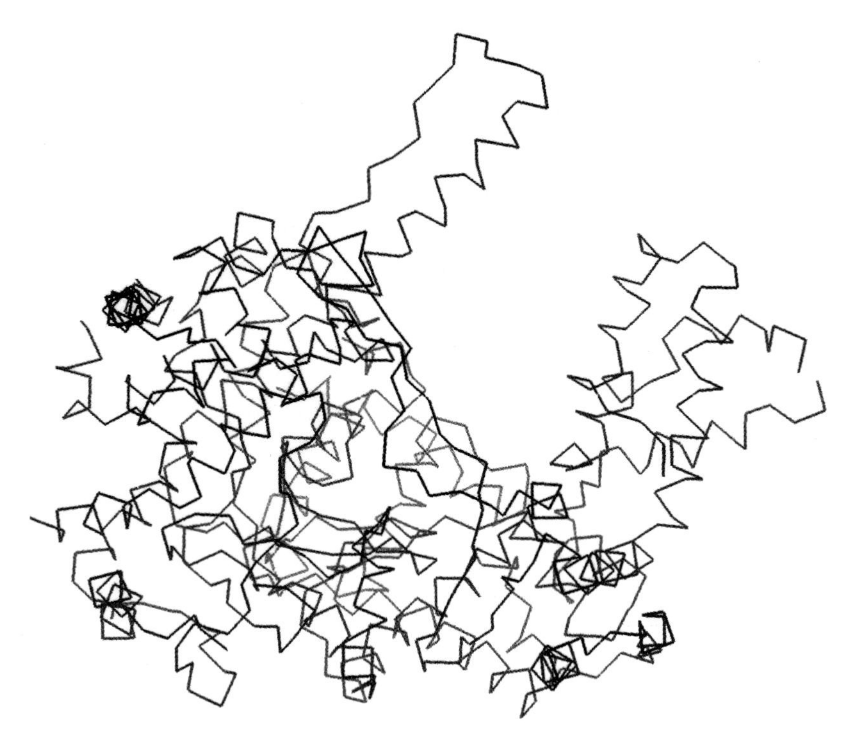

Figure 1.12. X-Ray Structure of T7 RNAP at 3.3Å Resolution.
Note the presence of a deep cleft in the single-polypeptide RNAP, that can accomodate double-helical template DNA. After Sousa *et al.* (1993). PDB Access No. 4RNP.

used by the two apparently divergent RNAP types, will be necessary to see whether the catalytic centers of single- and multisubunit RNAPs are organized along similar structural and mechanistic principles.

Mitochondrial RNAPs Are Homologous to Bacteriophage RNAPs

In addition to the three nuclear RNAPs, the vast majority of eukaryotic cells (with a few interesting exceptions; Cavalier-Smith, 1987; Martin and Muller, 1998) contain organelles (Figure 1.3.). Mitochondria harbor a small, independent genome and a distinct mitochondrial RNAP for transcribing it. Mitochondrial RNAPs are usually single-polypeptide enzymes, that are encoded by the nucleus (!), and display a high degree of sequence homology to the bacteriophage polymerases (Figure 1.13; Masters *et al.*, 1987; Cermakian *et al.*, 1996). This result was initially surprising, because mitochondrial RNAPs were expected to display a bacterial-like, multisubunit-type RNAP as predicted by the endosymbiont-hypothesis of mitochondrial evolution (e.g. Gray and Doolittle, 1982). A large-scale PCR-based study of a phylogenetically broad range of eukaryotic species by Cermiakan *et al.* (1996) essentially confirmed the view that the vast majority of eukaryotes, even evolutionarily highly-diverged species with unusually-structured mitochondria, contained such bacteriophage RNAP-like enzymes. The mitochondrial RNAPs are more homologous to each other as a group than to the bacteriophage enzymes, which is consistent with the idea that they originated from a common ancestral gene, that was already present in one of pre-symbiontic cell precursors before the diversification of the eukaryotic lineages.

Interestingly, after this question appeared to have beeen settled unanimously, a bacterial-like mitochondrial RNAP did turn up in the 'truly primitive' mitochondrion from the dinoflagellate *Reclinomonas americana* (Lang *et al.*, 1997; see also Palmer *et al.*, 1997). In addition to the presence of genes encoding bacterial-like core RNAP subunits, with a significant degree of homology to the *E.coli* α, β and β' proteins, this mitochondrial genome also contains a gene encoding a σ^{70} homolog. As we shall see in Chapter 2, σ^{70} is a crucial selectivity

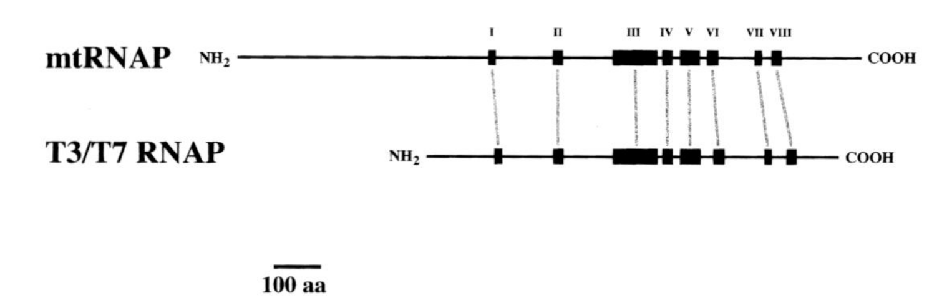

Figure 1.13. Mitochondrial RNAPs are Homologous to Bacteriophage RNAPs.
The yeast mitochondrial RNAP and the T3/T7 RNAPs are aligned for maximum overlap. Regions with a high degree of sequence similarities are indicated as black blocks and numbered in roman numerals (I-VIII). Note that the mitochondrial RNAP contains an extended N-terminal domain, that may interact with regulatory factors. After Masters *et al.*, 1987.

factor that allows sequence-specific recognition of promoter sequences by bacterial RNAPs. From these results it is likely that the *R. americana* mitochondrial genome is similarly transcribed as the well-studied bacterial genomes.

The widespread occurrance of a (nuclear-encoded) bacteriophage-type RNAP in most mitochondria, and the presence of a bacterial-type RNAP in at least one species, is compatible with the endosymbiontic origin hypothesis. The original bacteria-like mitochondrial precursor probably contained a genome, similar to that found in *R. americana*, that encoded a multi-subunit RNAP responsible for the transcription of all mitochondrial genes. The presence of an additional bacteriophage-like RNAP, possibly already present in the nuclear host genome, allowed at some point a functional substitution of the two RNAPs (Lang *et al.*, 1997). Since mitochondrial genomes contain only a small number of promoters, such a takeover event could have proceeded rather gradually and without major disruption of the overall mitochondrial gene expression program. Once transcription by the bacteriophage-like RNAP was comprehensively established, the genes encoding the bacterial-like subunits were lost from the mitochondrial genome due to the absence of further selective pressure favouring their continued maintainance. As we will see in the next section, a similar process may be currently going on in the chloroplasts of higher plants.

Higher Plant Chloroplasts Contain Two Distinct RNAPs

Chloroplasts, like mitochondria, are thought to be originally of bacterial origin and were acquired by eukaryotic algae and plants through endosymbiosis. Consistent with this view is the presence of genes encoding a bacterial-like RNAP subunits, encoding α, β, β' within the plastid genome (Ohyama *et al.*, 1986; Shinozaki *et al.*, 1986; Sijben-Muller *et al.*, 1986). The necessary σ factor to complete this enzyme is separately encoded by a nuclear gene and imported into the chloroplast (reviewed in Igloi and Kossel, 1992; Link, 1996). In addition to the plastid-encoded polymerase system (PEP), Allison *et al.* (1996) found evidence for a second nuclear encoded polymerase (NEP). Interestingly, many of the plastid genes have a dual promoter arrangement and are transcribed by both PEP and NEP, and only a small number of genes seem to depend exclusively on only one of the two RNAP systems (Hajdukiewicz *et al.*, 1997; Pfannschmidt and Link, 1997). Evidence from *Arabidopsis* suggests that the NEP, similar to the mitochondrial enzymes, is a bacteriophage-like RNAP (Hedtke *et al.*, 1997). We have already seen earlier, that the mitochondrial genome was probably originally transcribed by bacteria-like RNAP, that was replaced in most eukaryotes by nuclear-encoded bacteriophage-like enzyme later in evolution. The presence of two independent RNAP systems in higher plant chloroplasts, with substantially overlapping functions (!), may represent a similar attempt by the NEP system to gradually replace the PEP system.

1.4. Conclusions

Considering that RNA polymerases play such a central role in transcription, relatively little information is known about their general functions and structural organization. In all cellular/nuclear RNAPs many of the important regions, such as the sites involved in nucleotide binding and the RNA catalysis, are localized in the two large subunits. These large RNAP subunits are the only polypeptides of the transcriptional machinery that are universally conserved across all three evolutionary domains (bacteria, archaea and eukaryotes). In addition to the large subunits, cellular/nuclear RNAPs contain small subunits with mostly unidentified functions. Many of these small subunits are highly conserved between eukaryotes and archaea, but are absent in bacterial RNAPs.

The single-polypeptide RNAPs found in bacteriophages and mitochondria display a significant degree of structural homology to DNA polymerases and reverse transcriptases, suggesting that they are derived from a different evolutionary precursor enzyme than the cellular/nuclear RNAPs.

References

Internet Sites of Interest

http://www.rockefeller.edu/labheads/darst/structures.htm
Electron microscopic images of pro- and eukaryotic RNAPs.

Research Literature

Allison, L.A., Moyle, M., Shales, M., and Ingles, C.J. (1985). Extensive homology among the largest subunits of eukaryotic and prokaryotic RNA polymerases. *Cell* 42, 599–610.

Allison, L.A., Simon, L.D., and Maliga, P. (1996). Deletion of *rpoB* reveals a second distinct transcription system in plastids of higher plants. *EMBO J.* 15, 2802–2809.

Beese, L.S., Derbyshire, V., and Steitz, T.A. (1993). Structure of DNA polymerase I Klenow fragment bound to duplex DNA. *Science* 260, 352–355.

Brown, J.R., and Doolittle, W.F. (1997). Archaea and the prokaryote-to-eukaryote transition. *Microbiol. Mol. Biol. Rev.* 61, 456–502.

Bult, C. J., *et al.* (1996). Complete genome sequence of the methanogenic archaeon, *Methanococcus jannaschii*. *Science* 273, 1058–1073.

Carles, C., Treich, I., Bouet, F., Riva, M., and Sentenac, A. (1991). Two additional common subunits, ABC10α and ABC10β, are shared by yeast RNA polymerases. *J. Biol. Chem.* 266, 24092–24906.

Carroll, S.B., and Stollar, B.D. (1983). Conservation of a DNA-binding site in the largest subunit of eukaryotic RNA polymerase II. *J. Mol. Biol.* 170, 777–790.

Cavalier-Smith, T. (1987). Eukaryotes with no mitochondria. *Nature* 326, 332–333.

Cermiakan, N., Ikeda, T.M., Cedergren, R., and Gray, M.W. (1996). Sequences homologous to yeast mitochondrial and bacteriophage T3 and T7 RNA polymerases are widespread throughout the eukaryotic lineage. *Nucl. Acids Res.* 24, 648–654.

Chamberlain, M., McGrath, J., and Waskoll, L. (1970). New RNA polymerase from *Escherichia coli* infected with bacteriophage T7. *Nature* 228, 227–230.

Cho, J.M., and Kimball, A.P. (1982). Probes of eukaryotic DNA-dependent RNA polymerase II - I. Binding of 9-β-D-arabinofuranosyl-6-mercaptopurine to the elongation subsite. *Biochem. Pharmacol.* 31, 2575–2581.

Darst, S.A., Kubalek, E.W., and Kornberg, R.D. (1989). Three-dimensional structure of *Escherichia coli* RNA polymerase holoenzyme determined by electron crystallography. *Nature* 340, 730–732.

Darst, S.A., Edwards, A.M., Kubalek, E.W., and Kornberg, R.D. (1991). Three-dimensional structure of yeast RNA polymerase at 16Å resolution. *Cell* 66, 121–128.

Davies, J.F., Almassy, R.J., Hotomska, Z., Ferre, R.A., and Hostomsky, Z. (1994). 2.3Å crystal structure of the catalytic domain of DNA polymerase β. *Cell* 76, 1123–1133.

Delarue, M., Poch, O., Tordo, N., Moras, D., and Argos, P. (1990). An attempt to unify the structure of polymerases. *Protein Engineering* 3, 461–467.

Dequard-Chablat, M., Riva, M., Carles, C., and Sentenac, A. (1991). RPC19, the gene for a subunit common to yeast RNA polymerases A (I) and C (III). *J. Biol. Chem.* 266, 15300–15307.

Eloranta, J.J., Kato, A., Teng, M.S., and Weinzierl, R.O.J. (1998). *In vitro* assembly of an archaeal D-L-N RNA polymerase subunit complex reveals a eukaryote-like structural arrangement. *Nucl. Acids Res.* 26, 5562–5567.

Furter-Graves, E.M., Hall, B.D., and Furter, R. (1994). Role of a small RNA pol II subunit in TATA to transcription start site spacing. *Nucl. Acids. Res.* 22, 4932–4936.

Gray, M.W., and Doolittle, W.F. (1982). Has the endosymbiont hypothesis been proven? *Microbiol. Rev.* 46, 1–42.

Hajdukiewicz, P.T.J., Allison, L.A., and Maliga, P. (1997). The two RNA polymerases encoded by the nuclear and the plastid compartments transcribe distinct groups of genes in tobacco plastids. *EMBO J.* 16, 4041–4048.

Hedtke, B., Borner, T., and Weihe, A. (1997). Mitochondrial and chloroplast phage-type RNA polymerases in *Arabidopsis*. *Science* 277, 809–811.

Heyduk, T., Heyduk, E., Severinov, K., Tang, H., and Ebright, R. (1996). Determinants of RNA polymerase α subunit for interaction with β, β' and σ subunits: hydroxyl-radical protein footprinting. *Proc. Natl. Acad. Sci. USA* 93, 10162–10166.

Hull, M.W., McKune, K., and Woychik, N.A. (1995). RNA polymerase II subunit RPB9 is required for accurate start site selection. *Genes Dev.* 9, 481–490.

Igloi, G.L., and Kossel, H. (1992). The transcriptional apparatus of chloroplasts. *Crit. Rev. Plant Sci.* 10, 525–558.

Khazak, V., Sadhale, P.P., Woychik, N.A., Brent, R., and Golemis, E.A. (1995). Human RNA polymerase II subunit hsRPB7 functions in yeast and influences stress survival and cell morphology. *Mol. Biol. Cell* 6, 759–775.

Klenk, H.-P., and Doolittle, W.F. (1994). Archaea and eukaryotes versus bacteria? *Curr. Biol.* 4, 920–922.

Kolodziej, P.A., Woychik, N., Liao, S.M., and Young, R.A. (1990). RNA polymerase II subunit composition, stoichiometry, and phosphorylation. *Mol. Cell. Biol.* 10, 1915–1920.

Korobko, I.V., Yamamoto, K., Nogi, Y., and Muramatsu, M. (1997). Protein interaction cloning in yeast of the mouse third largest RNA polymerase II subunit, mRPB31. *Gene* 185, 1–4.

Krapp, S., Kelly, G., Reischl, J., Weinzierl, R.O.J., and Matthews, S. (1998). Eukaryotic RNA polymerase subunit RPB8 is a new relative of the OB family. *Nature Struct. Biol.* 5, 110–114.

Krummel, B., and Chamberlain, M. (1989). RNA chain initiation by *Escherichia coli* RNA polymerase. Structural transitions of the enzyme in early ternary complexes. *Biochemistry* 28, 7829–7842.

Lalo, D., Carles, C., Sentenac, A., and Thuriaux, P. (1993). Interactions between three subunits of yeast RNA polymerases I and III. *Proc. Natl. Acad. Sci. USA* 90, 5524–5528.

Lang, B.F., Burger, G., O'Kelly, C.J., Cedergren, R., Golding, G.B., Lemieux, C., Sankoff, D., Turmel, M., and Gray, M.W. (1997). An ancestral mitochondrial DNA resembling a eubacterial genome in miniature. *Nature* 387, 493–497.

Lanzendorfer, M., Langer, D., Hain, J., Klenk, H.P., Holz, I., Arnold-Ammer, I., and Zillig, W. (1994). Structure and function of the DNA-dependent RNA polymerase of *Sulfolobus*. *Systematic Appl. Microbiol.* 16, 656–664.

Link, G. (1996). Green life: control of chloroplast gene transcription. *BioEssays* 18, 465–471.

Mann, C., Buhler, J.-M., Treich, I., and Sentenac, A. (1987). *RPC40*, a unique gene for a subunit shared between yeast RNA polymerases A and C. *Cell* 48, 627–637.

Martin, W., and Muller, M. (1998). The hydrogen hypothesis for the first eukaryote. *Nature* 392, 37–41.

Masters, B.S., Stohl, L.L., and Clayton, D.A. (1987). Yeast mitochondrial RNA polymerase is homologous to those encoded by bacteriophages T3 and T7. *Cell* 51, 89–99.

McKune, K., Moore, P.A., Hull, M.W., and Woychik, N.A. (1995). Six human RNA polymerase subunits functionally substitute for their yeast counterparts. *Mol. Cell. Biol.* 15, 6895–6900.

Memet, S., Saurin, W., and Sentenac, A. (1988). RNA polymerases B and C are more closely related to each other than to RNA polymerase A. *J. Biol. Chem.* 263, 10048–10051.

Ohyama, K., Fukuzawa, H., Kohchi, T., Shirai, H., Sano, T., Sano, S., Umesono, K., Shiki, Y., Takeuchi, M., Chang, Z., Aota, S., Inokuchi, H., and Ozeki, H. (1986). Chloroplast gene orgnization deduced from complete sequence of liverwort *Marchantia polymorpha* chloroplast DNA. *Nature* 322, 572–574.

Polyakov, A., Severinova, E., and Darst, S.A. (1995). Three-dimensional structure of *E.coli* core RNA polymerase: promoter binding and elongation conformation of the enzyme. *Cell* 83, 365–373.

Palmer, J.D. (1997). The mitochondrion that time forgot. *Nature* 387, 454–455.

Pfannschmidt, T., and Link, G. (1997). The A and B forms of plastid DNA-dependent RNA polymerase from mustard (*Sinapis alba L.*) transcribe the same genes in a different developmental context. *Mol. Gen. Genet.* 257, 35–44.

Puhler, G., Leffers, H., Gropp, F., Palm, P. K., Klenk, H.-P., Lottspeich, F., Garrett, R. A., and Zillig, W. (1989). Archaebacterial DNA-dependent RNA polymerases testify to the evolution of the eukaryotic nuclear genome. *Proc. Natl. Acad. Sci. USA* 86, 4569–4573.

Riva, M., Schaffner, A.R., Sentenac, A., Hartmann, G.R., Mustaev, A.A., Zaychikov, E.F., and Grachev, M.A. (1987). Mapping the active site of yeast RNA polymerase B (II). *J. Biol. Chem.* 262, 14377–14380.

Ross, W., Gosink, K.K., Salomon, J., Igarashi, K., Zou, C., Ishihama, A., Severinov, K., Gourse, R.L. (1993). A third recognition element in bacterial promoters: DNA binding by the α subunit of RNA polymerase. *Science* 262, 1407–1413.

Schultz, P., Celia, H., Riva, M., Sentenac, A., and Oudet, P. (1993). Three-dimensional model of yeast RNA polymerase I determined by electron microscopy of two-dimensional crystals. *EMBO J.* 12, 2601–2607.

Severinov, K., Mustaev, A., Kukarin, A., Muzzin, O., Bass, I., Darst, S.A., and Goldfarb, A. (1996). Structural modules of the large subunits of RNA polymerase. Introducing archaebacterial and chloroplast split sites in the beta and beta' subunits of *Escherichia coli* RNA polymerase. *J. Biol. Chem.* 271, 27969–27974.

Shinozaki, K., *et al.* (1986). The complete nucleotide sequence of the tobacco chloroplast genome: its gene organization and expression. *EMBO J.* 5, 2043–2049.

Shpakovski, G.V., Acker, J., Wintzerith, M., Lacroix, J.F., Thuriaux, P., and Vigneron, M. (1995). Four subunits that are shared by the three classes of RNA polymerase are functionally interchangeable between *Homo sapiens* and *Saccharomyces cerevisiae*. *Mol. Cell. Biol.* 15, 4702–4710.

Sentenac, A. (1985). Eukaryotic RNA polymerases. *CRC Crit. Rev. Biochem.* 18, 31–91.

Sijben-Muller, G., Hallick, R.B., Alt, J., Westhoff, P., and Herrmann, R.G. (1986). Spinach plastid genes coding for initiation factor IF-1, ribosomal protein S11 and RNA polymerase α-subunit. *Nucl. Acids Res.* 14, 1029–1044.

Sousa, R., Chung, Y.J., Rose, J.P., and Wang, B.-C. (1993). Crystal structure of bacteriophage T7 RNA polymerase at 3.3Å resolution. *Nature* 364, 593–599.

Sweetser, D., Nonet, M., and Young, R.A. (1987). Prokaryotic and eukaryotic RNA polymerases have homologous core subunits. *Proc. Natl. Acad. Sci. USA* 84, 1192–1196.

Thuriaux, P., and Sentenac, A. (1992). Yeast nuclear RNA polymerases. In 'The Molecular Biology of the Yeast *Saccharomyces*: Gene Expression'. Cold Spring Harbor Press, New York.

Treich, I., Riva, M., and Sentenac, A. (1991). Zinc-binding subunits of yeast RNA polymerases. *J. Biol. Chem.* 266, 21971–21976.

Treich, I., Carles, C., Riva, M., and Sentenac, A. (1992a). RPC10 encodes a new mini subunit shared by yeast nuclear RNA polymerases. *Gene Express.* 2, 31–37.

Treich, I., Carles, C., Sentenac, A., and Riva, M. (1992b). Determination of lysine residues affinity labeled in the active site of yeast RNA polymerase II (B) by mutagenesis. *Nucl. Acids Res.* 20, 4721–4725.

Ulmasov, T., Larkin, R.M., and Guilfoyle, T.J. (1996). Association between 36- and 13.6-kDa alpha-like subunits of *Arabidopsis thaliana* RNA polymerase II. *J. Biol. Chem.* 271, 5085–5094.

Wang, B., Jones, D.N., Kaine, B.P., and Weiss, M.A. (1998). High-resolution structure of an archaeal zinc ribbon defines a general architectural motif in eukaryotic RNA polymerases. *Structure* 6, 555–569.

Werner, M., Hermann-Le Denmat, S., Treich, I., Sentenac, A., and Thuriaux, P. (1992). Effects of mutations in a zinc-binding domain of yeast RNA polymerase C (III) on enzyme function and subunit association. *Mol. Cell. Biol.* 12, 1087–1095.

Woese, C. R., Kandler, O., and Wheelis, M. L. (1990). Towards a natural system of organisms: proposals for the domains Archea, Bacteria and Eukarya. *Proc. Natl. Acad. Sci. USA* 87, 4576–4579.

Woychik, N.A., Liao, S.-M., Kolodziej, P.A., and Young, R.A. (1990). Subunits shared by eukaryotic nuclear RNA polymerases. *Genes Dev.* 4, 313–323.

Young, R.A. (1991). RNA polymerase II. *Ann. Rev. Biochem.* 60, 689–715.

Zaychikov, E., Martin, E., Denissova, L., Kozlov, M., Markovtsov, V., Kashlev, M., Heumann, H., Nikiforov, V., Goldfarb, A., and Mustaev, A. (1996). Mapping of catalytic residues in the RNA polymerase active center. *Science* 273, 107–109.

Zillig, W., Stetter, K.O., and Janekowic, D. (1979). DNA-dependent RNA polymerase from *Sulfolobus acidocaldarius. Eur. J. Biochem.* 96, 597–604.

Zillig, W., Palm, P., Klenk, H.-P., Langer, D., Hudepohl, U., Hain, J., Lanzendorfer, M., and Holz, I. (1993). Transcription in archaea. In 'The Biochemistry of Archaea (archaebacteria)'. M. Kates, D. J. Kushner and A. T. Matheson (Eds). Elsevier, Amsterdam.

Chapter 2

Basal Factors Recognize Promoters and Assemble the Pre-Initiation Transcription Complexes

All bacterial, archaeal and eukaryotic RNA polymerases are complex enzymes consisting of many different subunits (Chapter 1). In spite of this high degree of complexity, they are unable to locate by themselves the specific start sites ('promoters') that are located upstream of the coding regions of transcribed genes. This task is carried out by the 'basal' factors, which are a functionally defined family of diverse proteins that assist RNAPs in the task of finding promoters and establishing active pre-initiation transcription complexes on them. Basal factors cooperate intimately with RNAPs at numerous stages to enable them to go through a distinct sequence of steps that are required before a transcript can be specifically initiated (Figure 2.1). These steps either constitute kinetic 'bottle-necks' or involve sequence-specific molecular recognition events (e.g. of core promoter elements) that RNAPs themselves are incapable of. Furthermore, especially in eukaryotic systems, the transition from the transcript initiation to elongation mode of $RNAP_{II}$ is highly controlled (see Chapter 5) and several basal factors play an important role in allowing RNAP to 'escape' efficiently from the initiation complex. An increasing amount of evidence also suggests that the eukaryotic basal factors

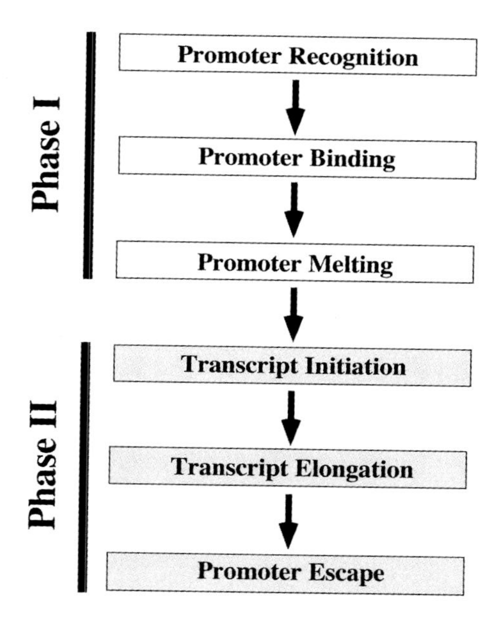

Figure 2.1. Basal Transcription Factors Enable and Facilitate Many Different Steps During Transcript Initiation by Cellular RNA Polymerases.
This diagram illustrates the various steps of the transcription initiation process. In bacteria, the σ factors participate in all three stages of phase I, before RNAP takes over and completes all the remaining steps. In eukaryotic systems, basal transcription factors participate in each of the stages in both phase I and phase II.

play an important role in continously maintaining high levels of gene transcription from activated genes through the formation of a stable initiation complexes that favour high rates of repeated re-initiation of new transcripts.

Before focusing on the various aspects of the complex eukaryotic $RNAP_{II}$ pre-initiation complex, we will briefly review the differences and similarities of the various basal factors using a variety of specific examples derived from all three evolutionary domains. This will allow the common themes to emerge and establish the evolutionary relationships of the different types of basal transcriptional machineries in a range of organisms.

2.1. σ Factors Recruit Bacterial RNA Polymerases to Different Promoters and Assist Transcript Initiation

σ Factors Determine the Promoter-Specific Binding of Bacterial RNA Polymerases

Bacteria thrive under many different environmental conditions, that either change gradually or sometimes quite rapidly. Such changes require the constant adjustment of the expression rate of numerous genes to allow the cells to utilize the available environmental resources as efficiently as possible. This is usually based on controlling the rate of transcription of operons encoding metabolic enzymes through gene-specific transcription factors.

In addition to such highly-specialized systems, there are several different genome-wide ('global') gene expression programs, that are implemented through controlling the promoter-specific recognition abilities of the bacterial RNAP initiation complex. The bacterial 'core' enzyme (α_2-β-β') has essentially no ability to recognize specific promoter sequences, but there are a number of polypeptides, the σ factors, that temporarily associate with the core enzyme to form the 'holoenzyme.' The σ factors present in such a holoenzyme configuration guide the core RNAP to promoters through their ability to recognize particular sequence motifs in a sequence-specific manner. Although σ factors have been regarded as *bona fide* bacterial RNAP subunits for historical reasons, they fulfill essentially a similar role as many of the other basal factors that we will encounter at a later stage in the description of archaeal and eukaryotic transcription systems.

One of the most commonly used σ factors in *E. coli* is σ^{70} (the superscript indicates its approximate molecular size in kDa). σ^{70} is the dominant σ factor in cells during exponential growth in rich medium and is therefore one of the biochemically and structurally best-understood σ factors (reviewed in Helmann and Chamberlin, 1988). In addition to σ^{70}, there are several other, often structurally substantially diverged σ factors with more specialized roles that direct transcription of specific gene sets under particular environmental conditions (summarized in Figure 2.2; Strauss *et al.*, 1987; Helman and Chamberlin, 1988; Arnosti and Chamberlin, 1989; Erickson and Gross, 1989; Merrick, 1993; Lowen and Hengge-Aronis, 1994; Angerer *et al.* 1995; Jishage *et al.*, 1996).

σ Factors Facilitate the Formation of Initiation-Competent RNA Polymerases in Bacteria

High resolution electron microscopic studies of *E.coli* core and holoenzyme RNAPs have convincingly shown that binding of σ^{70} causes a substantial conformational change in the overall appearance of the enzyme (Figure 2.3;

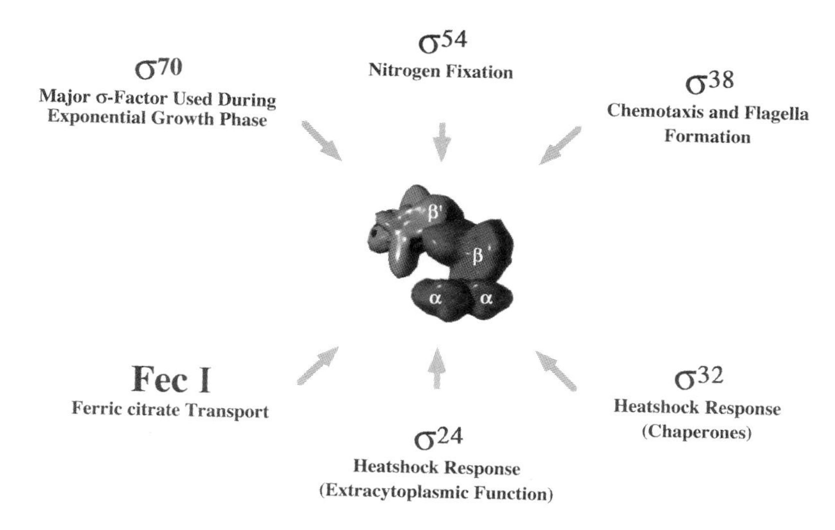

Figure 2.2. Different σ Factors Change the Promoter Specificity of Bacterial RNAPs. Under different environmental conditions the bacterial core RNA polymerase (α_2-β-β′) associates with a variety of distinct σ factors that influence the ability of the enzyme to recognize specific types of promoters.

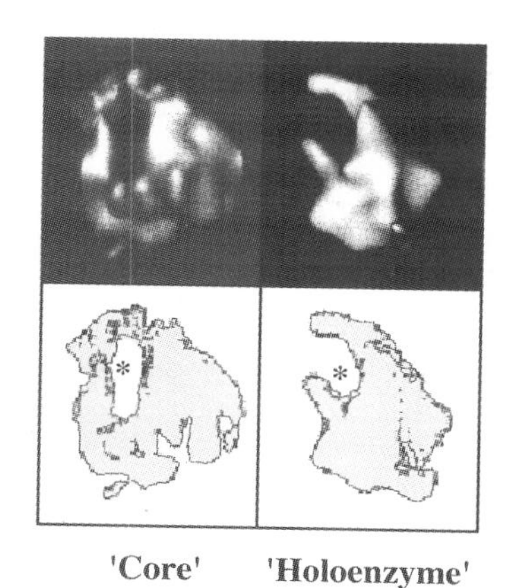

Figure 2.3. Electron Micrographs of the 'Core' and 'Holoenzyme' Configurations of *E. coli* RNA Polymerase.
The presence of σ^{70} in the holoenzyme (top and bottom right) causes a substantial conformational change in the overall appearance of *E. coli* RNAP. The 'open hand' conformation of the holoenzyme may facilitate its binding to DNA while searching for promoter sequences. Once σ^{70} dissociates, the thumb structure closes and locks the RNAP core enzyme onto DNA (top and bottom left). Note that there are also substantial changes in other parts of the molecule.
The bottom part of the figure has been processed to visually enhance the crucial structural features. The asterisk symbol indicates the channel through which the DNA template is thought to be threaded during transcription.
Original (top) photograph from Polyakov *et al.* (1995).

Polyakov *et al.*, 1995). While the core enzyme has an overall ring-like structure with a central channnel, through which DNA could be threaded during the elongation process, binding of σ^{70} causes this channel to open and form a structure reminiscent of a human hand with a 'thumb'-like projection. These structures immediately suggest that σ^{70} allows *E. coli* RNAP sterical access to double-stranded DNA and to 'feel' for the sequence-specific motifs that constitute bacterial promoters. In addition to inducing such a conformational change, σ^{70} is also directly involved in the recognition of important *E.coli* promoter elements through protein-DNA interactions (Figure 2.4). Although the complete structure of σ^{70} has yet to be determined, X-ray crystallographic studies of the domain involved in the recognition of the 'Pribnow' box (an A/T-rich region close to the transcript initiation site) have revealed many fascinating insights into σ^{70} function involved in specific recognition of promoter sequences and the subsequent 'melting' of the double-helical DNA template in preparation for transcription initiation (Figure 2.5; Malhotra *et al.*, 1996).

Once σ^{70} has fulfilled its role, it dissociates from the RNAP core enzyme and allows it to revert to its 'closed hand' conformation (Figure 2.3). It is very likely that the high degree of processivity associated with elongating bacterial

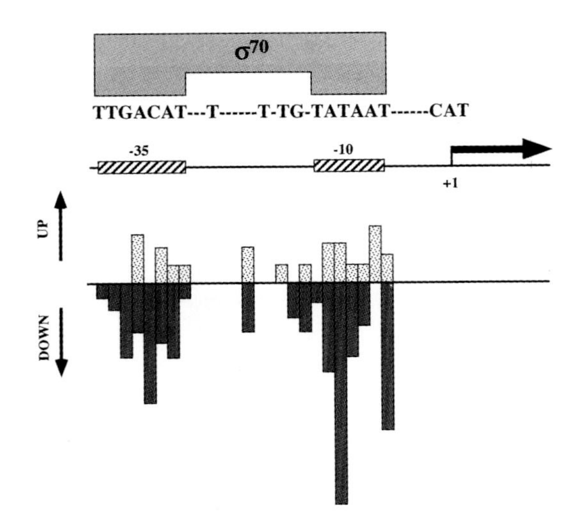

Figure 2.4. Basal Sequence Motifs of Bacterial Promoters Recognized by σ^{70}.
The major *E. coli* σ factor, σ^{70}, specifically recognizes both consensus motifs centered around −10 and −35 (note the transcription start site at +1). Most mutations in the consensus sequences cause a substantial decrease in promoter 'strength' ('down' mutations; shown in bottom part of figure).

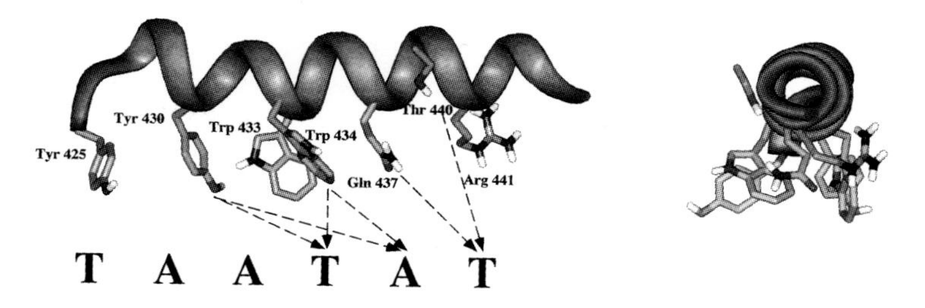

Figure 2.5. DNA Interaction Surface of σ^{70}.
Left: Helix 14 of σ^{70} contains many of the amino acid side chains that are involved in promoter recognition and melting. The dashed lines indicate known DNA-protein interactions that have been defined by biochemical and/or genetic means. Mutations in the aromatic residues tyr^{425}, tyr^{430}, trp^{433} and trp^{434} interfere with promoter 'melting' and may stack with the exposed bases through hydrophobic interactions.
Right: View down the axis of the recognition helix. All the crucial amino acid side chains involved in the DNA interaction face into the same direction and are solvent-exposed.
After Malhotra *et al.*, 1996.

core RNAPs is a direct consequence of this conformational change because it physically prevents the RNAP from dissociating from its template prematureley (Polyakov *et al.*, 1995).

σ Factors are Highly Specialized and do not have Obvious Similarities to Basal Factors in Other Transcription Systems

We have already seen in Chapter 1 that bacterial RNAPs, although they contain the highly conserved large subunits β and β′, are highly diverged from the archaeal/eukaryotic RNAPs with respect to their organization of the 'small' RNAP subunits. σ is no exception to this rule. Although many researchers have looked for sequence similarities between bacterial σ factors and the various archaeal and eukaryotic basal transcription factors, the search has up to now been essentially unsuccessful. Some intriguing localized sequence homologies between the archaeal/eukaryotic TATA-binding protein and σ^{70} initially led to the idea that there might be some structural (and functional) similarities between these proteins (Horikoshi *et al.*, 1989), but this idea was refuted when the three-dimensional structures of both proteins were compared (Nikolov *et al.*, 1992; Malhotra *et al.*, 1996). Similar to the bacterial α RNAP subunits, we have to regard the σ factors as highly specialized bacterial adaptations that do not have readily recognizable homologs in the other evolutionary domains.

2.2. Simple Beginnings: The Archaeal TATA-Binding Protein (TBP)-TFB-RNAP Initiation Complex

From the fact that bacterial RNAPs contain a very different organization of small subunits in comparison to the archaeal/eukaryotic enzymes, it perhaps does not come as a big surprise that the σ factors are also structurally very different from the basal factors found in the other evolutionary domains. We have previously seen that the archaea contain an RNAP that is very closely related to the eukaryotic RNAPs. Research during the last few years has shown that this similarity between the archael and eukaryotic transcriptional machineries also encompasses, at least to a certain extent, the basal factors. Archaeal RNAPs require two additional basal factors for accurately initiating transcription on most promoters under *in vitro* conditions, and both of these factors are also present and highly conserved in all eukaryotes. Before we will analyze the requirement for basal factors on archaeal promoters in more detail, we will briefly review some of the general technical approaches that have been used to identify the basal factors from numerous archaeal/eukaryotic species, using the archaeal transcription system as a simple example to illustrate the underlying principles.

Basic Philosophy of 'Fractionated' Transcription Systems for Identification and Purification of Basal Transcription Factors

Archeal transcription has been studied in detail using *in vitro* transcription systems derived from several different thermophilic species, including *Pyrococcus furiosus* (Hethke *et al.*, 1996) and *Sulfolobus shibatae* (Qureshi *et al.*, 1997). Generally, the high temperatures at which the cells grow best is also reflected in the optimal temperatures at which archaeal *in vitro* transcription extracts work best (60 to 85°C; Figure 2.6), with little or no activity at 'normal' temperatures at which transcription extracts from the vast majority of bacterial and eukaryotic systems perform well (usually around 20–30°C). The establishment of *in vitro* transcription systems, that carry out accurately initiated

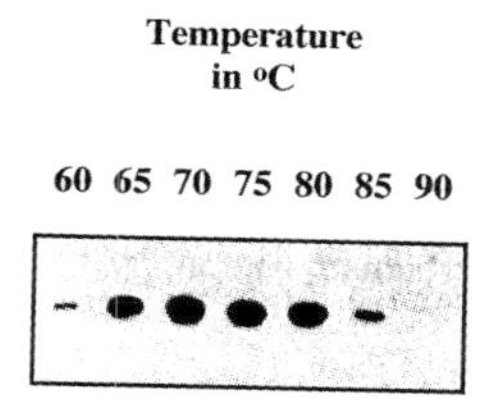

Figure 2.6. Temperature Dependance of Archaeal *In Vitro* Transcription Reactions.
Cell-free transcription reactions derived from hyperthermophilic archaea work optimally at high temperatures (70–80°C). Comparable *in vitro* systems derived from other model organisms (such as yeast, *Drosophila* and humans) would rapidly loose activity at temperatures above 45–50°C. Data from Hethke *et al.* (1996).

transcription of promoter-containing DNA templates, is a crucial precondition for any attempts to identify and biochemically characterize the participating proteins. Crude cellular (or eukaryotic nuclear) extracts usually contain thousands of different types of proteins in varying degrees of abundance, most of which are not involved in transcription. To identify the transcription factors that actively participate in the transcription process, it becomes necessary to purify them with various biochemical techniques, that usually include various types of protein chromatography and differential precipiations steps (Figure 2.7). Since it is very likely (and desirable for purification purposes) that the individual transcription factors required for the *in vitro* transcriptional activity end up in different fractions, it then becomes necessary to test the various fractions obtained either individually or in various combinations to identify the ones containing functionally active components. Further continuation of the fractionation procedures allows the essential proteins to be highly enriched, or even to be purified to homogeneity. The availability of all of the proteins required for reconstituting transcription accurately *in vitro* in purified form allows their functional contributions to be analyzed in detail on many different promoters, and in the presence/absence of gene-specific transcription factors (Chapter 3). Furthermore, once various basal factors are highly purified, it becomes possible to obtain (partial) amino acid sequence data that allows the genes encoding them to be cloned and molecularly characterized.

Ultimately the availability of such clones provides an opportunity to produce the encoded basal factors as recombinant proteins in various bacterial and eukaryotic expression system for reconstituting complete *in vitro* transcription systems and for structural studies. Many of the components of the transcriptional machineries from different organisms are nowadays already available in

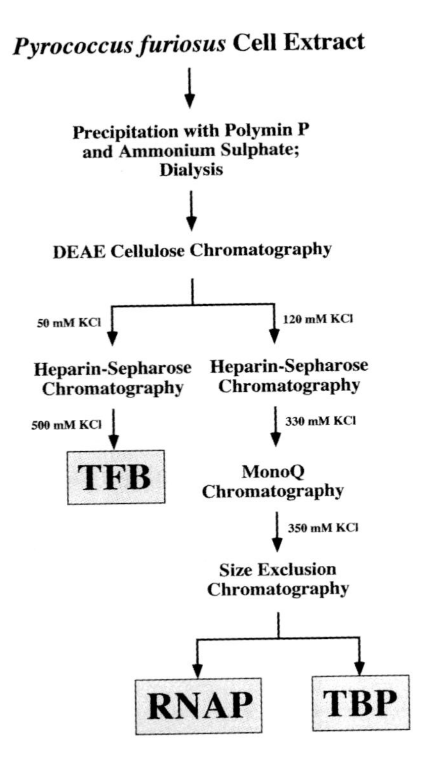

Figure 2.7. Biochemical Fractionation of a Cell-Free Archaeal *In Vitro* Transcription System.
The three active protein components (TBP, TFB, and RNAP; shown as grey boxes) are partially purified from a crude cellular extract through a series of precipiations and chromatographic steps. This example represents a relatively simple purification scheme. In the more complex eukaryotic systems purification of each basal factor and RNAPs through a dozen chromatographic steps is not unusual.
After Hethke *et al.,* 1996.

recombinant form, and it is likely that a complete system, reconstituted entirely from recombinant proteins, will become available in the forseeable future.

Archaeal Promoters Require Two Basal Factors, TBP and TFB, in Addition to RNA Polymerase for Accurate Transcription *In Vitro*

After this brief excursion into some of the technical details underlying much of the experimental work with basal factors, we are now returning to look at the biological insights that have been gained from studying various archaeal transcription systems *in vitro* and *in vivo*. Detailed mutagenesis experiments of several archaeal promoters revealed TATA-like motifs present near the transcription initiation site, that were shown to be essential for start site selection and for specifying the overall efficiency of transcription (Reiter *et al.*, 1988; Reiter *et al.*, 1990; Hausner *et al.*, 1991). Such TATA-motifs are a general feature of many eukaryotic promoters and provided the first hint about the underlying similarities between the archaeal and eukaryotic transcriptional machineries. Nevertheless, because the eukaryotic basal factor (TATA-binding protein, TBP), that binds to the eukaryotic TATA sequence, was only molecularly characterized during the late 1980s and early 1990s, the significance of this similarity between archaeal and eukaryotic promoter motifs was not fully appreciated until the (somewhat unexpected!) discovery of TBP homologs in archaea (Marsh *et al.* 1994; Rowlands *et al.*, 1994). With hindsight, we now realize that TBP is one of the most highly conserved basal transcription factors present in all archaeal and eukaryotic species (but not in bacteria), where it plays many fundamental roles (Figure 2.8; see e.g. Hernandez, 1993).

From the biochemical fraction scheme of an archaeal extract illustrated in the previous section we can see that another factor, TFB, is required in addition to TBP and RNAP to reconstitute transcription from archaeal promoters *in vitro* (Figure 2.7; Hethke *et al.*, 1996; Qureshi *et al.,* 1997; Qureshi and Jackson, 1998). The binding of TFB to the TBP/TATA-box binary complex stabilizes its overall conformation greatly and enables the specific recruitment of archaeal RNAP to the the promoter (Rowlands *et al.*, 1994). TFB is another highly

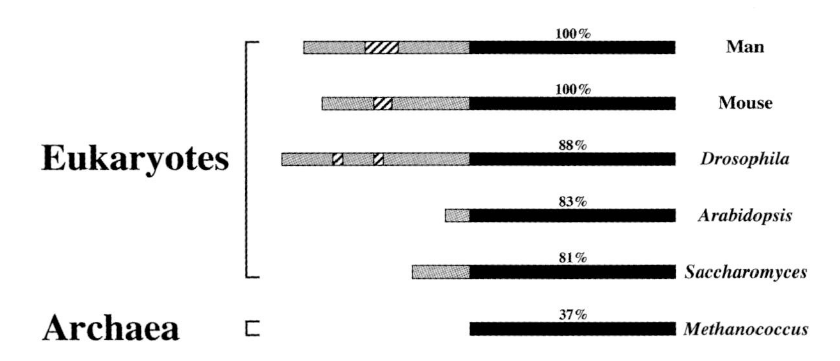

Figure 2.8. Conservation of TBP Structure During Evolution.
The highly conserved C-terminal domain is shown in dark grey and the more divergent (or substantially reduced) N-terminal domains are shown in light grey. The degree of sequence conservation of TBP from different species is indicated as a % relative to the human sequence. The position and extent of glutamine-rich regions in the N-terminal sequences of the human, mouse and *Drosophila* TBPs are shown as hatched boxes.

conserved basal factor that is structurally very similar to the important eukaryotic $RNAP_{II}$ basal transcription factor TFIIB (see below; Ouzonis and Sander, 1992). We will see at a later stage that these observations in the archeal system are highly relevant for understanding the fundamental role of their eukaryotic homologs, TBP and TFIIB, in the eukaryotic $RNAP_{II}$ pre-initiation complex.

Significance of the Archaeal System for Understanding Other Transcription Systems

The high degree of structural similarity between the archeal and eukaryotic RNAPs and the two basal factors, TBP and TFB/TFIIB, points to the fact that these proteins have arisen at a very early stage of evolution of life on earth and have subsequently remained essentially invariant in the archaeal/eukaryotic lineages over billions of years. From an archaeal point of view, many of the additional eukaryotic basal factors used by the three eukaryotic RNAP transcription systems ($RNAP_{I}$, $RNAP_{II}$ and $RNAP_{III}$), that we will encounter later in this chapter and in other parts of the book (Chapter 6), are 'merely' additions and extensive elaborations of this fundamental theme. The combination of two basal factors (TBP-TFB/TFIIB) and RNAP constitutes the ultimate functional core of the archaeal/eukaryotic machineries, that combines the ability to recognize promoter motifs in a sequence-specific manner (mostly via TBP) with the catalytic activity necessary to initate a specific transcript (RNAP). Such an archaeal/eukaryotic minimal complex, consisting of at least 12–15 different polypeptides, is functionally (but by no means structurally!) comparable to the much less complex bacterial holoenzyme, consisting of only four different polypeptides.

2.3. Eukaryotic $RNAP_{II}$ Promoters: TATA-Boxes, Initiator Elements and TBP/TBP-Associated Factors Responsible for Sequence-Specific Recognition

The three different types of eukaryotic nuclear RNAPs use distinct sets of basal factors to assemble specific transcription preinitiation complexes (PICs) on the relevant promoters (Figure 2.9). In the remaining part of this chapter

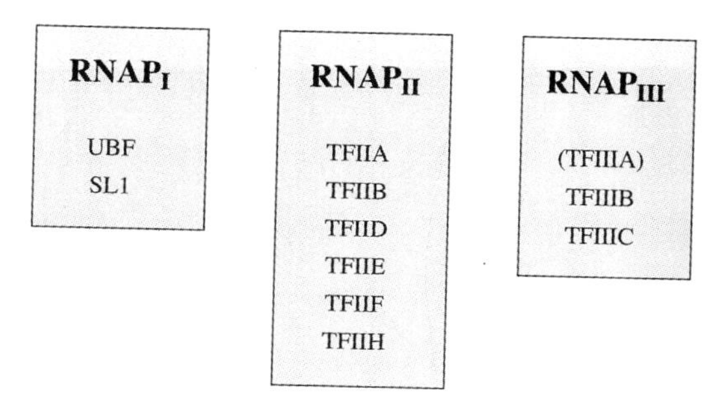

Figure 2.9. Basal Factors Associated with the Three Eukaryotic RNA Polymerase Systems. All three nuclear eukaryotic RNA polymerases are associated with a number of basal factors that assist them in finding their correct transcript initiation sites and facilitate the communication of the preinitiation complexes with various regulatory gene-specific transcription factors.

we will focus on the eukaryotic $RNAP_{II}$ transcriptional machinery to illustrate the structure and function of many of the basal factors involved in this system. A detailed knowledge of the functional properties of the $RNAP_{II}$ transcriptional machinery is essential for understanding many molecular aspects of differential gene expression during embryonic development and cell differentiation. For this reason, and because of its high degree of complexity, the $RNAP_{II}$ transcriptional machinery has been the focus of the intense research efforts of many scientists working in the field of eukaryotic gene regulation. We will return at a later stage to the basal factors that assist $RNAP_I$ and $RNAP_{III}$ in their tasks (Chapter 6), so that we can compare the many similarities and differences shared by all three eukaryotic nuclear transcription systems.

Overview: The Basal Factors of the $RNAP_{II}$ Transcriptional Machinery Assemble in a Distinct Order On Promoters

Most $RNAP_{II}$ promoters require at least 5 distinct biochemically-defined basal factors in addition to $RNAP_{II}$ to achieve a low (basal) level of specifically-initiated mRNA. An *in vitro* analysis of the assembly pathway of the basal factors TFIIB, TFIID, TFIIE, TFIIF and $RNAP_{II}$ into a 'Pre-Initiation Complex' ('PIC') has revealed a distinct order of recruitment (Figures 2.10 and 2.11; Van Dyke *et al.*, 1988; Buratowski *et al.*, 1989). The

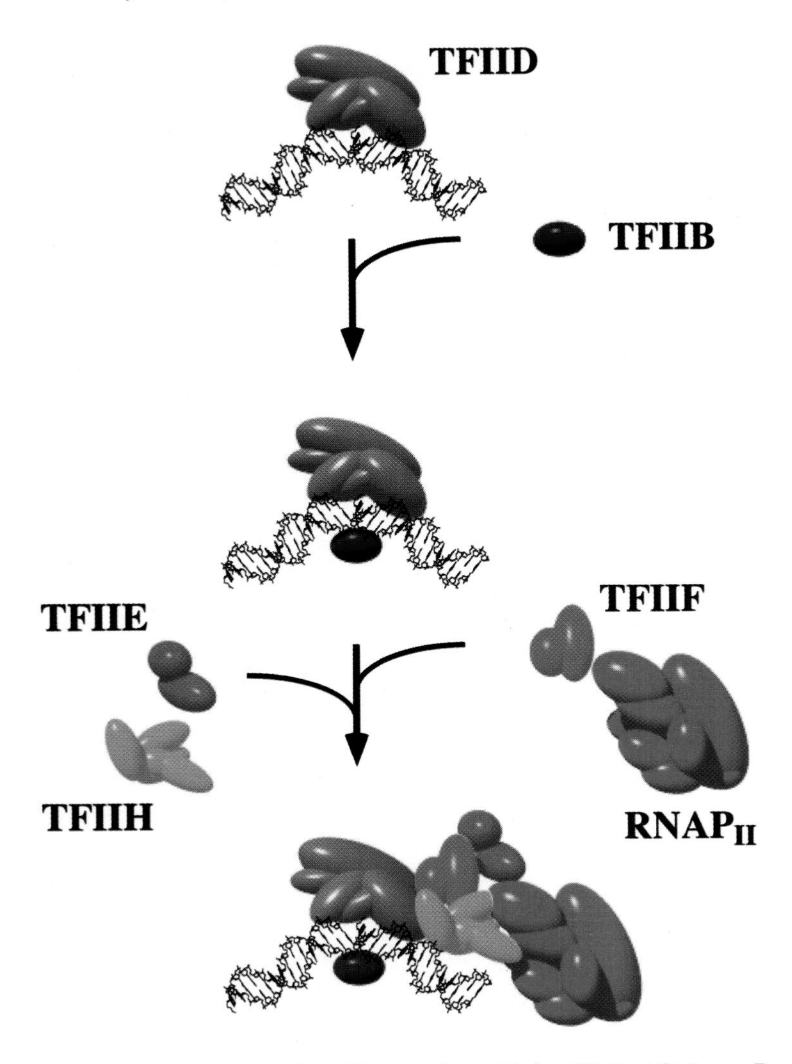

Figure 2.10. Eukaryotic RNAP$_{II}$ Basal Factors Assemble in a Distinct Order on Promoter Elements (Stepwise Assembly Model).
The sequence-specific binding of TFIID to the TATA-box and other promoter elements intitates the assembly sequence. TFIIB stabilizes this TFIID interaction through binding on the opposite site of the kink introduced into the DNA by the TFIID component TBP. This pre-assembled TFIID-TFIIB complex recruits RNAP$_{II}$/TFIIF and the release factors TFIIE and TFIIH. See text for more details.

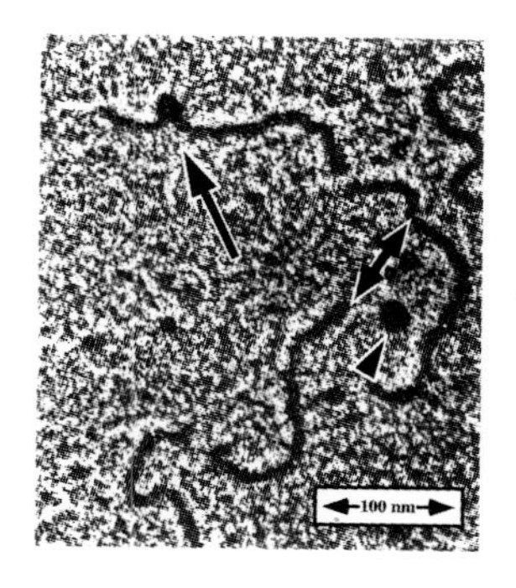

Figure 2.11. Electron Micrograph of the Eukaryotic Pre-Initaion Complex Assembled on Promoter-Containing DNA Molecules.

A pre-initiation complex consisting of purified TBP, TFIIB, TFIIE, TFIIF and $RNAP_{II}$ was assembled on the adenovirus major late promoter. The large arrow indicates a DNA-bound pre-initiation complex, the arrowhead points at an unbound complex, and the double-headed arrow shows protein-free DNA molecules. The relative position of the DNA-bound PIC complex along the DNA molecule is consistent with the location of the TATA-box (160 nucleotides from one end of the 630 nucleotide long promoter fragment). From Forget *et al.*, 1997.

only currently known exception is the human immunoglobulin heavy chain (IgH) promoter, which solely requires TBP, TFIIB and $RNAP_{II}$ if the template DNA is supercoiled, but a full complement of basal factors if the template DNA is linearized (Parvin and Sharp, 1993). This is not a general phenomenon, because many other frequently-used model promoters, such as the adenovirus major late promoter (AdMLP), can not function without a full set of basal factors, regardless of the superhelicity status of the the DNA. If only initiation of the RNA transcript is measured with specialized assays, TFIIE and TFIIH are dispensable because they are not involved in transcript initiation, but in promoter clearance (i.e. the separation of RNA polymerase from the initiation complex during transcript elongation; e.g. Goodrich and Tjian, 1994).

The TFIID multiprotein complex initiates the process through one or several specific DNA-binding events, that essentially determine the precise position of PIC assembly along a piece of template DNA. Binding of TFIID to DNA induces a distinct kink, that is energetically stabilized through the recruitment of another basal factor, TFIIB, to the TFIID-DNA complex. The TFIID-TFIIB 'platform' then allows $RNAP_{II}$ and TFIIF to enter the 'growing' PIC, followed finally by the specific recruitment of TFIIE and TFIIH. Once the PIC is assembled in its final form, specific 'escape' of $RNAP_{II}$ from the PIC occurs at a low, 'basal' rate. This rate can be either substantially stimulated by gene-specific activators, or reduced in the presence of repressors (see Chapter 3).

From such studies it has become clear that approximately 20–25 different individual polypeptides (most of them organized into multiprotein complexes) are required for basal transcription on most promoters. This is, however, only the minimal configuration and does not the reveal the complete picture. Taking into account the actual number of proteins present in a PIC under *in vivo* conditions, we end up with an estimate containing at least twice that figure and possibly substantially more, considering recent reports that the eukaryotic $RNAP_{II}$ transcriptional machinery may be closely associated with DNA-repair and chromatin remodelling activities (the '$RNAP_{II}$ holoenzyme'; see below).

The large number of proteins involved in the PIC enables numerous molecular interaction to take place that will exert a substantial influence on the regulation of gene expression. It is currently thought that specific allosteric changes, and activation of inherent enzymatic activities present in some basal factors, are the major biochemical mechanisms used by the PIC to process the transcriptional activation and repression signals emanating from promoter-specific transcription factors. From this point of view the PIC constitutes a complex analog computing device and future studies will be required to investigate its information-processing properties in more detail. At the current stage most researchers are still occupied with identifying the components present and assigning specific functional roles to them.

Because of the central role of basal factors in the regulation of $RNAP_{II}$-transcribed genes, much more is known about the the process than summarized above. Below we will review some of the most important structural and functional studies that have helped us to understand the various components of the $RNAP_{II}$ basal machinery in detail.

Assembly of a TBP-Containing Complex (TFIID) on Core Promoter Elements

Similar to archaeal genes, many eukaryotic promoters contain a TATA-box located close to the transcription start site. This motif is usually present approximately 25–30 nucleotides upstream of the transcript initiation site in higher eukaryotes and at a more variable distance (–30 to –120) in yeast (Figure 2.12; Corden *et al.*, 1980; Benoist and Chambon, 1981; Struhl, 1987). The TATA motifs present close to the initiation sites of both pro- and eukaryotic promoters have probably been selected during evolution for facilitating the

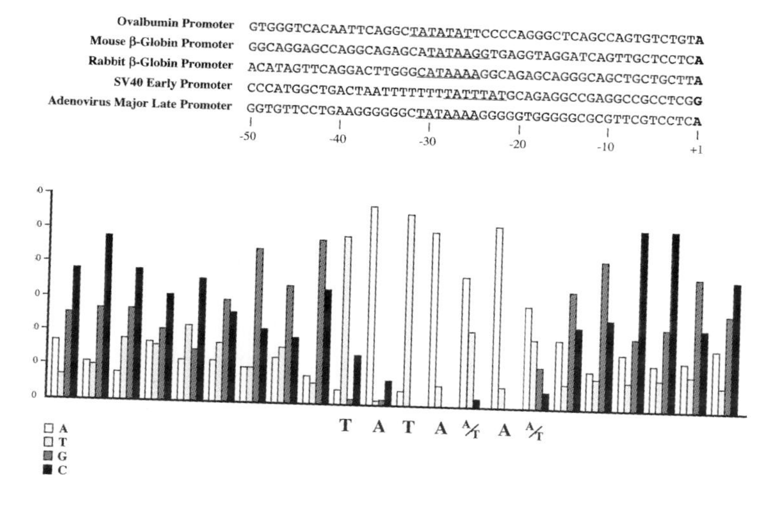

Figure 2.12. The Eukaryotic TATA-Box.
Top: Examples of TATA-boxes located upstream of various cellular and viral promoters. The TATA-motif is underlined and is usually located in the vicinity of the –30 region. The first nucleotide (position +1) of the transcribed region is shown at the right end side of each promoter in bold.
Bottom: Nucleotide composition of TATA-boxes (aligned around the almost invariant second 'A' residue of the TATA-box consensus) and their surrounding sequences from 60 different eukaryotic genes. The number each nucleotide occurs in a particular position is shown as differently shaded bars. Note the high degree of sequence conservation within the TATA-box sequence and the high average G/C content of the surrounding upstream promoter sequences. Data from Mathis and Chambon, 1981.

unwinding of the double-stranded promoter DNA, which is of course an essential precondition for initiating RNA synthesis on the single-stranded template strand. Of the 16 possible dinucleotide combinations T-A steps in DNA are thermally the least stable due to weak stacking interactions (Gotoh and Tagashira, 1981) and sequences containing several consecutive TA motifs 'melt' easily in supercoiled DNA at room temperature (Drew *et al.*, 1985).

The TATA-box represents one of the major eukaryotic promoter 'core' elements used to position the rest of the PIC accuarately over the transcription initiation site and is sequence-specifically bound by the TATA-binding protein

(TBP). Not all genes transcribed by the $RNAP_{II}$ system contain TATA-boxes. Many of the 'house-keeping' genes (i.e. those that encode commonly used metabolic enzymes and other cellular components) do not have any recognizable TATA-sequences. Promoters of this type are therefore usually referred to as 'TATA-less'. Transcription from TATA-less promoters is still dependent on the presence of TBP (Pugh and Tjian, 1991), and it has been shown that on such promoters TBP contacts the DNA in the –30 region, regardless of the sequence present at that position. It is thought that in TATA-less promoters TBP is not recruited through its sequence-specific DNA-binding ability, but is brought into the pre-initiation complex via other mechanisms (e.g. via 'initiator' elements, see below) and subsequently interacts with DNA by relatively 'non-specific' DNA contacts (Zenzie-Gregory *et al.*, 1992; 1993; Martinez *et al.*, 1995).

These findings, together with the results of many other studies, indicate that TBP plays a fundamental role in the transcription of all $RNAP_{II}$ promoters (and also all $RNAP_{I/III}$ promoters; see Chapter 6). The reason why TBP is still required, even for transcription of TATA-less promoters, is that (unlike in archaea) eukaryotic TBP does not function as a single polypeptide, but is part of the multiprotein TFIID complex. TFIID contains TBP and, depending on the species, an additional 10–12 subunits, the 'TBP-Associated Factors' ('TAFs'). Although many of the TAFs are probably involved in conveying regulatory signals from gene-specific transcription factors to the $RNAP_{II}$ basal transcriptional machinery (Chapter 4), several of them are involved in the specific recognition of additional promoter elements (Figure 2.13; Kaufmann and Smale, 1994; Purnell *et al.,* 1994; Sypes and Gilmour, 1994). The 'Initiator' motif (often abbreviated as 'Inr') is a pyrimidine-rich element located around the transcription start site whose existence was not adequately recognized for a long time, because of its rather low degree of sequence conservation (reviewed in Kaufmann *et al.*, 1996; Verrijzer and Tjian, 1996). One of the TAFs isolated from the *Drosophila* TFIID complex, $TAF_{II}150$ (the subscript indicates that it is a component of the $RNAP_{II}$ machinery) is directly implicated in recognition of Inr sequences (Verrijzer *et al.*, 1994; 1995). There are also other, less-defined sequence motifs, such as the 'Downstream Promoter Element' ('DPE'; Burke and Kadonaga, 1996) that are actually located within the transcribed region

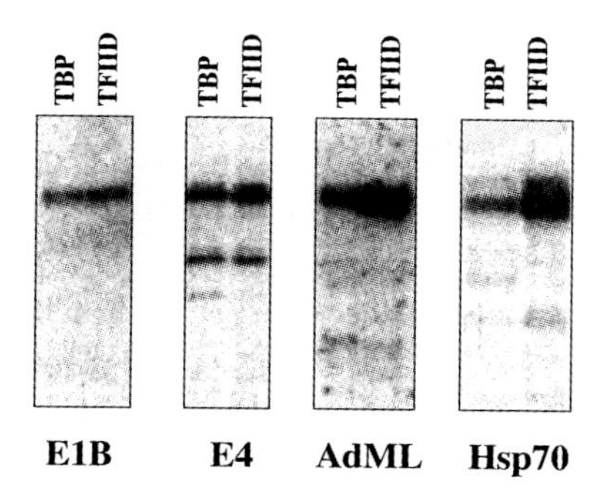

Figure 2.13. TBP and TFIID Direct Different Levels of Basal Transcription. Four different promoters were tested in a reconstituted *in vitro* transcription system for their ability to support basal levels of transcription in the presence of either TBP or TFIID (TBP+TAFs). Note that the basal levels of transcription from the adenovirus major late promoter (AdMLP) and the heatshock promoter (hsp70) are higher in the presence of TFIID than with TBP, indicating specific recognition of additional core elements by TFIID components. From Verrijzer *et al.*, 1995.

and also make a significant contribution to the overall promoter 'strength,' presumably through direct contacts with other TAFs, basal factors or possibly $RNAP_{II}$ itself (Figure 2.12).

Taken together, the various combinations of TATA-boxes, Inr motifs and DPEs are also known as 'core promoter elements' because they constitute the essential sequence configuration necessary for the $RNAP_{II}$ system to yield a low, but specific level of transcription under *in vitro* conditions, even in absence of gene-specific transcription factors (Roeder, 1996).

Structural Aspects of TBP-TATA Complex

The universal role of TBP in transcription processes has stimulated great interest in the structural aspects of TBP, either in isolation (Nikolov *et al.*, 1992; Chasman *et al.*, 1993; Nikolov and Burley, 1994), or bound to DNA (Kim *et al.*, 1993a, b; Kim and Burley, 1994; Nikolov *et al.*, 1996). The evolutionarily conserved C-terminal domain TBP bears a striking resemblance to a saddle-like structure with the DNA-binding surface facing the concave, lower part of the saddle (Figure 2.14). The TATA-motif is recognized via the

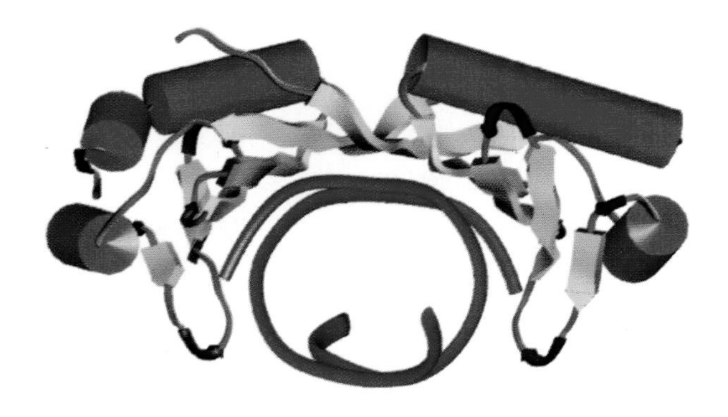

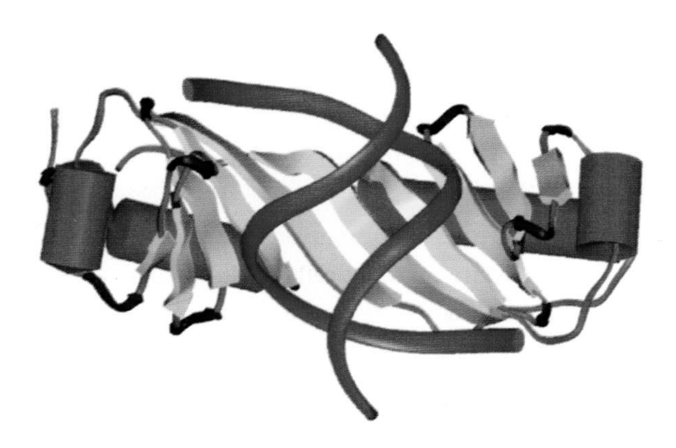

Figure 2.14. High-Resolution Structure of a TBP-DNA Complex.
The highly-conserved C-terminal domain of TBP forms a saddle-like structure that contacts the surface of DNA through an extensive β-sheets surface (shown in yellow). The top part of TBP is mainly α-helical (shown in red) and contains several distinct interaction surfaces for various TBP-associated factors. The TATA box-containing DNA molecule (purple) is severely distorted from the normal B-form of DNA through the interaction with TBP. Note that this structure does not show the extensive N-terminal domains (of essentially unknown functions) present in many TBPs from higher eukaryotic species. PDB access No. 1CDW.

minor groove (see Chapter 3 for more information about DNA-protein interactions via minor and major grooves; Lee *et al.*, 1991) and causes two distinct and severe bends in the DNA double helix resulting in an overall kink of approximately 95° (Kim *et al.*, 1993a, b) These bends are specifically induced through the insertion of two phenylalanine residues between adjacent nucleotides near the beginning and end of the TATA motif. Nucleosomes use a very similar strategy to bend DNA sharply around them in order to package it efficiently into chromatin structures (Chapter 7). The energetic cost of such a severe distortion of the double-helical axis almost certainly accounts for the rather weak and not particularly specific binding of TBP to TATA-elements under *in vitro* conditions. Furthermore, results revealing the presence of a rate-limiting isomerization step during the early stages of the formation of a TBP-TATA complexes are also consistent with the structural data obtained from the crystallographic studies. Once TBP overcomes this initial kinetic 'bottle-neck', the resulting TBP-TATA complexes are remarkably stable (Hoopes *et al.*, 1992). The functional significance of the kink that TBP introduces into DNA is currently not at all understood, but may indicate that TBP plays an important role in modifying the overall 'architecture' of promoters (see below).

TFIID/TATA Complex is Stabilized Through Interactions with TFIIA and TFIIB

The stability of specific DNA-TBP complexes is substantially increased in the presence of either TFIIA and/or TFIIB (this effect can be seen most clearly in footprinting and gelshift assays; Figures 2.15 and 2.16). The binding of TFIIA involves direct contact with TBP and promoter DNA, but TFIIA is not essential for accurately reconstituting transcription from most promoters under *in vitro* conditions (Cortes *et al.*, 1992; Lee *et al.*, 1992; Geiger *et al.*, 1996; Oelgeschlager *et al.*, 1996; Tan *et al.*, 1996). In contrast, TFIIB is universally required and, together with TBP and RNAP, is a components of the PIC that is highly conserved between the archaeal and eukaryotic transcriptional machineries (see above). TFIIB makes specific DNA contacts both upstream and downstream of the TATA-element (when present) and interacts with a defined portion of TBP through protein interactions (Figure 2.17; Lee and Hahn, 1995; Nikolov *et al.*, 1995). Specific contacts between the archaeal homolog

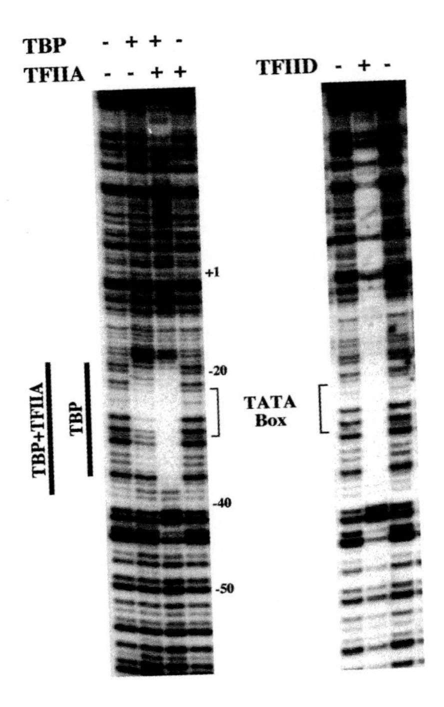

Figure 2.15. The Binding Specificity of TBP is Enhanced by TFIID Components and TFIIA .

Left Gel Image: In this experiment the adenovirus major late promoter (AdMLP) was incubated with DNAase I either in absence of any transcription factors, or with various combinations of TBP and TFIIA. In the presence of TBP a weak, but distinct footprint over the TATA-box is discernible when compared with unprotected DNA. In the presence of TBP and TFIIA this footprint does not extend significantly in size, but stabilization of the TBP/DNA complex by TFIIA is evident through the enhanced protection and sharpening of the footprint borders. TFIIA on its own does not bind to DNA to produce a detectable footprint.

Right Gel Image: In the presence of TFIID (TBP+TAF_{II}s) an extensive footprint is detected on the AdMLP promoter extending significanly into the transcribed region (beyond the +1 position; towards the top of the gel image). Original data generously provided by Dr. Peter Verrijzer.

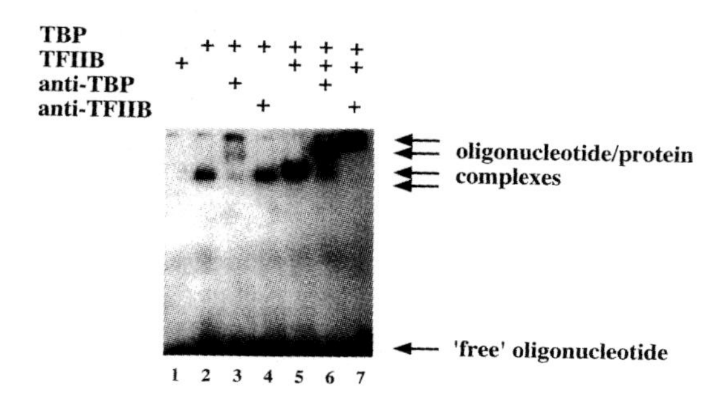

Figure 2.16. Application of EMSA to Study the Binding of TBP and TFIIB to a TATA-Box Oligonucleotide.
A ^{32}P-labelled doublestranded oligonucleotide containing a TATA-motif was incubated with various combinations of TBP, TFIIB and specific antibodies. The unbound oligonucleotide migrates quickly and has been electrophoresed all the way to the bottom of the gel. In presence of TBP a distinct shift can be detected (lane 2), whereas TFIIB does not bind to the oligonucleotide probe in absence of TBP (lanes 1 and 3). The stabilization of TBP binding in presence of TFIIB results in a slightly 'supershifted' and stronger band (compare lanes 2 and 5). The presence of TBP and/or TFIIB in the shifted complexes is confirmed by 'supershifting' them with antibodies specific for these transcription factors (lanes 3, 6 and 7). Photograph kindly provided by Promega UK.

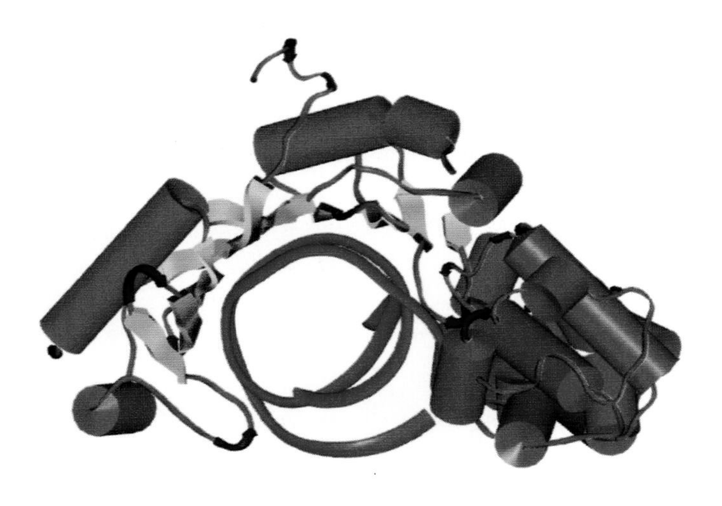

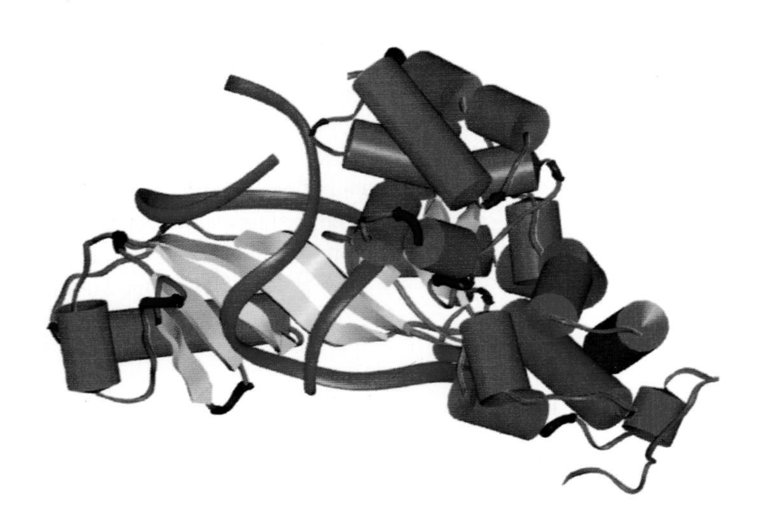

Figure 2.17. High Resolution Structure of a TBP-TFIIB-DNA Ternary Complex. The binding of TFIIB to the underside of the TBP saddle greatly stabilizes the ternary complex through additional protein-protein and protein-DNA contacts. PDB access No. 1VOL.

of TFIIB, TFB, and DNA have a substantial influence on promoter strength (Qureshi and Jackson, 1998) and it is likely that the eukaryotic TFIIBs are also (possibly to a lesser degree) involved in sequence-specific DNA recognition (see also Huet *et al.*, 1997). Interestingly, although the two distinct cyclin-like domains in the C-terminal domain of TFIIB initially suggested that the presence of a matching cdk-like kinase somewhere among the other basal factors, this prediction has not materialized up to now (there are, however, two additional *bona fide* cyclin/cdk pairs present in TFIIH and among the SRB components of the holoenzyme mediator complex; see below).

2.4. Recruitment of $RNAP_{II}$/TFIIF

Specific Contacts between $RNAP_{II}$ and TFIIB

Once recruited to the TFIID-DNA complex, TFIIB has a crucial function in initiating the recruitment of $RNAP_{II}$ to the PIC and positioning the enzyme over the start site at +1; (Buratowski *et al.*, 1989; Pinto *et al.*, 1992; Tschochner *et al.*, 1992; Malik *et al.*, 1993; Li *et al.*, 1994; Pinto *et al.*, 1994). Especially the recent studies of Leuther *et al.* (1996) have shed new light on the possible spatial arrangement of TFIIB relatively to $RNAP_{II}$ in the PIC. Electron microscopic imaging of $RNAP_{II}$/TFIIB complexes lends strong support to the view that direct protein contacts between the N-terminal domain of TFIIB and the large $RNAP_{II}$ subunit (RPB1; Tschochner *et al.*, 1992; Xiao *et al.*, 1994) account for many aspects of start selection by $RNAP_{II}$. These results have made it possible to establish a possible model of the initiation complex and places the catalytically active site of $RNAP_{II}$ approximately 30 basepairs away from the TATA-box. This fits well with the known position of TATA-boxes relative to mRNA start sites and suggests that DNA minimally follows a straight path from the TBP/TFIIB complex across the 'palm' of a groove present in $RNAP_{II}$ (Leuther *et al.*, 1996). In the cases where the TATA-box is further away from the initiation site (such as in yeast; Struhl, 1987), it is possible that $RNAP_{II}$ 'scans' for suitable initiation sites and loops out the intervening DNA sequences. This explanation requires that either $RNAP_{II}$ or some other associated basal factor displays sequence-specific DNA interaction abilities. Although conclusive evidence for such sequence-specifc DNA-protein interactions is currently not yet available, there are preliminary indications that this could be the case.

In a later chapter we will see that the transcription by the two other eukaryotic polymerases, i.e. $RNAP_I$ and $RNAP_{III}$, also use TBP-containing complexes for setting up and spatially orientating their initiation complexes (Chapter 6). Furthermore, the $RNAP_{III}$ system also contains a protein with a high degree of sequence similarity to TFIIB, suggesting that there may be a significant degree of similarity in the assembly and function of these otherwise quite differently structured $RNAP_{II}$/$RNAP_{III}$ PICs.

TFIIF: An Intimate Associate of $RNAP_{II}$

Although TFIIF is definded as a biochemically distinct basal factor, it associates with $RNAP_{II}$ with high affinity. The two human TFIIF subunits, Rap30 and Rap70, were initially cloned precisely because they were found to be tighthly associated with $RNAP_{II}$ (Rap = RNA polymerase-associated protein). $RNAP_{II}$ can potentially interact via direct protein-protein interactions with all (except TFIIA?) basal factors (Flores *et al.*, 1989; Gerard *et al.*, 1991; McCracken and Greenblatt, 1991; Usheva *et al.*, 1992; Ha *et al.*, 1993; Maxon *et al.*, 1994; Leuther *et al.*, 1996). The interaction between $RNAP_{II}$ and TFIIF is, however, likely to be a very significant for regulating recruitment of the polymerase to the PIC, and TFIIF is also crucially involved in controlling $RNAP_{II}$ elongation (Chapter 5) and recycling (Lei *et al.*, 1998). Photo-crosslinking studies revealed that TFIIF has an influence on the overall structure of the PIC that results in the formation of new contacts between $RNAP_{II}$ and promoter DNA sequences (Figure 2.18; Robert *et al.*, 1996; Forget *et al.*, 1997). These findings have interesting, and yet unexplored, implications for the initiation process and may contribute to substantial topological changes of promoter DNAs.

2.5. Promoter Melting, Transcript Initiation and the Role of TFIIE and TFIIH

The two additional basal factors TFIIE and TFIIH are not thought to be directly involved in promoter recognition, but work in conjunction with $RNAP_{II}$

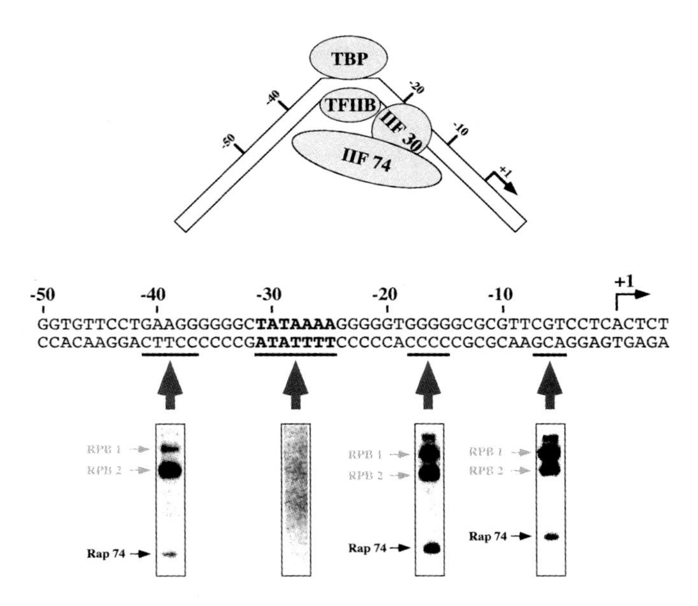

Figure 2.18. The Large Subunit of TFIIF (RAP74) is in Close Contact to Promoter DNA Upstream and Downstream of the TBP/TFIIB Complex.

Cross-linking studies with specifically-placed photoprobes at various positions along the adenovirus major late promoter reveal that the large TFIIF subunit can be crosslinked to positions upstream (around −40) and downstream (−18 → −15 and −7 → −5) of the TBP/TFIIB complex centered around position −28. Note the absence of Rap74 crosslinking to the region immediately around the TATA-element (center and bottom parts of the figure). The schematic diagram on top shows the proposed arrangement of Rap 74 (TFIIF 74) relative to some other basal factors (TBP and TFIIB), the small TFIIF subunit (TFIIF 30) and the promoter DNA. Data from Forget *et al.*, 1997.

to open the double-stranded DNA template in an ATP-dependent manner and subsequently allow $RNAP_{II}$ to 'escape' from the PIC (Wang *et al.*, 1992; Goodrich and Tjian, 1994; Ohkuma and Roeder, 1994; Zawel and Reinberg, 1995). TFIIE and TFIIH contact each other physically and cooperate functionally during this critical step (Li *et al.*, 1994; Maxon *et al.*, 1994). No strong, direct interactions between TFIIE and $RNAP_{II}$ have been detected with conventional biochemical methods (Sayre *et al.*, 1992), but a specific TFIIE-$RNAP_{II}$ complex has been visualized by electron microscopy (Leuther *et al.*, 1997).

Analysis of the energy requirements of the transcript initiation stage revealed a series of energetically and kinetically distinct stages in transcription initiation that can be experimentally detected by high resolution footprinting and abortive initiation assays (Holstege *et al.*, 1996 and 1997; Yan and Gralla, 1997). The first step is formation of an 'open complex' which requires an ATP-dependent DNA helicase activity present in TFIIH (Schaefer *et al.*, 1993 and 1994; Holstege *et al.*, 1996; reviewed in Svejstrup *et al.*, 1996). The promoter region 'melts' around the start site (+1) and most of the single-stranded region includes 'upstream' sequences (positions –9 to +2; i.e. nucleotides that immediately preceed the initiation site at +1). The open complex is intrinsically unstable and has a half-life of only 45 seconds (although this might be promoter-specific: measurements on the adenovirus E4 promoter suggest a longer half-life of approximately 5 minutes; Yan and Gralla, 1997). Transcript initiation has to occur within this time period, but the promoter can be sustained in the open complex configuration for a longer time by further ATP-driven TFIIH-helicase activity. Formation of the first RNA dinucleotide stabilizes the open complex by doubling its half-life to 90 seconds. Only after the synthesis of the first three nucleotides of mRNA (step 2) does DNA helicase activity become expendable for maintaining DNA strand separation. Afterwards the transcription bubble enlarges continuously until the growing transcript reaches a size of 11 nucleotides. At this stage $RNAP_{II}$ has the option to switch from abortive to productive RNA synthesis, and once this occurs further transcription continues with a high degree of processivity (step 3). This step is usually referred to as 'promoter clearance' or 'promoter escape' because $RNAP_{II}$ physically detaches at this point from the transcription initiation complex (Dvir *et al.*, 1996). The precise mechanism of promoter clearance is probably different for different promoter types and may constitute a major control point for the regulation of gene expression (Chapter 5; Bentley, 1995). From studies of the elongation properties of $RNAP_{II}$ it has become clear that phosphorylation of the C-terminal domain (CTD) of the largest $RNAP_{II}$ subunit through a TFIIH kinase activity greatly influences the ability of $RNAP_{II}$ to read through 'difficult' DNA sequences. Such a CTD phosphorylation event is required for promoter clearance from the AdE4 and DHFR promoters (Jiang *et al.*, 1996; Akoulitchev *et al.*, 1995), but not for AdMLP in a highly purified *in vitro* transcription system (Holstege *et al.*, 1997).

In summary, the experimental data obtained in various studies indicates that at least one ATP molecule needs to be hydrolyzed to open the promoter for the initiation of mRNA synthesis and, on some promoters, more ATP molecules are utilized to phosphorylate the CTD domain of $RNAP_{II}$. This is in stark contrast to bacterial systems where the σ^{70}-containing holoenzyme is capable of maintaining an open complex in absence of ATP hydrolysis merely through preferential binding of σ^{70} to the non-template DNA strand (Siebenlist *et al.*, 1980). This slow, rate-limiting step (compare to the TBP/DNA interaction described above!) results in a RNAP/promoter open complex that is essentially stable in the absence of nucleotide triphosphates. Interestingly, neither of the two other eukaryotic nuclear RNA polymerases, $RNAP_{I}$ and $RNAP_{III}$, require ATP-hydrolysis for RNAP recruitment to their respective initiation complexes, which strongly emphasizes the distinctiveness of the $RNAP_{II}$ system.

2.6. Re-initiation and Recycling of Basal Factors

The description of the PIC assembly process presented above illustrates the large number of molecules involved in setting up the PIC for the initiation of the synthesis of a single mRNA molecule. $RNAP_{II}$ (possibly as a complex with TFIIF) is the only PIC component that actually leaves the complex to transcribe the coding regions of genes, and for a long time the fate of the remaining basal factors was unclear. Do they all dissociate and start the whole assembly sequence all over again from scratch (e.g. Kadonaga, 1990)? Or does a stable complex remain behind that recruits new $RNAP_{II}$? Many important questions relating to this issue remain currently unanswered or are still controversial, but the experimental work by Zawel *et al.* (1995) has shed some light on these questions in an *in vitro* transcription system. Their studies suggest that many of the basal factors remain briefly associated with elongating $RNAP_{II}$ for varying distances into the transcribed region. TFIID remains bound to the promoter, TFIIB dissociates early, TFIIE between nucleotides 1–10, TFIIF between nucleotides 10–68 and TFIIH between 30 and 68. These observations suggest that TFIID, once bound to a promoter, can sustain multiple rounds of transcription initiation events. Further below we will encounter a different theory concerning the assembly of the PIC that postulates that many of the eukaryotic basal factors and $RNAP_{II}$ are present as a pre-assembled 'holoenzyme' (similar to the bacterial σ-containing holoenzyme). Interestingly, TFIID is not considered to be part of the holoenzyme, which fits well with the idea of a stable TFIID complex nucleating the recruitment of PICs over and over again on the same promoter without being displaced.

2.7. Basal Factors Reorganize Promoter Topology

Binding of TFIID to DNA May Cause Substantial Reorganization of Promoter Topology

We have seen earlier that the binding of TBP to DNA introduces a kink of 95° into the promoter DNA which may facilitate the formation of a more compact PIC. In addition to this deformation of DNA, additional evidence has emerged during the last few years suggesting that this may only be the tip of the iceberg. The binding of the TFIID complex to covalently-closed circular DNA templates introduces a distinct change into its superhelical density, implying that DNA may be wrapped around the TFIID complex (Figure 2.19; Oelgeschlager *et al.*, 1996; Hoffmann *et al.*, 1997). This radical idea is supported by several independent observations, including studies of the DNAase protection pattern conferred by TFIID on AdMLP (Zhou *et al.*, 1992) and photo-crosslinking of various TFIID components to distinct promoter sections (Oelgeschlager *et al.*, 1996). Crystallographic studies of two TFIID components (dTAF40 and dTAF60; Hoffmann *et al.*, 1996) revealed a histone-octamer type organization (see Chapter 7), which was originally used to support the TFIID wrapping model. More recent high resolution structural studies of nucleosome cores have shown, however, that the specific residues present in histones for contacting DNA are absent in the TAFs (Luger *et al.*, 1997). It is therefore unlikely that a comparable nucleosome-like packaging mechanism accounts for the contact points between TFIID and DNA.

TFIIF as an Additional Wrapping Factor

Photo-crosslinking studies of a partial PIC consisting of TFIID-TFIIB-$RNAP_{II}$-TFIIF-TFIIE have revealed that the large subunit of TFIIF, Rap74, is specifically crosslinked to a large portion of the AdML promoter (from positions

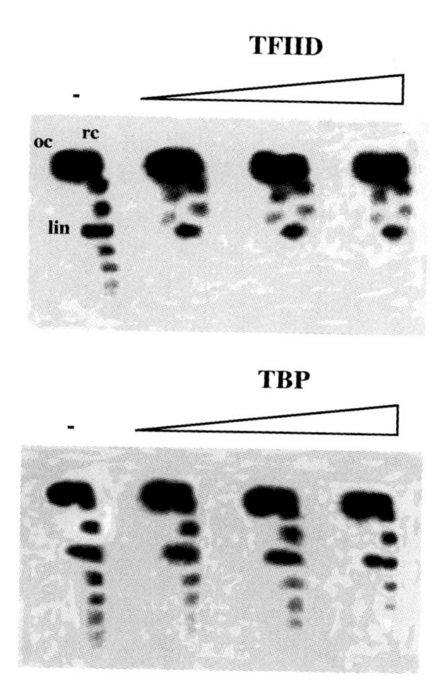

Figure 2.19. Binding of TFIID, but not TBP, to Circular DNA Causes Changes in Its Superhelicity State.

Changes in the superhelicity state of a circular plasmid containing a TATA-box are detected by two-dimensionl agarose gel electrophoresis. The position of linearized DNA (lin), open circular (oc) and relaxed circular (rc) DNA are indicated in the top left diagram. Addition of increasing amount of TFIID causes a substantial change, even at the lowest concentration used, in the superhelicity state of the plasmid compatible with the concept of TFIID inducing DNA to wrap around it. In contrast, the addition of increasing amounts of TBP to the same plasmid preparations causes only minor detectable topological changes. From Oelgeschlager *et al.*, 1996.

–61 to +26 [ie. a distance of almost 90 nucleotides!] Robert *et al.*, 1996; Forget *et al.*, 1997; see also Lei *et al.*, 1998). Since Rap74 is closely associated with $RNAP_{II}$, the simplest model to explain such multiple contacts with distant parts of the promoter is that RAP74 helps to wrap DNA around $RNAP_{II}$ (and possibly other basal factors). This second postulated instance of DNA wrapping within the PIC raises a number of fascinating questions regarding the path of DNA through the assembled $RNAP_{II}$ transcriptional machinery. Clearly, some of the older models picturing basal factors neatly lined up on top of a linear DNA molecule (such as shown in Figure 2.10) will probably have to be substantially revised in the foreseeable future. At this stage the purpose of the DNA wrapping is unclear, but it seems to offer the advantage of bringing a substantial part of the linear promoter sequence into close physical proximity to each other, thus allowing numerous sequence-specific (and nonspecific) DNA-protein contacts to be made. We will also see in the next chapter that many gene-specific transcription factors, such as Sp1, are also very likely participating in large scale DNA looping events that help to bring distantly bound Sp1 molecules into the vicinity of the PIC so that they can exert a stimulatory function through direct physical contacts with basal factors and associated coactivators. Finally, in Chapters 7 and 8 we will encounter the ultimate topological organization machineries, chromatin and nuclear 'transcription factories', that determine large-scale gene expression events entirely through changing the spatial accessibility of DNA sequences within the nucleus.

2.8. The $RNAP_{II}$ Holoenzyme

During the 1980s and the early 1990s most of the research concerning the mechanisms of transcription initiation by $RNAP_{II}$ focused on identifying various individual components of the basal transcriptional machinery and defining their functional contribution to the transcription initiation process. The isolation of each of the basal facors as a distinct biochemical entity, and the experimental demonstration of a distinct sequence of entry of basal factors into the preinitiation complex, led to the it general assumption that such a stepwise-assembly process also occurs on promoters under *in vivo* conditions. More recent experimental evidence, demonstrating that $RNAP_{II}$ and many basal factors often occur in a pre-assembled 'holoenzyme' configuration, stimulated a careful re-investigation of many previously-held assumptions on the *in vivo* assembly of the $RNAP_{II}$ transcriptional machinery on the promoters of different genes. Genetic studies carried out in 'simple' eukaryotic model systems, such as yeast, provided many of the initial hints of these additional unexpected complexities of the basal $RNAP_{II}$ transcriptional machinery.

Genetic Identification of SRB Proteins in *Saccharomyces cerevisiae*

Some of the key experiments leading to the discovery of the eukaryotic $RNAP_{II}$ holoenzyme are based on a genetic screen carried almost a decade ago (Nonet and Young, 1992; Koleske *et al.*, 1992; Thompson *et al.*, 1993). This group was looking for genes that compensated for a the phenotype caused by a mutation in largest subunit, RPB1, of $RNAP_{II}$. Briefly, RPB1 contains a unique C-terminal domain consisting of a series of near-perfect repeats varying in length in different organisms 26 to 52 copies. Generally, there are fewer repeats in RPB1 from lower eukaryotes than the ones found in higher eukaryotes (Figure 2.20; e.g. Allison *et al.*, 1985; Corden *et al.*, 1985). Also, the C-terminal repeats (CTDs) are absolutely specific for eukaryotic $RNAP_{II}$ and even the archaeal RNAPs, that are in many other respects very similar in structure and function to eukaryotic RNAPs, do not contain this unique sequence motif.

The RPB1 subunit from *S. cerevisiae* $RNAP_{II}$ contains 26 repeats and partial deletions of repeat units within the the CTD cause phenotypes ranging from mild temperature-sensitivity to lethality, in direct proportion to the number of missing heptad repeats (Nonet and Young, 1992; Koleske *et al.*, 1992; Thompson *et al.*, 1993). It was thus possible to design a genetic screen to identify mutations in other genes that would compensate for CTD-defect by allowing the cells to grow at a temperature which would ordinarily kill them. The reasoning behind this approach was that such a strategy would facilitate the discovery of genes encoding proteins that are in close contact with the $RNAP_{II}$-CTD and thus provide a lead for elucidating the functional role of this curious structure. A number of mutations in a relatively large group of genes, the Suppressors of RNA polymerase B (SRB), was indeed identified using this approach (Figure 2.21). Most of the SRBs characterized to date (SRB2, SRB4, SRB5, SRB6, SRB8, SRB9, SRB10 and SRB11) encode proteins that do not contain any detectable homology to other proteins or display recognizable sequence motifs in their primary amino acid sequence. Mutations in SRB genes affect the transcription of many genes (Thompson and Young, 1995), but at least some SRBs are not absolutely essential for cell survival (Thompson *et al.*, 1993). Two of them, SRB10 and SRB11, encode a cyclin/cyclin-dependent kinase pair that may play a (possibly functionally redundant) role in the phosphorylation of the CTD domain (Liao *et al.*, 1995).

	Mouse	*Drosophila*	*Arabidopsis*	*Caenorhabditis*	Yeast	*Plasmodium*
1	YSPTSPA	YSPTSPN	YSPSSPA	LSPRTPS	FGVSSPG	YSPS---
2	YEPRSPGG	YTASSPG	YSPTSPG	YGGMSPGV	FSPTSPT	YSPTSPT
3	YSPTSPS	--GASPN	YSPTSPG	YSPSSPQ	YSPTSPA	YNANNAY
4	YSPTSPS	YSPSSPN	YSPTSPG	FSMTSPH	YSPTSPS	YSPTSPK
5	YSPTSPS	YSPTSPL	YSPTSPT	YSPSSPQ	YSPTSPS	NQNDQMNVNSQ
6	YSPTSPN	YA--SPR	YSPSSPG	YSPTSPS	YSPTSPS	YNVMSPV
7	YSPTSPS	YASTTPN	YSPTSPA	YSPTSPAAG	YSPTSPS	YSVTSPK
8	YSPTSPS	FNPNSTG	YSPTSPS	QSPVSPS	YSPTSPS	YSPTSPK
9	YSPTSPS	YSPSSSG	YSPTSPS	YSPTSPS	YSPTSPS	YSPTSPK
10	YSPTSPS	YSPTSPV	YSPTSPS	YSPTSPS	YSPTSPS	YSVTSPK
11	YSPTSPS	YSPTVQ-	YSPTSPS	YSPTSPS	YSPTSPS	YSVTSPK
12	YSPTSPS	FGSSPS-	YSPTSPS	YSPTSPS	YSPTSPS	YSVTSPK
13	YSPTSPS	FAGSGSNI	YSPTSPS	YSPTSPS	YSPTSPS	YSVTSPK
14	YSPTSPS	YSPGN-A	YSPTSPS	YSPTSPS	YSPTSPS	YSVTSPK
15	YSPTSPS	YSPSSSN	YSPTSPA	YSPTSPS	YSPTSPS	YSPTSPVA
16	YSPTSPS	YSPNSPS	YSPTSPA	YSPTSPS	YSPTSPS	QNIASPN
17	YSPTSPS	YSPTSPS	YSPTSPA	YSPTSPR	YSPTSPS	YSP----
18	YSPTSPS	YSPSSPS	YSPTSPS	YSPTSPT	YSPTSPA	YSITSPK
19	YSPTSPS	YSPTSPC	YSPTSPS	YSPTSPT	YSPTSPS	FSPTSPA
20	YSPTSPS	YSPTSPS	YSPTSPS	YSPTSPT	YSPTSPS	YSISSPV
21	YSPTSPS	YSPTSPN	YSPTSPS	YSPTSPT	YSPTSPS	
22	YSPTSPN	YTPVTPS	YSPTSPS	YSPTSPS	YSPTSPS	(+68aa)
23	YSPTSPN	YSPTSPN	YSPTSPA	YESGGG-	YSPTSPN	
24	YSPTSPS	YS-ASPQ	YSPTSPG	YSPSSPK	YSPTSPS	
25	YSPTSPS	YSPASPA	YSPTSPS	YSPSSPT	YSPTSPG	
26	YSPTSPN	YSQTGVK	YSPTSPS	YSPTSPS	YSPGSPA	
27	YSPTSPN	YSPTSPT	YSPTSPS	YSPTSPQ	YSPKQDEQKHNENENSR	
28	YSPTSPS	YSPPSPSDG	YNPQSAK	YSPTSPQ		
29	YSPTSPS	YSPGSPQ	YSP-SIA	YSPSSPT		
30	YSPTSPS	YTPGSPQ	YSPSNAR	YSPTSPT		
31	YSPSSPR	YSPASPK	LSPASP-	YSPTSPS		
32	YTPQSPT	YSPTSPL	YSPTSPN	YTPSSPQ		
33	YSPSSPS	YSPSSPQ	YSPTSPS	YSPTSPT		
34	YSPSSPS	HSPS-SQ	YSPTSPS	YTPSPSEQPGTSAQVDFFYSKLNF		
35	YSPTSPK	YSPTGST	YSPTSPT			
36	YTPTSPS	YSPTSPR	YSPTSPY			
37	YSPSSPE	YSPNMSI	SSGASPD			
38	YTPASPK	YSPSSTK	YSPSAG-			
39	YSPTSPK	YSPTSPT	YSPTLPG			
40	YSPTSPK	YTPTARN	YTPSSTGQ			
41	YSPTSPT	YSPTSPM	YTPHEGDKKDKTGKKDASKDDKGNP			
42	YSPTTPK	YSPTAPSH				
43	YSPTSPT	YSPTSPA				
44	YSPTSPV	YSPSSPTFEESED				
45	YTPTSPK					
46	YSPTSPT					
47	YSPTSPK					
48	YSPTSPT					
49	YSPTSPKGST					
50	YSPTSPG					
51	YSPTSPT					
52	YSLTSPAISPDDSDEEN					

Figure 2.20. C-Terminal Domains (CTD) of the Largest Eukaryotic $RNAP_{II}$ Subunits Contain An Array of Heptapeptide Repeats.

The CTDs from a number of eukaryotic species across the evolutionary range is shown. More 'advanced' eukaryotic $RNAP_{II}$s (e.g. mouse) contain longer CTDs than the ones from more 'primitive' species, such as yeast and *Plasmodium*. Note that many of the repeats conform to the overall consensus sequence YSPTSPA.

Modified from Young (1991).

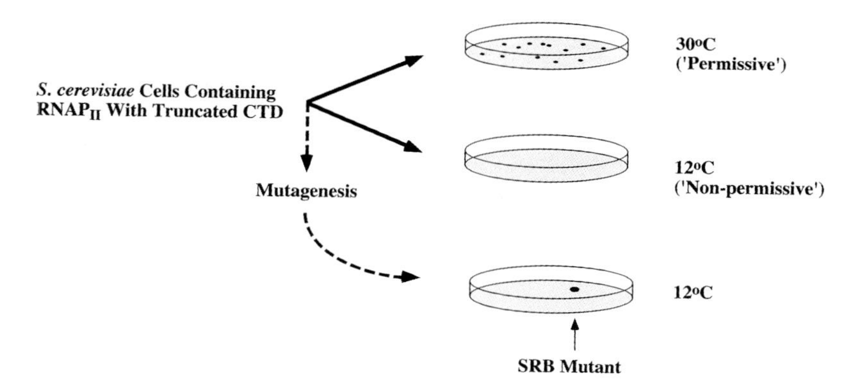

Figure 2.21. Schematic Diagram of the Genetic Screen Leading to the Identification of SRB Proteins.
See text for a description of the experimental strategy.

SRBs are Part of a Multiprotein Complex Containing RNAP$_{II}$ and Basal Factors

After raising antibodies directed against several SRB proteins, it became possible to analyze their biochemical properties further. It was found that all the SRBs co-purified with each other through a series of chromatographic steps, suggesting that they were subunits of a single complex. More surprisingly, purification of the SRB-containing complex to homogeneity showed that it did not only contain an SRB complex, but also stoichiometric amounts of RNAP$_{II}$, the basal factors TFIIB, TFIIF, TFIIH and several additional polypeptides in a stable configuration (Thompson *et al.*, 1993). When supplemented with TBP and TFIIE, this 'holoenzyme' complex was able to support basal, as well as activated levels of transcription *in vitro*. The result proved that the holoenzyme contained all the necessary basal activities for specific initiation of transcripts and that it also contained coactivators allowing it to respond to gene-specific activators (see Chapter 4; Koleske and Young, 1994).

Independent biochemical investigations in different laboratories have shed more light on the polypeptide composition of the yeast RNAP$_{II}$ holoenzyme. Kim *et al.* (1994b) identified a SRB-containing multiprotein complex, the 'mediator' complex, which was found to be stably associated with up to 50% of enogenous $RNAP_{II}$ and certain basal factors in a holoenzyme-like configuration. The presence of mediator in reconstituted *in vitro* transcription reactions stimulated the level of basal transcription and increased the rate phosphorylation of the RPB-1 CTD by enhancing the kinase activity present in TFIIH (see above). Analysis of the protein composition of the mediator complex revealed more than 16 distinct polypeptides, including most of the SRBs and other proteins that had been previously identified in various genetic screens as general transcriptional regulators, such as GAL11, SIN4, RGR1, ROX3 and several newly-identified 'MED' proteins (Figure 2.22; Li *et al.*, 1995; Gustafsson *et al.*, 1997; Myers *et al.*, 1998). A substantial proportion of

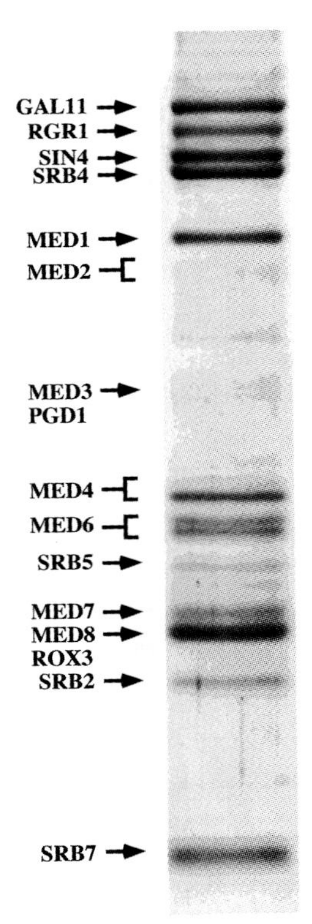

Figure 2.22. Protein Composition of the Yeast Mediator Complex.
Highly purified mediator complex was analyzed by SDS-gel electrophoresis and Coomassie Blue gel staining. The identified proteins are indicated on the left (Myers *et al.*, 1998).

cellular $RNAP_{II}$ is therefore not, as previously thought, present as a separate enzyme, but is tightly associated with several basal factors and a regulatory mediator complex. The total number of proteins in the yeast $RNAP_{II}$-holoenzyme has been conservatively estimated to be more than 30, resulting in a combined molecular weight in excess of 1,200 kDa. Growth of yeast cells harboring a temperature-sensitive SRB 4 mutation at restrictive temperatures abolishes transcription from all $RNAP_{II}$-dependent genes, suggesting that the holoenzyme configuration is physiologically relevant and required for transcription from most, if not all $RNAP_{II}$ promoters (Thompson and Young, 1995).

Yeast $RNAP_{II}$ Holoenzyme Variants

During the last years it has become apparent that there may be alternative yeast holoenzyme versions that completely lack SRB proteins (Shi *et al.*, 1997). Immunopurification of yeast $RNAP_{II}$ with an antibody recognizing the CTD domain yielded more than 20 $RNAP_{II}$-associated proteins (RAPs) that co-purified with polymerase under these conditions (Shi *et al.*, 1996; Wade *et al.*, 1996). This collection contained some known general factors (TFIIB, TFIIF and TFIIS), but also a number of previously uncharacterized proteins, such as CDC73 and PAF1. A more detailed biochemical characterization of $RNAP_{II}$ complexes containing CDC73/PAF1 showed that they also contained GAL11, a characteristic ingredient of the 'classical' holoenzyme, but no SRB proteins. The complementary observation, that the SRB-containing holoenzyme does not contain CDC73 or PAF1, supports the interpretation that there are at least two physically distinct forms of holoenzymes in yeast that share certain common features (presence of $RNAP_{II}$, TFIIB, TFIIF and GAL11), but differ substantially in the composition of other proteins. The fact that both complexes can be simultanously isolated using an affinity-tagged TFIIF subunit suggests that alternative holoenzyme versions coexist in the cell (Figure 2.23; Shi *et al.*, 1997).

Interestingly, the results of these experiments propel GAL11 into the center stage as the only subunit that is permanently associated with both known holoenzyme versions. Genetic studies have up to now failed to provide clear-cut answers about the *in vivo* role of GAL11 (see also Nishizawa *et al.*, 1994).

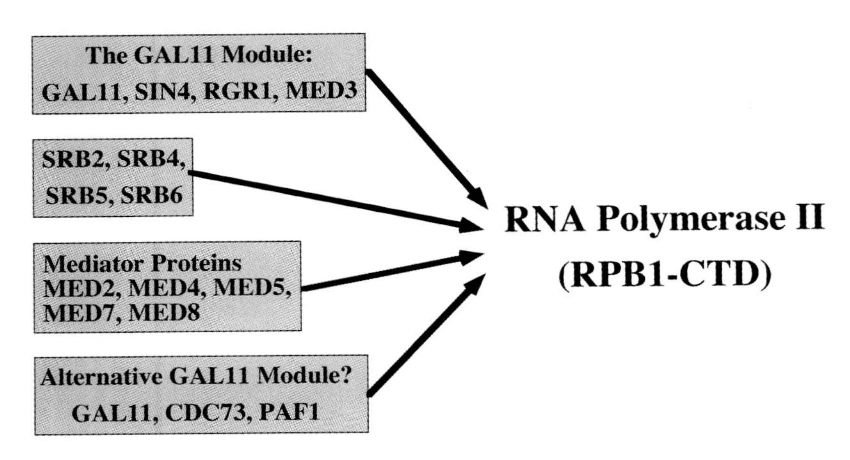

Figure 2.23. General and Alternative Mediator Components.
The purified mediator complex (as defined by Myers *et al.*, 1998) consists of SRBs, the mediator proteins (MED) and the 'GAL11 module'. The existence of an alternative GAL11-containing complex which lacks SRBs, but contains the CDC73 and PAF1 subunits suggests that there may be several distinct $RNAP_{II}$ holoenzyme versions in yeast cells. The precise relationship of several other proposed holoenzyme subunits (including SUG1, SRB8-SRB11 and the SWI/SNF chromatin remodelling complex) to the mediator complex is currently either unclear or disputed.

Although the expression of some genes is substantially affected, GAL11 mutations often have little or no effect on the viability of yeast strains under laboratory conditions. The phenotypical consequences of GAL11 mutations in the presence of mutation in other genes (e.g. PAF1; Shi *et al.*, 1996) suggests that GAL11 may participate in a functionally redundant genetic pathway where other proteins can at least partially substitute for GAL11 function. This view is supported by recent reports suggesting that GAL11 and TFIIE act in a common regulatory pathway to stimulate the kinase activity of TFIIH (Sakurai and Fukasawa, 1997; 1998).

Although the PAF1/CDC73-containing holoenzyme version is currently the best-characterized example that supports the notion of structurally and functionally distinct holoenzymes, there are also a number of hints in the current research literature suggesting that there is more to come in the near future. It is quite possible that one or more of the partially characterized yeast transcription complexes containing the ADA2/ADA3/GCN5 proteins (Marcus *et al.* 1994; Barlev *et al.*, 1995; Horiuchi *et al.*, 1995), the NOT1/NOT2/NOT3/NOT4 proteins (Collart and Struhl, 1994) and the SPT4/SPT5/SPT6 proteins (Swanson and Winston, 1992) participate in additional holoenzyme variants with more specialized roles in the transcription of particular gene subsets. The recent discovery of tissue-specific variants of TFIID complexes (see Chapter 4), combined with the existence of holoenzyme variants in yeast, makes it very likely that this pattern will be repeated (possibly in an even more elaborate form) in the cells of higher eukaryotes.

Holoenzymes in Higher Eukaryotes

The discovery of the $RNAP_{II}$ holoenzyme in yeast sparked off a search for similar holoenzymes in higher eukaryotes. Although SRBs are the 'hallmarks' of one of the yeast holoenzymes, they are not very highly conserved among higher eukaryotes and up to now only a single human homolog of SRB7 has been clearly identified (Chao *et al.*, 1996; it is possible, however, that cdk8 and cyclin C are homologs of the yeast SRB10 and SRB11 proteins: Tassan *et al.*, 1995; Rickert *et al.*, 1996). Rescue of yeast SRB7 mutants with the human SRB7 were initially not successful, but a hybrid version (57% human/ 43% yeast) compensated for a deletion of the yeast SRB7 gene *in vivo*, thus indicating at least partial functional conservation. The limited structural and functional conservation of SRB7 between yeast and humans and the failure to discover homologs of other SRBs through sequence similarities indicates that SRBs have rapidly evolved in the eukaryotic evolutionary lineage.

Antibodies directed against the human SRB7 homolog provided a diagnostic tool for the detection of a human holoenzyme and led to the purification of a complex with an estimated size of around 2,000 kDa. This complex was found to contain $RNAP_{II}$, SRB7, the basal factors TFIIE and TFIIH and a number of unidentified polypeptides (Chao *et al.*, 1996; Maldonado *et al.*, 1996). Studies in several laboratories have also revealed evidence for the presence of DNA-repair enzymes (Maldonado *et al.*, 1996), the tumor supressor protein BRCA1 (Scully *et al.*, 1997), the coactivators CBP/p300 (see Chapter 4; Nakjima *et al.*, 1997; Neish *et al.*, 1998) and chromatin-remodelling functions (see Chapter 7; Neish *et al.*, 1998) in mammalian $RNAP_{II}$ holoenzymes purified from different sources. Some of these studies yielded conflicting results, especially regarding the presence of CBP and DNA-repair enzymes in mammalian holoenzymes (see e.g. Neish *et al.*, 1998), and it is difficult at this stage to clearly establish the protein composition of mammalian holoenzymes. The evidence indicating the presence of alternative holoenzymes in yeast (see above) supports the view that many of the observed differences may, at least to a certain extent, reflect a high degree of diversity of higher eukaryotic holoenzyme configurations in different cell- and tissue types. Alternatively, it is possible that some of the experimental discrepancies are due to differences in experimental procedures leading either to the artefactual partial dissociation of large and fragile holoenzyme complexes, or non-specific association of various nuclear proteins with holoenzymes. A substantial amount of experimental work will be required in the forseeable future to define the structure and function of eukaryotic holoenzymes in more detail in a variety of cell types.

The Significance of the Holoenzyme Concept for Developing Theoretical Models of Transcriptional Regulation

The question, whether the eukaryotic transcriptional machineries are assembled from individual components on the promoter start site, or arrive as preassembled complexes, almost appears to be trivial. To the casual observer it may seem that it does not matter, because the end product, an assembled $RNAP_{II}$ transcription initiation complex located over the transcription start site is identical, regardless how it got there. There are, however, potentially important kinetic factors to consider.

It has been postulated that certain types of gene-specific transcription factors increase the frequency of transcription from a promoter by increasing the assembly rate of the basal initiation complex through an increased rate of basal factor recruitment (e.g. Choy and Green, 1993; Roberts and Green, 1994). In the sequential assembly model (Buratowski *et al.*, 1989), with many different basal factors entering the increasingly growing initiation complex in a defined sequence, there are many potentially rate-limiting steps where transcriptional activators could play an important recruitment role (reviewed in Buratowski, 1994). In contrast, the $RNAP_{II}$ holoenzyme model suggests that there may be far fewer recruitment steps and thus less opportunity for transcriptional activators to play an important kinetic role in speeding up the different stages of the assembly process. This point was specifically addressed in a series of experiments performed by Ptashne and coworkers. These researchers demonstrated that the transcription of a selected target gene could be stimulated *in vivo* by creating just one protein-protein interaction between a gene-specific transcription factor and the universal holoenzyme component GAL11 (Barberis *et al.*, 1995; Farrell *et al.*, 1996). The result is in excellent agreement with the hypothesis that recruitment of a large portion of the basal transcriptional machinery happens as a single rate-limiting 'all-or-nothing' event, rather than as a series of small steps as proposed by the stepwise assembly model. Taking into account some of the recent evidence documenting that the $RNAP_I$ and $RNAP_{III}$ transcriptional machineries do also occur in a holoenzyme configuration (Saez-Vasquez and Pikaard, 1997; Wang *et al.*, 1997), it is likely that this finding will have widespread implications for our understanding of the fundamental principles underlying the transcriptional control mechanisms operating in these diverse systems.

2.9. Conclusions

Basal factors assist cellular RNAPs in promoters recognition and and various stages of the transcript initiation. In bacteria the σ factors fulfill all these functions and switching among different members of distinct σ factors is used as a regulatory strategy to implement large-scale changes in the promoter-specificity of RNAP. In the archaea we can already discern a simple, eukaryote-like basal machinery consisting of TBP, TFB (TFIIB homolog) and RNAP. These factors are also present at the core of the substantially more complex eukaryotic $RNAP_{II}$ PIC. All of the identified PIC components play clearly defined roles during promoter recognition, melting, transcript initation and elongation. Currently we still do not understand many of the structural and functional aspects of the interaction of the basal factors with each other and with $RNAP_{II}$ in great detail. Bacterial RNAPs do not require NTP hydrolysis for the formation stable open promoter complexes, wheras eukaryotic open complexes have a significantly shorter half life and require NTP hydrolysis for their formation and maintenance. Several instances of severe topological distortion of DNA in the eukaryotic PIC have been reported, which may facilitate the initiation process. In eukaryotes many of the basal factors and $RNAP_{II}$ are present as a large complexes which may represent the physiologically relevant configuration of the $RNAP_{II}$ basal transcriptional machinery.

References

Internet Sites of Interest

http://tfiib.med.harvard.edu/transcription/basaltx.html
Information about eukaryotic basal factors.

Research Literature

Akoulitchev, S., Makela, T.P., Weinberg, R.A., and Reinberg, D. (1995). Requirement of TFIIH kinase activity in transcription by RNA polymerase II. *Nature* 377, 577–560.

Allison, L.A., Moyle, M., Shales, M., and Ingles, C.J. (1985). Extensive homology among the largest subunits of eukaryotic and prokaryotic RNA polymerases. *Cell* 42, 599–610.

Angerer, A., Enz, S., Ochs, M., and Braun, V. (1995). Transcription regulation of ferric citrate transport in *Escherichia coli* K12. FecI belongs to a new subfamily of σ^{70}-type factors that respond to extracytoplasmic stimuli. *Mol. Microbiol.* 18, 163–174.

Arnosti, D.N., and Chamberlin, M.J. (1989). Secondary σ factor controls transcription of flagellar and chemotaxis genes in *Escherichia coli. Proc. Natl. Acad. Sci. USA* 86, 830–834.

Barberis, A., Pearlberg, J., Simkovitch, N., Farrell, S., Reinagel, P., Bamdad, C., Sigal, G., and Ptashne, M. (1995). Contact with a component of the polymerase II holoenzyme suffices for gene activation. *Cell* 81, 359–368.

Barlev, N.A., Candau, R., Wang, L., Darpino, P., Silverman, N., and Berger, S.L. (1995). Characterization of physical interactions of the putative transcriptional adaptor ADA2 with acidic activation domains and TATA-binding protein. *J. Biol. Chem.* 270, 19337–19344.

Benoist, C., and Chambon, P. (1981). *In vivo* sequence requirements of the SV40 early promoter region. *Nature* 290, 304–310.

Bentley, D. (1995). Regulation of transcriptional elongation by RNA polymerase II. *Curr. Opin. Genet. Dev.* 5, 210–216.

Buratowski, S., Hahn, S., Guarente, L., and Sharp, P.A. (1989). Five intermediate complexes in transcription initiation by RNA polymerase II. *Cell* 56, 549–561.

Buratowski, S. (1994). The basics of basal transcription by RNA polymerase II. *Cell* 77, 1–3.

Burke, T.W., and Kadonaga, J.T. (1996). *Drosophila* TFIID binds to a conserved downstream promoter element that is present in many TATA-box deficient promoters. *Genes Dev.* 10, 711–724.

Chao, D.M., Gadbois, E.L., Murray, P.J., Anderson, S.F., Sonu, M.S., Parvin, J.D., and Young, R.A. (1996). A mammalian SRB protein associated with an RNA polymerase II holoenzyme. *Nature* 380, 82–84.

Chasman, D., Flaherty, K., Sharp, P. and Kornberg, R. (1993). Crystal structure of yeast TATA-binding protein and model for interaction with DNA. *Proc. Natl. Acad. Sci. USA* 90, 8174–8178.

Choy, B., and Green, M.R. (1993). Eukaryotic activators function during multiple steps of preinitiation complex assembly. *Nature* 366, 531–536.

Collart, M.A., and Struhl, K. (1994). NOT1/CDC39, NOT2/CDC36, NOT3 and NOT4 encode a global-negative regulator of transcription that differentially affect TATA-element utilization. *Genes Dev.* 8, 525–537.

Corden, J., Wasylyk, B., Buchwalder, A., Corsi, P.S., Kedinger, C., and Chambon, P. (1980). Promoter sequences of eukaryotic protein-coding genes. *Science* 209, 1406–1414.

Corden, J.L., Cadena, D.L., Ahearn, J.M., and Dahmus, M.E. (1985). A unique structure at the carboxyl terminus of the largest subunit of eukaryotic RNA polymerase II. *Proc. Natl. Acad. Sci. USA* 82, 7934–7938.

Cortes, P., Flores, O., and Reinberg, D. (1992). Factors involved in specific transcription by mammalian polymerase II: purification and analysis of transcription factor IIA and identification of transcription factor IIJ. *Mol. Cell. Biol.* 12, 413–421.

Drew, H. R., Weeks, J. R., and Travers, A. A. (1985). Negative supercoiling induces spontaneous unwinding of a bacterial promoter. *EMBO J.* 4, 1025–1032.

Dvir, A., Conaway, R.C., and Conaway, J.W. (1996). Promoter escape by RNA polymerase II. *J. Biol. Chem.* 271, 23352–23356.

Erickson, J.W., and Gross, C.A. (1989). Indentification of the σ^E subunit of *Escherichia coli* RNA polymerase: a second alternative σ factor involved in high-temperature gene expression. *Genes Dev.* 3, 1462–1471.

Farrell, S., Simkovich, N., Wu, Y., Barberis, A., and Ptashne, M. (1996). Gene activation by recruitment of the RNA polymerase II holoenzyme. *Genes & Dev.* 10, 2359–2367.

Flores, O., Maldonado, E., and Reinberg, D. (1989). Factors involved in specific transcription by mammalian RNA polymerase II. Factors IIE and IIF independently interact with RNA polymerase II. *J. Biol. Chem.* 264, 8913–8921.

Forget, D., Robert, F., Grondin, G., Burton, Z.F., Greenblatt, J., and Coulombe, B. (1997). RAP74 induces promoter contacts by RNA polymerase II upstream and downstream of a DNA bend centered on the TATA box. *Proc. Natl. Acad. Sci. USA* 94, 7150–7155.

Geiger, J.H., Hahn, S., Lee, S., and Sigler, P.B. (1996). Crystal structure of the yeast TFIIA/TBP/DNA complex. *Science* 272, 830–836.

Gerard, M., Fischer, L., Moncollin, V., Chpoulet, J. M., Chambon, P., and Egly, J. M. (1991). Purification and interaction properties of the human RNA polymerase b (II) general transcription factor BTF2. *J. Biol. Chem.* 266, 20940–20945.

Goodrich, J.A., and Tjian, R. (1994). Transcription factors IIE and IIH and ATP hydrolysis direct promoter clearance by RNA polymerase II. *Cell* 77, 145–156.

Gotoh, O., and Tagashira, Y. (1981). Stabilities of nearest-neighbour doublets in double-helical DNA determined by fitting calculated melting profiles to observed profiles. *Biopolymers* 20, 1033–1042.

Gustaffson, C.M., Myers, L.C., Li, Y., Redd, M.J., Lui, M., Erdjument-Bromage, H., Tempst, P., and Kornberg, R.D. (1997). Identification of Rox3 as a component of mediator and RNA polymerase II holoenzyme. *J. Biol. Chem.* 272, 48–50.

Ha, I., Roberts, S., Maldonado, E., Sun, X., Green, M. R., and Reinberg, D. (1993). Multiple functional domains of human transcription factor IIB: distinct interactions with two general transcription factors and RNA polymerase II. *Genes Dev.* 7, 1021–1032.

Hausner, W., Frey, G., and Thomm, M. (1991). Control regions of an archaeal gene. A TATA box and an initiator element promote cell-free transcription of the $tRNA^{Val}$ gene of *Methanococcus vanielli. J. Mol. Biol.* 222, 495–508.

Hausner, W., Frey, G., and Thomm, M. (1993). Purification and characterization of a general transcription factor, aTFB, from the Archaeon *Methanococcus thermolithotrophicus. J. Biol. Chem.* 268, 24047–24052.

Helman, J.D., and Chamberlin, M.J. (1988). Structure and function of bacterial sigma factors. *Annu. Rev. Biochem.* 57, 839–872.

Hernandez, N. (1993). TBP, a universal eukaryotic transcription factor? *Genes Dev.* 7, 1291–1308.

Hethke, C., Geerling, A. C. M., Hausner, W., de Vos, W. M., and Thomm, M. (1996). A cell-free transcription system for the hyperthermophylic archaeon *Pyrococcus furiosus. Nucl. Acids Res.* 24, 2369–2376.

Hoffmann, A., Chiang, C.-M., Oelgeschlager, T., Xie, X., Burley, S.K., Nakatani, Y., and Roeder, R.G. (1996). A histone octamer-like structure within TFIID. *Nature* 380, 356–359.

Hoffmann, A., Oelgeschlager, T., and Roeder, R.G. (1997). Considerations of transcriptional control mechanisms: do TFIID-core promoter complexes recapitulate nucleosome-like functions? *Proc. Natl. Acad. Sci. USA* 94, 8928–8935.

Holstege, F.C.P., van der Vliet, P.C., and Timmers, H.T.M. (1996). Opening of an RNA polymerase II promoter occurs in two distinct steps and requires the basal factors IIE and IIH. *EMBO J.* 15, 1666–1677.

Holstege, F.C.P., Fiedler, U., and Timmers, H.T.M. (1997). Three transitions in the RNA polymerase II transcription complex during initiation. *EMBO J.* 16, 7468–7480.

Hoopes, B., LeBlanc, J., and Hawley, D. (1992). Kinetic analysis of yeast TFIID-TATA box complex formation suggests a multistep pathway. *J. Biol. Chem.* 267, 11539–11546.

Horikoshi, M., Wang, C.K., Fujii, H., Cromlish, J.A., Weil, P.A., and Roeder, R.G. (1989). Cloning and structure of a yeast gene encoding a general transcription initiation factor TFIID that binds to the TATA box. *Nature* 341, 299–303.

Horiuchi, J., Silverman, N., Marcus, G.A., and Guarente, L. (1995). ADA3, a putative transcriptional adapter, consists of two separable domains and interacts with ADA2 and GCN5 in a trimeric complex. *Mol. Cell. Biol.* 15, 1203–1209.

Huet, J., Conesa, C., Carles, C., and Sentenac, A. (1997). A cryptic DNA-binding domain at the COOH terminus of TFIIIB70 affects formation, stability and function of preinitiation complexes. *J. Biol. Chem.* 272, 18341–18349.

Jiang, Y., Yan, M., and Gralla, J.D. (1996). A three-step pathway of transcription initiation leading to promoter clearance at an activated RNA polymerase II promoter. *Mol. Cell. Biol.* 16, 1614–1621.

Jishage, M., Iwata, A., Ueda, S., and Ishihama, A. (1996). Regulation of RNA polymerase sigma subunit synthesis in *Escherichia coli*: intracellular levels of four species of sigma subunit under various growth conditions. *J. Bacteriol.* 178, 5447–5451.

Kadonaga, J.T. (1990). Assembly and disassembly of the *Drosophila* RNA polymerase II complex during transcription. *J. Biol. Chem.* 260, 2624–2631.

Kaufmann, J., and Smale, S.T. (1994). Direct recognition of initiator elements by a component of the transcription factor IID complex. *Genes Dev.* 8, 821–829.

Kaufmann, J., Verrijzer, C.P., Shao, J., and Smale, S.T. (1996). CIF, an essential cofactor for TFIID-dependent initiator function. *Genes Dev.* 10, 873–886.

Kim, J.L., and Burley, S.K. (1994a). 1.9Å resolution refined structure of TBP recogninzing the minor groove of TATAAAAG. *Nature Struct. Biol.* 1, 638–653.

Kim, J.L., Nikolov, D.B., and Burley, S.K. (1993a). Cocrystal structure of TBP recognizing the minor groove of a TATA element. *Nature* 365, 520–527.

Kim, Y., Geiger, J.H., Hahn, S., and Sigler, P.B. (1993b). Crystal structure of a yeast TBP/TATA complex. *Nature* 365, 512–520.

Kim, Y.J., Bjorklund, S., Li, Y., Sayre, M.H., and Kornberg, A.D. (1994b). A multiprotein mediator of transcriptional activation and its interaction with the C-terminal repeat domain of RNA polymerase II. *Cell* 77, 599–608.

Koleske, A.J., and Young, R.A. (1994). An RNA polymerase II holoenzyme responsive to activators. *Nature* 368, 466–469.

Koleske, A.J., Buratowski, S., Nonet, M., and Young, R.A. (1992). A novel transcription factor reveals a functional link between the RNA polymerase II CTD and TFIID. *Cell* 69, 883–894.

Lee, S., and Hahn, S. (1995). Model for binding of transcription factor TFIIB to the TBP-DNA complex. *Nature* 376, 609–612.

Lee, D. K., Horikoshi, M., and Roeder, R. G. (1991). Interaction of TFIID in the minor groove of the TATA element. *Cell* 67, 1241–1250.

Lee, D. K., Dejong, J., Hashimoto, S., Horikoshi, M., and Roeder, R. G. (1992). TFIIA induces conformational changes in TFIID via interactions with the basic repeat. *Mol. Cell. Biol.* 12, 5189–5196.

Lei, L., Ren, D., Finkelstein, A., and Burton, Z. F. (1998). Function of the N- and C-terminal domains of RAP74 in transcription initiation, elongation and recycling of RNA polymerase II. *Mol. Cell. Biol.* 18, 2130–2142.

Leuther, K.K., Bushnell, D.A., and Kornberg, R.D. (1996). Two-dimensional crystallography of TFIIB- and IIE-RNA polymerase II complexes: implications for start site selection and inititiation complex formation. *Cell* 85, 773–779.

Li, Y., Flanagan, P.M., Tschochner, H., and Kornberg, R.D. (1994). RNA polymerase II initiation factor interactions and transcription start site selection. *Science* 263, 805–807.

Li, Y., Bjorklund, S., Jiang, J.W., Kim, Y.J., Lane, W.S., Stillman, D.J., and Kornberg, R.D. (1995). Yeast global transcriptional regulators Sin4 and Rgrl are components of mediator complex-RNA polymerase II holoenzyme. *Proc. Natl. Acad. Sci. USA* 92, 10864–10868.

Liao, S.-M., Zhuang, J., Jeffery, D.A., Koleske, A.J., Thompson, C.M., Chao, D.M., Viljoen, M., van Vuuren, H.J.J., and Young, R.A. (1995). A kinase-cyclin pair in the RNA polymerase II holoenzyme. *Nature* 374, 193–196.

Lowen, P.C., and Hengge-Aronis, R. (1994). The role of the sigma factor σ^S (*katF*) in bacterial global regulation. *Annu. Rev. Microbiol.* 48, 53–80.

Luger, K., Mader, A.W., Richmond, R.K., Sargent, D.F., and Richmond, T.J. (1997). Crystal structure of the nucleosome at 2.8Å resolution. *Nature* 389, 251–260.

Maldonado, E., Shiekhattar, R., Sheldon, M., Cho, H., Drapkin, R., Rickert, P., Lees, E., Anderson. C.W., Linn, S., and Reinberg, D. (1996). A human RNA polymerase II complex associated with SRB and DNA-repair proteins. *Nature* 381, 86–89 (+ erratum: *Nature* 384, 384).

Malhotra, A., Severinova, E., and Darst, S.A. (1996). Crystal structure of a σ^{70} subunit fragment from *E.coli* RNA polymerase. *Cell* 87, 127–136.

Malik, S., Lee, D. K., and Roeder, R. G. (1993). Potential RNA polymerase II-induced interactions of transcription factor TFIIB. *Mol. Cell. Biol.* 13, 6253–6259.

Marcus, G.A., Silverman, N., Berger, S.L., Horiuchi, J., and Guarente, L. (1994). Functional similarity and physical association between GCN5 and ADA2: putative transcriptional adaptors. *EMBO J.* 13, 4807–4815.

Marsh, T.L., Reich, C.I., Whitelock, R.B., and Olsen, G.J. (1994). Transcription factor IID in the archaea: sequences in the *Thermococcus celer* genome would encode a product closely related to the TATA-binding protein in eukaryotes. *Proc. Natl. Acad. Sci. USA* 91, 4180–4184.

Martinez, E., Zhou, Q., L'Etoile, N., Oelgeschlager, T., Berk, A.J., and Roeder, R.G. (1995). Core promoter-specific function of a mutant transcription factor TFIID defective in TATA-box binding. *Proc. Natl. Acad. Sci. USA* 92, 11864–11868.

Mathis, D.J., and Chambon, P. (1981). The SV40 early region TATA box is required for accurate *in vitro* initiation of transcription. *Nature* 290, 310–315.

Maxon, M. E., Goodrich, J. A., and Tjian, R. (1994). Transcription factor IIE binds preferentially to RNA polymerase IIa and recruits TFIIH: a model for promoter clearance. *Genes Dev.,* 8, 515–524.

McCracken, S., and Greenblatt, J. (1991). Related RNA polymerase-binding regions in human RAP30/74 and *Escherichia coli* sigma 70. *Science* 253, 900–902.

Merrick, M. (1993). In a class of its own — the RNA polymerase sigma factor σ^{54}. *Mol. Microbiol.* 10, 903–909.

Myers, L.C., Gustafsson, C.M., Bushnell, D.A., Lui, M., Erjument-Bromage, H., Tempst, P., and Kornberg, R.D. (1998). The Med proteins of yeast and their function through the RNA polymerase II carboxy-terminal domain. *Genes Dev.* 12, 45–54.

Nakajima, T., Uchida, C., Anderson, S.F., Parvin, J.D., and Montminy, M. (1997). Analysis of a cAMP-responsive activator reveals a two-component mechanism for transcriptional induction via signal-dependent factors. *Genes Dev.* 11, 738–747.

Neish, A.S., Anderson, S.F., Schlegel, B.P., Weo, W., and Parvin, J.D. (1998). Factors associated with the mammalian RNA polymerase II holoenzyme. *Nucl. Acids Res.* 26, 847–853.

Nikolov, D.B., and Burley, S.K. (1994). 2.1Å resolution refined structure of a TATA box-binding protein (TBP). *Nature Struct. Biol.* 1, 621–637.

Nikolov, D.B., Hu, S.H., Lin, J., Gasch, A., Hoffmann, A., Horikoshi, M., Chua, N.H., Roeder, R.G., and Burley, S.K. (1992). Crystal structure of TFIID-TATA box binding protein. *Nature* 360, 40–46.

Nikolov, D.B., Chen, H., Halay, E.D., Usheva, A.A., Hisatake, K., Lee, D.K., Roeder, R.G., and Burley, S.K. (1995). Crystal structure of a TFIIB-TBP-TATA-element ternary complex. *Nature* 377, 119–128.

Nikolov, D.B., Chen, H., Halay, E.D., Hoffmann, A., Roeder, R.G., and Burley, S.K. (1996). Crystal structure of a human TATA box-binding protein/TATA element complex. *Proc. Natl. Acad. Sci. USA* 93, 4956–4961.

Nishizawa, M., Taga, S., and Matsubara, A. (1994). Positive and negative transcriptional regulation by the yeast GAL11 protein depends on the structure of the promoter and a combination of *cis* elements. *Mol. Gen. Genet.* 245, 301–312.

Nonet, M.L., and Young, R.A. (1989). Intragenic and extragenic suppressors of mutations in the heptarepeat domain of *Saccharomyces cerevisiae* RNA polymerase II. *Genetics* 123, 715–724.

Oelgeschlager, T., Chiang, C.-M., and Roeder, R.G. (1996). Topology and reorganization of a human TFIID-promoter complex. *Nature* 382, 735–738.

Ohkuma, Y., and Roeder, R.G. (1994). Regulation of TFIIH ATPase and kinase activities by TFIIE during active initiation complex formation. *Nature* 368, 160–163.

Ossipow, V., Tassan, J.P., Nigg, E.A., and Schibler, U. (1995). A mammalian RNA polymerase II holoenzyme containing all components required for promoter-specific transcription initiation. *Cell* 83, 137–146.

Ouzonis, C., and Sander, C. (1992). TFIIB, an evolutionary link between the transcription machineries of archaebacteria and eukaryotes. *Cell* 71, 189–190.

Parvin, J.D., and Sharp, P.A. (1993). DNA topology and a minimal set of basal factors for transcription by RNA polymerase II. *Cell* 73, 533–540.

Pinto, I., Ware, D.E., and Hampsey, M. (1992). The yeast *sua7* gene encodes a homolog of human transcription factor TFIIB and is required for normal start site selection *in vivo. Cell* 68, 977–988.

Pinto, I., Wu, W.-H., Na, J.G., and Hampsey, M. (1994). Characterization of *sua7* mutations defines a domain of TFIIB involved in transcription start site selection in yeast. *J. Biol. Chem.* 269, 30569–30573.

Polyakov, A., Severinova, E., and Darst, S.A. (1995). Three-dimensional structure of *E.coli* core RNA polymerase: promoter binding and elongation conformation of the enzyme. *Cell* 83, 365–373.

Pugh, B.F., and Tjian, R. (1991). Transcription from a TATA-less promoter requires a multisubunit TFIID complex. *Genes Dev.* 5, 1935–1945.

Purnell, B.A., Emanuel, P.A., and Gilmour, D.S. (1994). TFIID sequence recognition of the initiator and sequences farther downstream in *Drosophila* class II genes. *Genes Dev.* 8, 830–842.

Qureshi, S.A., and Jackson, S.P. (1998). Sequence-specific DNA binding by the *S. shibatae* TFIIB homolog, TFB, and its effect on promoter strenght. *Mol. Cell* 1, 389–400.

Qureshi, S. A., Bell, S. D., and Jackson, S. P. (1997). Factor requirements for transcription in the archaeon *Sulfolobus shibatae. EMBO J.* 16, 2927–2936.

Reiter, W.-D., Palm, P., and Zillig, W. (1988). Analysis of transcription in the archaebacterium *Sulfolobus* indicates that archaebacterial promoters are homologous to eukaryotic pol II promoters. *Nucl. Acids Res.* 16, 1–19.

Reiter, W.-D., Hudepohl, U., and Zillig, W. (1990). Mutational analysis of an archaebacterial promoter: essential role of a TATA box for transcription efficiency and start site selection *in vivo. Proc. Natl. Acad. Sci. USA* 87, 9509–9513.

Rickert, P., Seghezzi, W., Shanahan, F., Cho, and Lees, E. (1996). Cyclin C/cdk8 is a novel CTD kinase associated with RNA polymerase II. *Oncogene* 12, 2631–2640.

Robert, F.D., Forget, D., Li, J., Greenblatt, J., and Coulombe, B. (1996). Localization of subunits of transcription factors IIE and IIF immediately upstream of the transcriptional initiation site of the adenovirus major late promoter. *J. Biol. Chem.* 271, 8517–8520.

Roberts, S.G., and Green, M.R. (1994). Activator-induced conformational change in general transcription factor TFIIB. *Nature* 371, 717–720.

Roeder, R.G. (1996). The role of general initiation factors in transcription by RNA polymerase II. *Trends Biochem. Sci.* 21, 327–335.

Rowlands, T., Baumeister, P., and Jackson, S. P. (1994). The TATA-binding protein: a general transcription factor present in both eukaryotes and archaebacteria. *Science* 264, 1326–1329.

Saez-Vasquez, J., and Pikaard, C.S. (1997). Extensive purification of a putative RNA polymerase I holoenzyme from plants that accurately initiates rRNA gene transcription in vitro. *Proc. Natl. Acad. Sci. USA* 94, 11869–11874.

Sakurai, H., and Fukasawa, T. (1997). Yeast GAL11 and transcription factor IIE function through a common pathway in transcriptional regulation. *J. Biol. Chem.* 272, 32663–32669.

Sakurai, H., and Fukasawa, T. (1998). Functional correlation among GAL11, transcription factor (TF) IIE, and TFIIH in *Saccharomyces cerevisiae*. GAL11 and TFIIE cooperatively enhance TFIIH-mediated phosphorylation of RNA polymerase II carboxyl-terminal domain sequences. *J. Biol. Chem.* 273, 9534–9538.

Schaeffer, L., Roy, R., Humbert, S., Moncollin, V., Vermeulen, W., Hoeijmakers, J.H.J., and Egly, J.-M. (1994). DNA-repair helicase: a component of BTF2 (TFIIH) basic transcription factor. *Science* 260, 58–63.

Schaeffer, L., Moncollin, V., Roy, R., Staub, A., Mezzina, M., Sarasin, A., Weeda, G., Hoeijmakers, J.H.J., and Egly, J.-M. (1994). The ERCC2 DNA repair protein is associated with the class II BTF2/TFIIH transcription factor. *EMBO J.* 13, 2388–2392.

Scully, R., Anderson, S.F., Chao, D.M., Wei, W., Ye, L., Young, R.A., Livingston, D.M., and Parvin, J.D. (1997). BRCA1 is a component of the RNA polymerase II holoenzyme. *Proc. Natl. Acad. Sci. USA* 94, 5605–5610.

Shi. X., Finkelstein, A., Wolf, P.A., Wade, P.A., Burton, Z.F., and Jaehning, J.A. (1996). Paf1p, an RNA polymerase II-associated factor in *Saccharomyces cerevisiae*, may have both positive and negative roles in transcription. *Mol. Cell. Biol.* 16, 669–676.

Shi, X., Chang, M., Wolf, A.J., Chang, C.-H., Frazer-Abel, A.A., Wade, P.A., Burton, Z.F., and Jaehning J.A. (1997). Cdc73p and Paf1p are found in a novel RNA polymerase II-containing complex distinct from the Srbp-containing holoenzyme. *Mol. Cell. Biol.* 17, 1160–1169.

Siebenlist, U., Simpson, R.B., and Gilbert, W. (1980). *E. coli* RNA polymerase interacts homologously with two different promoters. *Cell* 20, 269–281.

Strauss, D.B., Walter, W.A., and Gross, C.A. (1987). The heat shock response of *Escherichia coli* is regulated by changes in the concentration of σ^{32}. *Nature* 329, 348–351.

Struhl, K. (1987). Promoters, activator proteins, and the mechanism of transcriptional initiation in yeast. *Cell* 49, 295–297.

Svejstrup, J.Q., Vichi, P., and Egly, J.-M. (1996). The multiple roles of transcription/repair factor TFIIH. *Trends Biochem. Sci.* 21, 346–350.

Swanson, M.S., and Winston, F. (1992). SPT4, SPT5 and SPT6 interactions: effects on transcription and viability in *Saccharomyces cerevisiae*. *Genetics* 132, 325–336.

Sypes, M.A., and Gilmour, D.S. (1994). Protein/DNA crosslinking of a TFIID complex reveals novel interactions downstream of the transcription start. *Nucl. Acids Res.* 22, 807–814.

Tan, S., Hunziker, Y., Sargent, D.F., and Richmond, T.J. (1996). Crystal structure of a yeast TFIIA/TBP/DNA complex. *Nature* 381, 127–134.

Tassan, J.-P., Jaquenoud, M., Leopold, P., Schultz, S.J., and Nigg, E.A. (1995). Identification of human cyclin-dependent kinase 8, a putative protein kinase partner for cyclin C. *Proc. Natl. Acad. Sci. USA* 92, 8871–8875.

Thompson, C.M., Koleske, A.J., Chao, D.M., and Young, R.A. (1993). A multisubunit complex associated with the RNA polymerase II CTD and TATA-binding protein in yeast. *Cell* 73, 1361–1375.

Thompson, C.M., and Young, R.A. (1995). General requirement for RNA polymerase II holoenzymes *in vivo*. *Proc. Natl. Acad. Sci. USA* 92, 4587–4590.

Tschochner, H., Sayre, M.H., Flanagan, P.M., Feaver, W.J., and Kornberg, R.D. (1992). Yeast RNA polymerase II initiation factor e: isolation and identification as the functional counterpart of human transcription factor IIB. *Proc. Natl. Acad. Sci. USA* 89, 11292–11296.

Usheva, A., Maldonado, E., Goldring, A., Lu, H., Houbavi, C., Reinberg, D., and Aloni, Y. (1992). Specific interaction between the nonphosphorylated form of RNA polymerase II and the TATA-binding protein. *Cell* 69, 871–881.

Van Dyke, M.W., Roeder, R.G., and Sawadogo, M. (1988). Physical analysis of transcription preinititation complex assembly on a class II gene promoter. *Science* 241, 1335–1338.

Verrijzer, C.P., and Tjian, R. (1996). TAFs mediate transcriptional activation and promoter selectivity. *Trends Biochem. Sci.* 21, 338–342.

Verrijzer, C.P., Yokomori, K., Chen, J.-L., and Tjian, R. (1994). *Drosophila* $TAF_{II}150$: similarity to yeast gene TSM-1 and specific binding to core promoter DNA. *Science* 264, 933–941.

Verrijzer, C.P., Chen, J.-L., Yokomori, K., and Tjian, R. (1995). Binding of TAFs to core elements direct promoter selectivity by RNA polymerase II. *Cell* 81, 1115–1125.

Wade, P.A., Werel, W., Fentzke, R.C., Thompson, N.E., Leykam, J.F., Burgess, R.R., Jaehning, J.A., and Burton, Z.F. (1996). A novel collection of accessory factors associated with yeast RNA polymerase II. *Protein Expr. Purif.* 8, 85–90.

Wang, A., Luo, T., and Roeder, R.G. (1997). Identification of an autonomously initiating RNA polymerase III holoenzyme containing a novel factor that is selectively inactivated during protein synthesis inhibition. *Genes Dev.* 11, 2371–2382.

Wang, W., Carey, M., and Gralla, J.D. (1992). Polymerase II promoter activation: closed complex formation and ATP-driven start site opening. *Science* 255, 450–453.

Xiao, H., Friesen, J.D., and Lis, J.T. (1994). A highly conserved domain of RNA polymerase II shares a functional element with acidic activation domains of upstream transcription factors. *Mol. Cell. Biol.* 14, 7507–7516.

Yan, M., and Gralla, J.D. (1997). Multiple ATP-dependent steps in RNA polymerase II promoter melting and initiation. *EMBO J.* 16, 7457–7467.

Zawel, L., and Reinberg, D. (1995). Common themes in the assembly of eukaryotic transcription complexes. *Annu. Rev. Biochem.* 64, 533–561.

Zawel, L., Kumar, P. and Reinberg, D. (1995). Recycling of the general transcription factors during RNA polymerase II transcription. *Genes Dev.* 9, 1479–1490.

Zenzie-Gregory, B., O'Shea-Greenfeld, A., and Smale, S. T. (1992). Similar mechanisms for transcription initiation mediated through a TATA box or an initiator element. *J. Biol. Chem.* 267, 2823–2830.

Zenzie-Gregory, B., Khachi, A., Garraway, I. P., and Smale, S. T. (1993). Mechanism of initiator-mediated transcription: evidence for a functional interaction between the TATA-binding protein and DNA in the absence of a specific recognition sequence. *Mol. Cell. Biol.* 13, 3841–3849.

Chapter 3

Gene-Specific Transcription Factors

Basal transcription factors assemble the $RNAP_{II}$ preinitiation complex on the core promoter elements and subsequently assist $RNAP_{II}$ with the various stages of transcript initiation and promoter clearance (Chapter 2). The overall outcome of all these activities is only a low level of uncontrolled 'basal' transcription. In order to specifically up- or down-regulate the activity of the $RNAP_{II}$ pre-initiation complex, the vast majority of eukaryotic genes contain additional *cis*-acting sequence motifs that recruit *trans*-acting, gene-specific transcriptional activators and repressors through their sequence-specific DNA-binding domains and direct protein-protein contacts (e.g. Figure 3.1; Banerji *et al.*, 1981; Gillies *et al.*, 1981). The different arrangement of such regulatory motifs within various promoters, together with the often cell-type specific expression pattern of gene-specific transcription factors interacting with them, leads to the regulation of numerous biological phenomena in eukaryotic cells through complex networks (Serfling *et al.*, 1985; Jones *et al.*, 1987). Some regulatory sequences influence transcription of a nearby gene in response to inducible physiological signals (such as hormones), whereas others respond to highly-specialized gene-specific transcription factors that are only present at certain stages of embryonic development or only in differentiated cell-types (e.g. Gillies *et al.*, 1981; Gerster *et al.*, 1987; Leonardo *et al.*, 1987). Particular

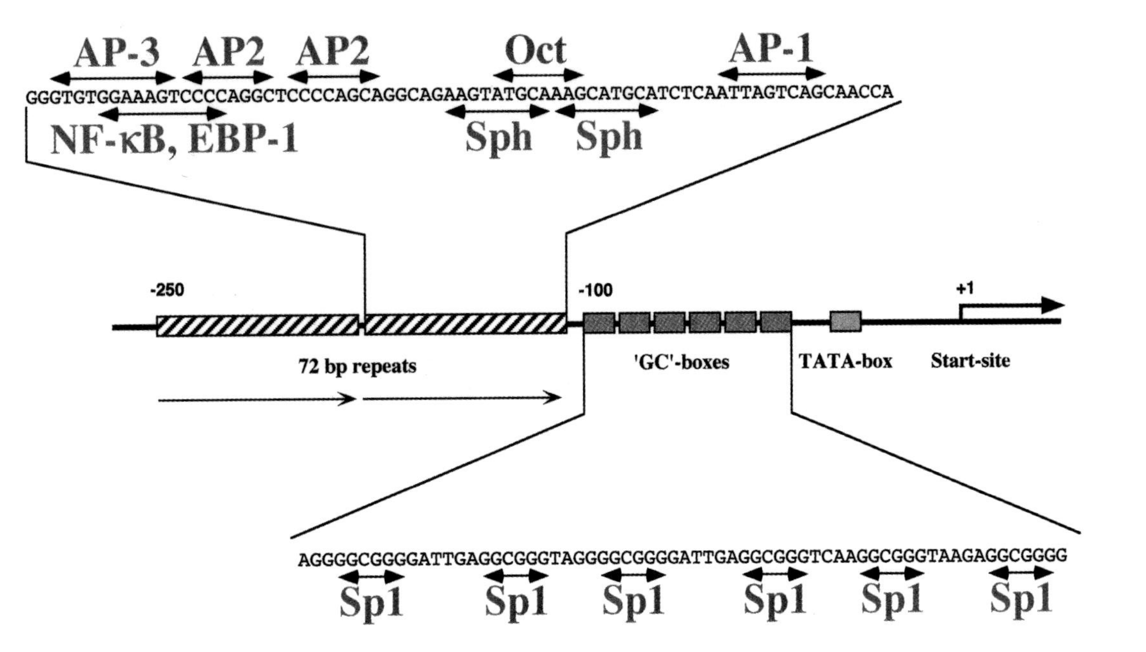

Figure 3.1. Eukaryotic Promoters Contain Multiple Sequence Motifs for a Diverse Range of Gene-Specific Transcription Factors.

For historic and technical reasons viral promoters have been intensely studied and a substantial amount of information about the regulatory motifs and their functions is available. The arrangement of motifs used to recruit gene-specific transcription factors to the SV40 'early' viral promoter is shown. mRNA transcription starts at position +1. Six 'GC' boxes binding the transcription factor Sp1 are tandemly arranged within 100 nucleotides of the initiation site. Further upstream there are two copies of a 72 basepair repeat that contain many additional sites for a variety of other gene-specific transcriptional activators, such as Oct-1, Sph, AP2 and NF-κB. Many of these transcription factors interact either synergistically or in a mutually exclusive manner with each other. Based on data from Dynan and Tjian, 1983; Lee *et al.*, 1987; Mitchell *et al.*,1987; Clark *et al.*, 1988; Macchi *et al.*, 1989; Mercurio and Karin, 1989.

sequence motifs also mediate negative regulatory effects through the action of gene-specific repressors (e.g. Borelli *et al.*, 1984; Gorman *et al.*, 1985; Goodbourn *et al.*, 1986; Mitchell *et al.*, 1987). Nowadays it is also clear that many of the regulatory properties of gene-specific promoter elements are due to their ability to direct, in conjunction with gene-specific transcription factors, the formation of transcriptionally active chromatin configurations (Chapter 7) and their influence on the attachment of chromatin fibres to 'transcription factories' in the nuclear matrix (Chapter 8).

In this chapter we will look in more detail at some of the fundamental properties of the *trans*-acting gene-specific transcription factors and the molecular structures responsible for them. The biochemical characterization of hundreds of gene-specific transcription factor/promoter systems has provided a wealth of information for understanding various aspects of regulated gene expression in eukaryotic cells. Many other aspects of the biological functions of gene-specific transcription factors will be discussed in the subsequent chapters to illustrate their wide influence on numerous steps in gene transcription.

3.1. Regulatory Roles of Gene-Specific Transcription Factors in Eukaryotes

Gene-specific Activators Stimulate Different Functions of the $RNAP_{II}$ Transcriptional Machinery

Bacterial transcription systems depend extensively on slowing down the transcription of particular genes with the help of gene-specific repressors, that prevent access of the bacterial RNAP holoenzyme to certain promoters (e.g. *lac*). Eukaryotic cells, on the other hand, rely very substantially on stimulating transcription of certain genes through the action of gene-specific activators bound to DNA motifs present in promoters (Figure 3.1). At first sight this point might seem obvious. The majority of eukaryotic DNA is densely packed into generally inaccessible chromatin structures, where it needs to be made available to the transcriptional machinery through a series of specific enzymatic steps before any active transcription can take place (Chapters 7 and 8). This is, however, not the complete explanation, because transcriptional activation is also largely due to the specific biochemical properties of the eukaryotic preinitiation complex (PIC). Specifically activated transcription can be observed as an *in vitro* phenomenon with highly purified transcription factors in the complete absence of chromatin, which proves that transcriptional activation is not solely due to relieving repressive effects of DNA packaging. Biochemical studies have established a wide variety of potential activation mechanisms that act on numerous independent stages of mRNA production. Some of the more obvious control points limiting the rate of transcription from a particular promoter have already been identified in Chapter 2 (see e.g. Figure 2.1). Many gene-specific transcriptional activators function by stimulating the recruitment of various basal factors during PIC assembly, counteracting the inhibitory function of other transcription factors, facilitating the rate of promoter clearance by $RNAP_{II}$ and increasing the overall recycling rate of basal factors (re-initiation on actively transcribed promoters). In addition to enhancing the rate of transcript initiation, the importance of converting $RNAP_{II}$ into a highly processive enzyme that can read through various transcription 'roadblocks' has become more and more obvious during the last few years (Chapter 5). Finally, in addition to the purely biochemically-defined stimulatory role on the various components of the basal transcriptional machinery, gene-specific activators do also play a major role in counteracting chromatin-mediated inhibition effects and in controlling the large-scale remodeling of chromatin and nuclear structure (discussed in more detail in Chapters 7 and 8).

It is already becoming very obvious from this list of proposed functions, that gene-specific activators have evolved a wide spectrum of different mechanisms that lead to the enhanced production of mRNAs from different genes. Although we are still quite far from understanding any of these activies in molecular detail, it is clear that no single explanation will suffice to explain the functions of these fascinating molecules. Below we will first investigate some of the technical approaches that have revealed various aspects of promoter organization and transcription factor function, and then consider the functional components of gene-specific transcription factors in more detail.

3.2. Mapping of Regulatory Motifs in Eukaryotic Promoters

Deletion Mutagenesis and Linker Scanning Identify DNA Sequences Responsible for Recruitment and Function of Gene-specific Transcription Factors

One of the first steps involved in the characterization of the functional role of gene-specific transcription factors on controlling the activity of a chosen promoter is the mapping of DNA sequences that are responsible for the recruitment and function of such proteins. It is estimated that an average eukaryotic promoter contains at least a dozen or so DNA motifs that are specifically recognized by gene-specific activator and repressor molecules. Many of these sites are usually clustered around the transcription initiation site ('+1' position), but at least some of them may also be spread over large regions of DNA, often several thousands of nucleotides away from the start site. Although it may be initially difficult to imagine how such distantly-located regulatory elements could influence events at the transcription inititation site, the effects of DNA looping and chromatin-packaging have made it easier to understand how gene-specific transcription factors bound to remote sites can be brought into a close physical proximity of the pre-initiation complex (see below for more details). Furthermore, although many gene-specific transcription factors bind to the 'upstream' control regions of a gene, there are also numerous examples of important control elements being located 'downstream' of the transcription start site, such as within coding regions, introns and even at positions beyond the 3' end of transcribed sequences. When studying promoter organization it is therefore important to initially define how much 'upstream' or 'downstream' sequence is required to faithfully reproduce all (or at least some of the more interesting) regulatory features of a particular promoter. Such features may include increased transcription in the presence of particular stimuli, cell type- and tissue-specific expression, or activation of transcription at a stage during embryonic development. The ideal test for having defined an 'intact' promoter is to show that the chosen promoter fragment directs the expression of a suitable reporter construct in transgenic animals in the same pattern as the expression pattern of the endogenous gene from which the promoter was originally derived.

Once a minimal promoter is defined, it becomes feasible to dissect the corresponding DNA fragment with a range of genetic engineering and mutagenesis techniques to localize the functionally active sequence motifs. In most cases a series of deletions (produced e.g. by PCR methods) can be used to estimate the location of regulatory sites, mostly because they allow the elimination of non-functional DNA sequences from further studies (Fig. 3.2). In addition, it is possible to use 'linker scanning' to attempt small localized disruptions of functionally active DNA motifs and targeted point mutations to define the binding sites of gene-specific transcription factors to a high degree of resolution. In all these cases the transcriptional activity of the mutated promoter constructs needs to be monitored using *in vitro* or *in vivo* transcription assays to check the effect of the introduced mutations on the level of specific mRNA produced. Ultimately, once the binding sites of particular gene-specific transcription factors is narrowed down to a few dozen nucleotides or so, footprinting and electrophoretic mobility shift assays (EMSA) allows precise definition of the DNA-protein complexes.

Biochemical Characterization and Purification of Gene-specific Transcription Factors

Once functionally important DNA motifs have been identified, a wide range of general and specialized biochemical methods can be applied to characterize the protein(s) that specifically interact with these sequences. Although it is quite feasible to use 'traditional' protein purification methods (e.g. various forms of protein chromatography) it is often quicker to take advantage of the DNA sequence-specific binding ability of the transcription factor of interest by the use of DNA-affinity chromatography. In this method short double-stranded oligonucleotides, containing the target sequence of the gene-specific transcription factor to be purified, are covalently linked to chromatography beads to generate a DNA affinity matrix. When an extract is passed over such a matrix, most proteins will not interact with the beads and therefore just flow through the column. Proteins that interact with the immobilized DNA fragments will, however, be specifically retained and can usually be eluted at a later stage in highly enriched form by increasing the ionic strength of the elution buffer (Kadonaga and Tjian, 1986). Since many proteins present in nuclear extracts

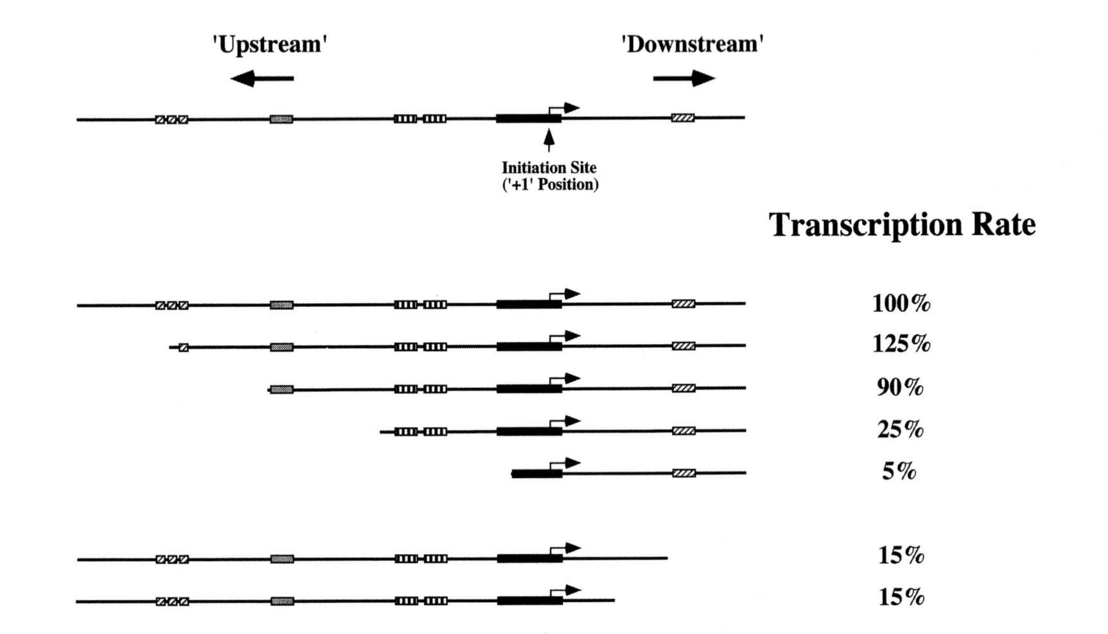

Figure 3.2. Promoter Deletion Study to Locate Binding Sites for Gene-specific Transcription Factors.
On the top a hypothetical eukaryotic promoter containing multiple binding sites for gene-specific activators and repressors (shaded boxes) is shown. Below various upstream and downstream deletion variants are displayed, together with the transcription rate from the mutated promoters (the full-length promoter is used as the standard). Note that in this example the deletion of the far left binding sites actually results in an increase in the transcription rate as compared to the full-length promoter (from 100 to 125%), indicating that they have an inhibitory function. Progressive deletion of further binding sites causes a substantial drop in transcription rate because these elements are sequence motifs bound by transcriptional activators that stimulate transcription from the start site. Also note that deletion of a sequence motif 'downstream' of the transcription initiation site causes a substantial drop in promoter activity in this particular example.

bind to DNA in general, it is usually necessary to mix the extract with competitor DNA (such as polydeoxyinosine-cytosine) before it is applied to the DNA-affinity column, to prevent such proteins from being non-specifically retained on the immobilized oligonucleotides (Figure 3.3). This method frequently allows a one-step purification of (at least some of the more abundant)

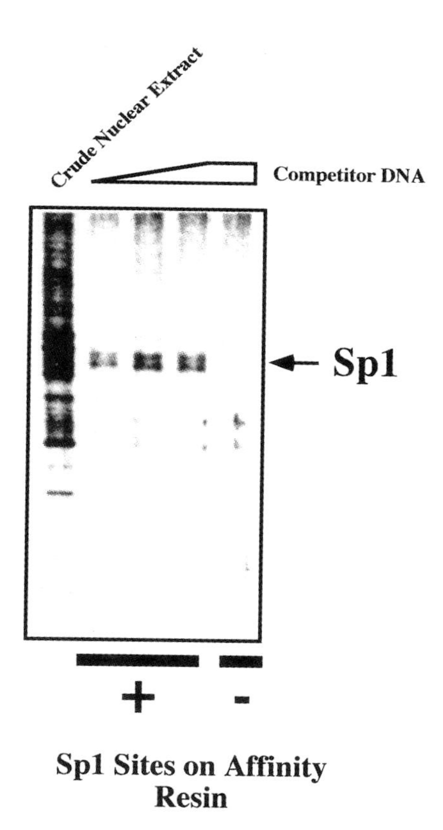

Figure 3.3. DNA-Affinity Chromatography for the Purification of Gene-Specific Transcription Factors.

The gene-specific transcription factor Sp1 can be purified from a nuclear extract fraction (left lane) to a high degree of homogeneity on a chromatography resin containing double-stranded oligonucleotides with Sp1 binding sites. From Kadonaga and Tjian, 1986.

transcription factors from nuclear extracts or provides at least a fraction that is substantially enriched in the desired transcription factor. During the biochemical purification the activity of the desired protein(s) is continuously monitored by DNA-binding studies (e.g. Footprinting and EMSA and, if at all possible, with *in vitro* transcription assays).

3.3. DNA-Binding Domains

Fundamental Aspects of Sequence-Specific DNA-Protein Interactions

Using the methodology outlined in the previous section, literally hundreds of eukaryotic promoters have been isolated and biochemically characterized to various degrees. A large number of transcription factors, mostly gene-specific transcriptional activators, have been identified and subjected to extensive mutagenesis studies to map the functionally important regions. From these studies several common features have emerged that apply to many (but definitely not all) types of gene-specific transcription factors. Many have a distinct modular structure and contain separate, individually folded domains that are either involved in the specific recognition of DNA sequence motifs, or interact with various proteins of the pre-initiation complex (PIC) to control their activities (Brent and Ptashne, 1985).

The overwhelming majority of gene-specific transcription factors contain at least one distinct DNA-binding domain, which is used to recruit the transcription factors to the promoter regions of distinct sets of genes within the genome (see further below for some notable exceptions!). The specificity of the DNA-binding domain is obviously of substantial significance, because it determines the biological effects of gene-specific transcription factors in the context of genetic regulatory networks. Transcription factors that bind only with a low degree of sequence-specificity could be used to regulate the activity of a wide range of genes, but may not be sufficiently accurate for controlling the expression of very specific subsets of genes involved in highly specialized cellular events. Many gene-specific transcription factors increase their specificity through the formation of dimeric and multimeric complexes, or by increasing the size or number of DNA-binding domains (see below for selected examples).

There are several salient features of double-stranded DNA molecules that play a fundamental role in the sequence-specific interactions between DNA and proteins. The 'normal' state of most of the DNA molecules present in cells is the standard ('type B') double-helical conformation. Transcription factors that bind to specific DNA sequences need to be able to recognize this DNA sequence from the surface features present on the *outside* of double-helical DNA because they can not unwind DNA to read the nucleotide sequence present on the 'inside' of the helix (this is of course different for RNA polymerases which need to unwind double-helical DNA to expose the 'sense' strand for synthesizing a complementary RNA sequence through specific nucleotide base pairing). Another consideration is that each individual base pair present in double-helical DNA can be physically read from either the major or the minor groove (Figure 3.4). Most DNA-binding transcription factors display a distinct selectivity for interacting with DNA either via the major or minor groove due

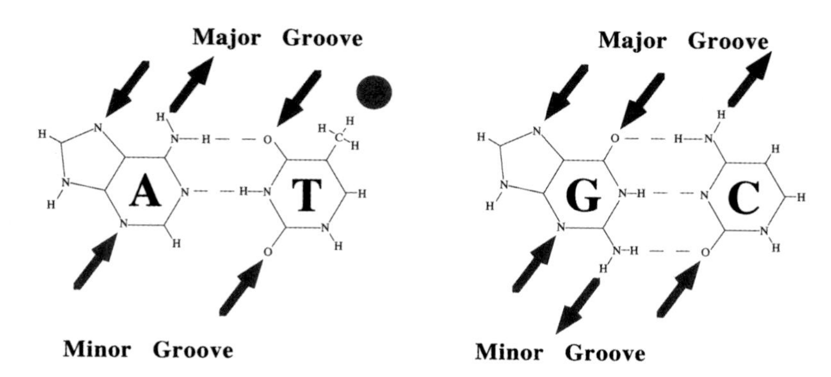

Figure 3.4. Contacts Available via the Major and Minor Grooves of Double-Stranded DNA for Sequence-Specific Interactions with Proteins.
Hydrogen-bonded A-T and G-C basepairs, as they occur in double-stranded 'B' form DNA. The different hydrogen bond donor and acceptor sites accessible via the major and minor grooves in double-stranded DNA are indicated with arrows. A 'bulky' methyl group that specifically identifies T from the major groove is labelled with a dot. Note the asymmetrical arrangement of accessible hydrogen donor and acceptor sites (indicated by the direction of the arrow), together with the presence of the T-specific methyl group in the major groove, which allows the unambigous identification of individual nucleotides by sequence-specific DNA-binding proteins.

to the physical structure of the DNA-binding domain and the specific contacts made with crucial bases present within their recognition sequence. This preference imposes important limits on the sequence-specificity of the resulting DNA-protein complex because full recognition of the nucleotide sequence is only possible for proteins that contact bases through the major groove. Transcription factors interacting with nucleotide sequences through the minor groove can only distinguish (A/T) base pairs from (G/C) base pairs, because the hydrogen bond donor and acceptor sites available for sequence-specific contacts are essentially symmetrical (Figure 3.5). The high vibration rate of hydrogen atoms involved in hydrogen bonding makes a precise recognition of the subtle differences found in differently oriented base pairs from the minor groove fundamentally impossible (Kearly *et al.*, 1994).

Consensus Sequences

Gene-specific transcription factors rarely (if ever) have an absolute requirement for a precise sequence motif in their target DNA. These proteins usually bind to a whole family of similar and inter-related sequences, often with similar binding kinetics and affinity. Comparison of individual members of the family of sequences bound by a particular transcription factor nevertheless often allows the identification of some form of consensus sequence that presents an idealized binding site. The consensus sequence is a DNA motif that the transcription factor, for which the consensus sequence was established, would predictably bind with very high affinity. Some transcription factors bind DNA via numerous weak contacts (as opposed to transcription factors that bind a smaller DNA sequence motif with fewer but strong contacts), which makes it often difficult or even impossible to define consensus sequences for such proteins in a conventional manner (Schneider, 1996).

A practical disadvantage of any type of consensus sequence is, however, that they are often not sufficiently well-defined to allow a precise definition of the binding sites of transcription factors along a DNA sequences. Moreover, although sequences close or identical to the consensus can be easily identified, there are many sequences that resemble the consensus sequence less distinctly and thus are difficult to categorize as relevant or non-relevant target sites for transcription factors. At the moment it is still impossible to sit down in front of a computer terminal and to accurately predict the number, positions and relative

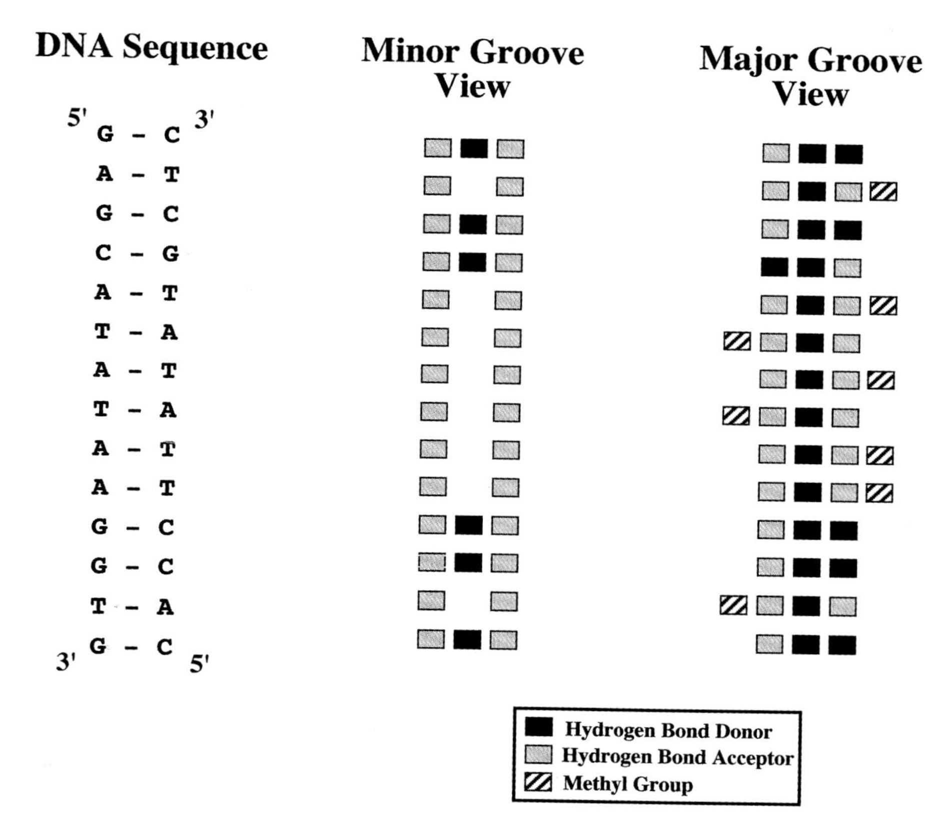

Figure 3.5. The 'Groove Code' for Sequence-Specific DNA-Protein Interactions. *Left*: An arbitrary DNA sequence is shown in the conventional double-stranded configuration. *Middle*: The same sequence as shown on the left is schematically represented as a view of hydrogen bond donor and acceptor sites as seen from the minor groove of the DNA double-helix. Note that it is possible to distinguish (A,T) from (G,C) basepairs, but it is not possible to read the sequence unambigously from the direction of the minor groove. *Right*: The same sequence as shown on the left is schematically represented as a view of hydrogen bond donor and acceptor sites, as seen from the major groove of the DNA double-helix. The asymmetric arrangement of hydrogen bond donor and acceptor sites, together with the presence of the distinctive methyl groups in T residues, allows the full sequence to be read via the 'groove code' without unwinding the double helix.

affinities of binding sites for a particular transcription factor within a stretch of known DNA sequence. Although there are no known fundamental reasons why this should not be eventually achievable once our understanding of DNA-protein interactions has reached a more sophisticated level, one ultimate problem still remains: a large proportion of the DNA sequences present in the nucleus is packaged into inaccessible chromatin strands that preclude either the binding of gene-specific transcription factors or prevent them from carrying out their functions properly (see Chapter 7 for more details). Even if it were possible to establish with absolute certainty the positions of strong target sites for particular transcription factors within a genome, it would still be difficult to assess whether they were functionally relevant and biologically active.

Helix-Turn-Helix Motif and its Variants

Several distinct types of DNA-binding domains are found in numerous transcription factors with diverse functions, indicating that such domains came into existence at a rather early stage in evolution. One of the best known DNA-binding motif is the 'helix-turn-helix' ('HTH') motif, which consists of two α-helices in a distinct configuration relative to each other. HTH-based DNA-binding domains are widespread among bacterial and eukaryotic gene-specific transcription factors, including bacterial repressors such as λ (Pabo and Lewis, 1982; Jordan and Pabo, 1988), *lac* (Kaptein *et al.*, 1985), *cro* (Wolberger *et al.*, 1988), *trp* (Schevitz et al., 1985; Mondragon et al., 1989) and the bacterial activator CAP (McKay and Steitz, 1981; Schultz *et al.*, 1991). The best-known eukaryotic representatives of HTH-containing transcription factors include the homeodomain proteins, a family of transcription factors that direct many important events during embryonic development and cellular differentiation (e.g. Desplan *et al.*, 1988; Kissinger *et al.*, 1990; Wolberger *et al.*, 1991; Dessain *et al.*, 1992).

In the HTH motif the C-terminal α-helix is often referred to as the recognition helix because it binds sequence-specifically to the individual nucleotides in the DNA target sequence via the major groove (Figure 3.6). This recognition helix is modified and/or elaborated on in many of the transcription factors containing variant HTH DNA-binding domains. One of the best-known HTH-variants is the 'POU domain', which was first identified as an extended homeodomain motif containing an additional 75 amino acid DNA binding

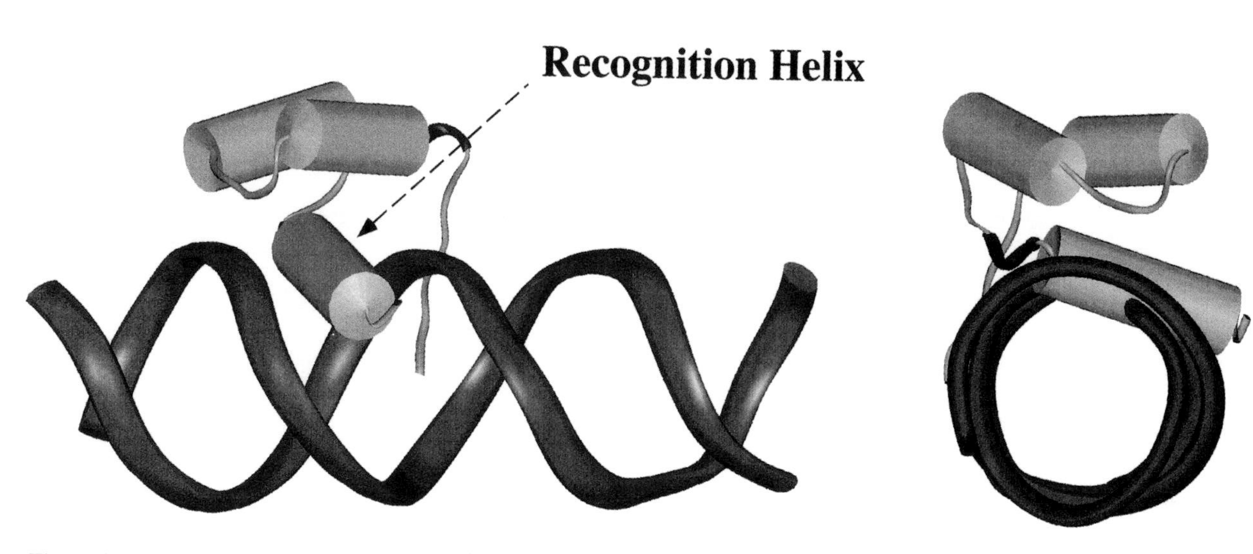

Figure 3.6. Structure of a Helix-Turn-Helix (Homeodomain)-DNA Complex.
Two views of a DNA-protein complex formed between the *engrailed* homeodomain from *Drosophila* and its cognate DNA are shown. The C-terminal α helix (the 'recognition helix') lies in the major groove of the double-stranded DNA motif allowing the amino acid sidechains to make a series of specific contacts with nucleotides present in the recognition sequence. PDB Access No. 1HDD.

module ('POU$_S$') 'tethered' to the carboxy-terminus of the HTH recognition helix. In the human transcription factor Oct-1 the presence of the POU$_S$ module expands the sequence-specific recognition capability of the transcription factor by four nucleotides in addition to the DNA target sequence recognized by the Oct-1 homeodomain (Fig. 3.7; Verrijzer et al., 1992; Assa-Munt *et al.*, 1993;

POU-Specific Domain

Helix 1 Helix 2 Helix 3 Helix 4

Pit-1	MDSPEIRELEQFANEFKVRRIKLGYTQTNVGEALAAVH----GSEFSQTTICRFENLQLSFKNACKLKAILSKWLEEAE
Oct-1	EEPSDLEELEQFAKTFKQRRIKLGFTQGDVGLAMGKLY----GNDFSQTTISRFEALNLSFKNMCKLKPLLEKWLNDAE
Oct-2	EEPSDLEELEQFARTFKQRRIKLGFTQGDVGLAMGKLY----GNDFSQTTISRFEALNLSFKNMCKLKPLLEKWLNDAE
Unc-86	DMDTDPRQLETFAEHFKQRRIKLGVTQADVGKALAHLKMPGV-S-LSQSTICRFESLTLSHNNMVALKPILHSWLEKAE

POU-Linker

Pit-1	QVAGALYNEKVGANER
Oct-1	NLSSDSSLSSPSALNSPGIEGLSR
Oct-2	TMSVDSSLPSPNQLSSPSLGFDGLPGR
Unc-86	EAMKQKDTIGDINGILPNTD

Homeo Domain

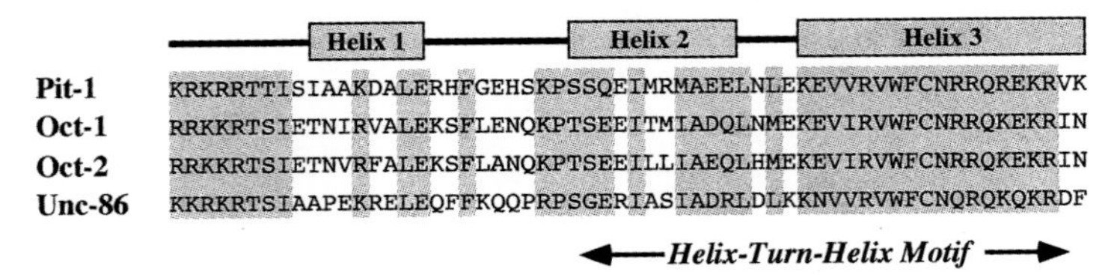

Figure 3.7. Alignment of Three POU Homeodomain-Containing Transcription Factors. The POU DNA-binding domain consists of the N-terminal POU-specific domain (top), a central unstructured linker region (middle) and a C-terminal homeodomain motif (bottom). The sequences of the three transcription factors, whose initials were used to name this DNA binding motif (POU=*P*it-1; *O*ct-1/2, *U*nc86) are aligned to each other to maximize their sequence similarities. Regions containing highly conserved or similar amino acid residues are highlighted with grey backgound bars. The homeodomain portion contains the typical helix-turn-helix motif, which is specifically marked. After Herr and Cleary, 1995.

Dekker *et al.*, 1993; Klemm *et al.*, 1994; Herr and Cleary, 1995). The Oct-1 HTH homeodomain motif and POU_S-module are separated by a highly flexible linker region, which has been shown to play an important role in determining the selectivity of DNA target site binding (Aurora and Herr, 1992; van Leeuwen *et al.*, 1997), and indirectly influences the interaction of Oct-1 with other transcription factors (Walker *et al.*, 1994; Gstaiger *et al.*, 1996). These studies show that DNA-binding domains are not merely involved in anchoring gene-specific transcription factors to particular promoter sequences, but are also responsible for conveying a significant amount of functional infomation from the DNA target site to the transcriptional machinery. The relative spacing of the DNA motifs recognized by the different parts of the Oct-1 POU domain almost certainly influences the three-dimensional structure of the DNA-bound protein greatly and thus affects the specific protein contacts made between Oct-1 and other transcription factors. While the Oct-1 homeodomain and POU_S-module are arranged on the opposite sides of the DNA-double helix, other POU-containing transcription factors, such as Pit-1, contain the two DNA-binding modules in a perpendicular arrangement (Figure 3.8; Klemm *et al.*, 1994; Jacobson *et al.*, 1997). These examples illustrate the extarordiary versatility and flexibility of POU DNA-binding domains as a direct consequence of their highly modular structure.

Other (less widely-spread) variants of the fundamental HTH domain include the addition of flexible wing-like structures in the heat shock transcription factors (Harrison *et al.*, 1994; Vuister *et al.*, 1994), specific alterations in the length and angle of the α-helices in LFB1/HNF1 (Ceska *et al.*, 1993; Leitin *et al.*, 1993), and unusual arrangements of the individual helices in PurR (Schumacher *et al.*, 1994).

Helix-Loop-Helix Motif

Many transcription factors contain a positively-charged region that interacts mainly through electrostatic forces with the negatively-charged sugar-phosphate backbone of DNA. Some particularly well-studied examples are the transcription factors containing a basic helix-*loop*-helix domain ('bHLH'; not to be confused with the helix-*turn*-helix domain described above; Murre *et al.*, 1989; Sun and Baltimore, 1991; Prendergast *et al.*, 1993; Ellenberger *et al.*,

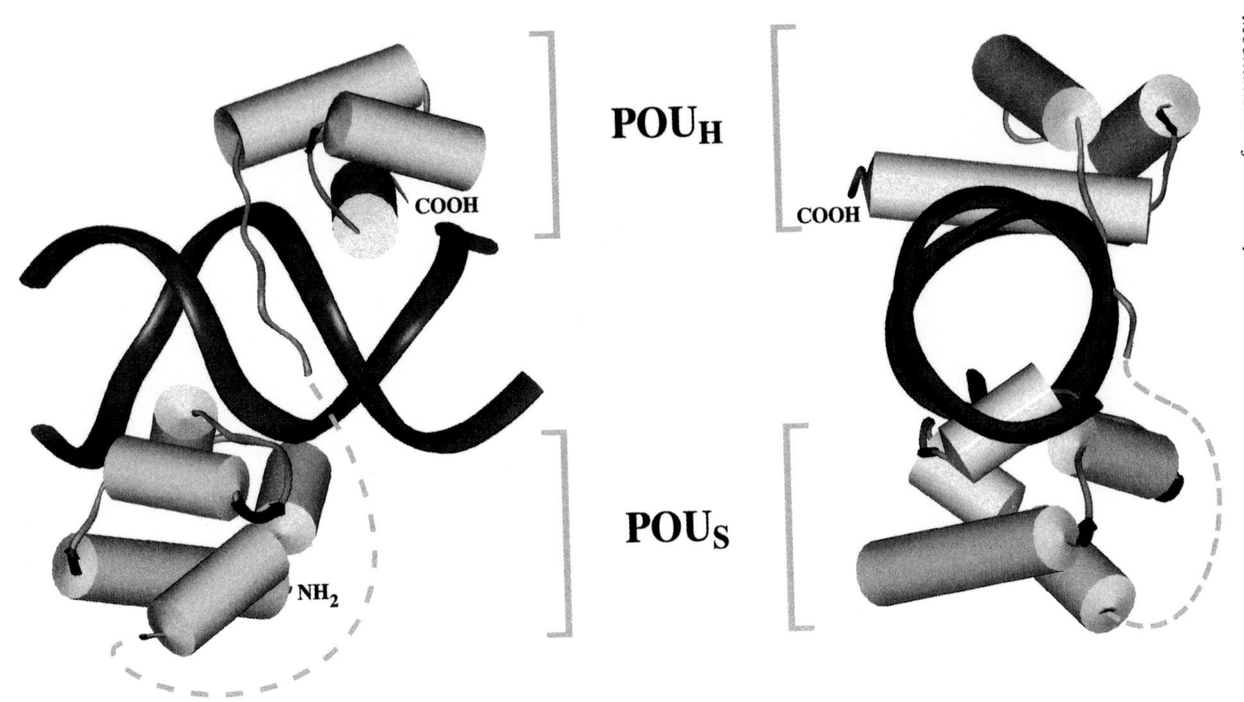

Figure 3.8. Structure of a POU Domain-DNA Complex.
Two views of a DNA-protein complex formed between the Oct-1 DNA Binding domain and its cognate DNA are shown. Note the separate contacts made by the Oct-1 homeo- and POU modules. The two domains are linked via a flexible linker region (shown as a dashed line). PDB Access No. 1OCT.

1994; Pognonec *et al.*, 1994). Several bHLH gene-specific transcription factors also contain an additional leucine-zipper dimerization domain and are referred to as bZIP proteins (Ferre-D'Amare *et al.*, 1994). Although the overall degree of primary sequence homology between HLH motifs from different transcription factors is not particularly high (Figure 3.9), X-ray crystallography studies have confirmed that different HLH motifs are very similarly structured on the three-dimensional level (Ellenberger *et al.*, 1992; Ferre-D'Amare *et al.*, 1993; Ma *et al.*, 1994). The two long α-helices present in the HLH motif participate in an amphipathic parallel helix bundle and are connected via a flexible loop. The helix bundle contacts the major groove of DNA sequence-specifically via the 'basic' α-helical domain (Figure 3.10). Since the helix bundle requires the participation of four α-helices, and each HLH motif contains only two, at least two HLH-containing transcription factors must dimerize with each other before a complete DNA-binding domain can be formed. This dimerization (or often even multimerization) event can either involve identical or different types of bHLH-containing molecules, and can therefore result in combinatorial hetero-oligomeric transcription factor complexes with unusual regulatory properties. One of the best known examples is the gene-specific transcription factor cMYC encoded by the cellular proto-oncogene *c-myc*. cMYC contains a well-conserved bHLH motif (Figure 3.9), but is incapable of binding to DNA by itself because it can not homodimerize to generate the four-helix bundle required the formation of a functional DNA-binding motif. It requires another bHLH-containing transcription factor, MAX, to form a cMYC/MAX complex to interact sequence-specifically with DNA (Blackwood and Eisenmann, 1991; Prendergast *et al.*, 1991). MAX also participates in hetero-oligomerization with other HLH partners and therefore plays a key role in the transcriptional regulation of numeorus cellular activities (e.g. Ayer *et al.*, 1993; Zervos *et al.*, 1993). Interestingly, some other HLH proteins present in eukaryotic cells (such as ID, EMC), lack a basic DNA-binding domain (see Figure 3.7), but are otherwise perfectly capable of hetero-oligmerization with other bHLH transcription factors (Benezra *et al.*, 1990; Ellis *et al.*, 1990; Pesce and Benezra, 1993). Such proteins are specific transcription inhitors that specifically 'sabotage' the function of their interaction partners through the formation of specific, but nonproductive transcription factor complexes.

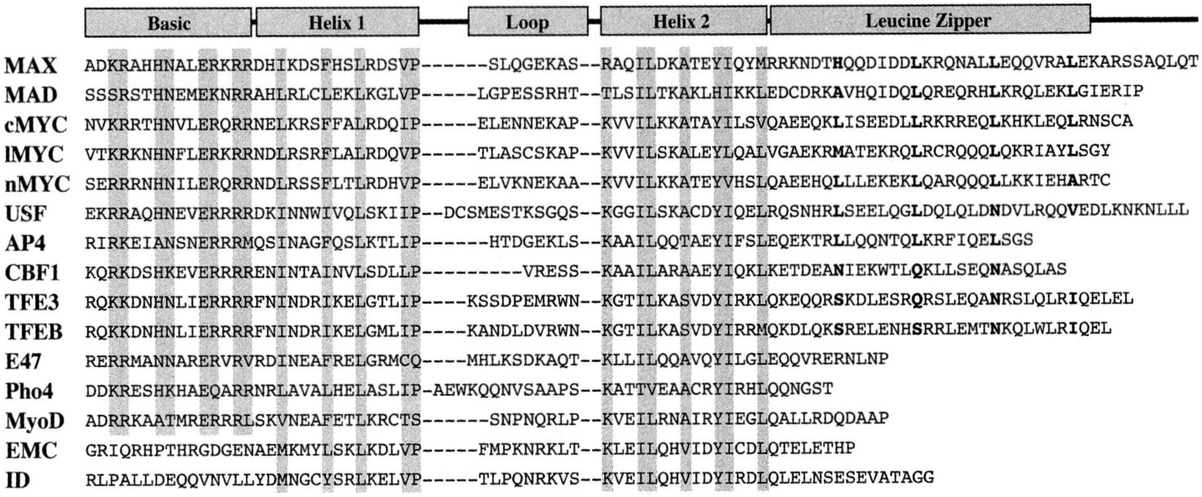

```
       Basic | Helix 1 | Loop | Helix 2 | Leucine Zipper
MAX    ADKRAHHNALERKRRDHIKDSFHSLRDSVP------SLQGEKAS--RAQILDKATEYIQYMRRKNDTHQQDIDDLKRQNALLEQQVRALEKARSSAQLQT
MAD    SSSRSTHNEMEKNRRAHLRLCLEKLKGLVP-----LGPESSRHT--TLSILTKAKLHIKKLEDCDRKAVHQIDQLQREQRHLKRQLEKLGIERIP
cMYC   NVKRRTHNVLERQRRNELKRSFFALRDQIP-----ELENNEKAP--KVVILKKATAYILSVQAEEQKLISEEDLLRKRREQLKHKLEQLRNSCA
lMYC   VTKRKNHNFLERKRRNDLRSRFLALRDQVP-----TLASCSKAP--KVVILSKALEYLQALVGAEKRMATEKRQLRCRQQQLQKRIAYLSGY
nMYC   SERRRNHNILERQRRNDLRSSFLTLRDHVP-----ELVKNEKAA--KVVILKKATEYVHSLQAEEHQLLLEKEKLQARQQQLLKKIEHARTC
USF    EKRRAQHNEVERRRRDKINNWIVQLSKIIP--DCSMESTKSGQS--KGGILSKACDYIQELRQSNHRLSEELQGLDQLQLDNDVLRQQVEDLKNKNLLL
AP4    RIRKEIANSNERRRMQSINAGFQSLKTLIP------HTDGEKLS--KAAILQQTAEYIFSLEQEKTRLLQQNTQLKRFIQELSGS
CBF1   KQRKDSHKEVERRRRENINTAINVLSDLLP---------VRESS--KAAILARAAEYIQKLKETDEANIEKWTLQKLLSEQNASQLAS
TFE3   RQKKDNHNLIERRRRFNINDRIKELGTLIP----KSSDPEMRWN--KGTILKASVDYIRKLQKEQQRSKDLESRQRSLEQANRSLQLRIQELEL
TFEB   RQKKDNHNLIERRRRFNINDRIKELGMLIP----KANDLDVRWN--KGTILKASVDYIRRMQKDLQKSRELENHSRRLEMTNKQLWLRIQEL
E47    RERRMANNARERVRVRDINEAFRELGRMCQ----MHLKSDKAQT--KLLILQQAVQYILGLEQQVRERNLNP
Pho4   DDKRESHKHAEQARRNRLAVALHELASLIP-AEWKQQNVSAAPS--KATTVEAACRYIRHLQQNGST
MyoD   ADRRKAATMRERRRLSKVNEAFETLKRCTS------SNPNQRLP--KVEILRNAIRYIEGLQALLRDQDAAP
EMC    GRIQRHPTHRGDGENAEMKMYLSKLKDLVP-----FMPKNRKLT--KLEILQHVIDYICDLQTELETHP
ID     RLPALLDEQQVNVLLYDMNGCYSRLKELVP-----TLPQNRKVS--KVEILQHVIDYIRDLQLELNSESEVATAGG
```

Figure 3.9. Alignment of Helix-Loop-Helix-(Leucine Zipper) Containing Transcription Factors.
15 different helix-loop-helix (HLH) are shown aligned to each other to maximize their similar sequence organization. Highly conserved or similar amino acids are highlighted with grey background bars in the basic region, helix1 and helix 2. The top ten HLH-containing transcription factors also contain a 'leucine-zipper' dimerization domain. The regularly spaced small apolar residues (usually leucine) located in every seventh position are shown in bold. Note that the two transcription inhibitors at the bottom of the figure (EMC and ID) contain the highly conserved residues in helix 1 and helix 2 necessary for interaction with other HLH factors, but lack the necessary basic amino acid residues necessary for DNA binding. After Ferre-D'Amare *et al.*, 1993.

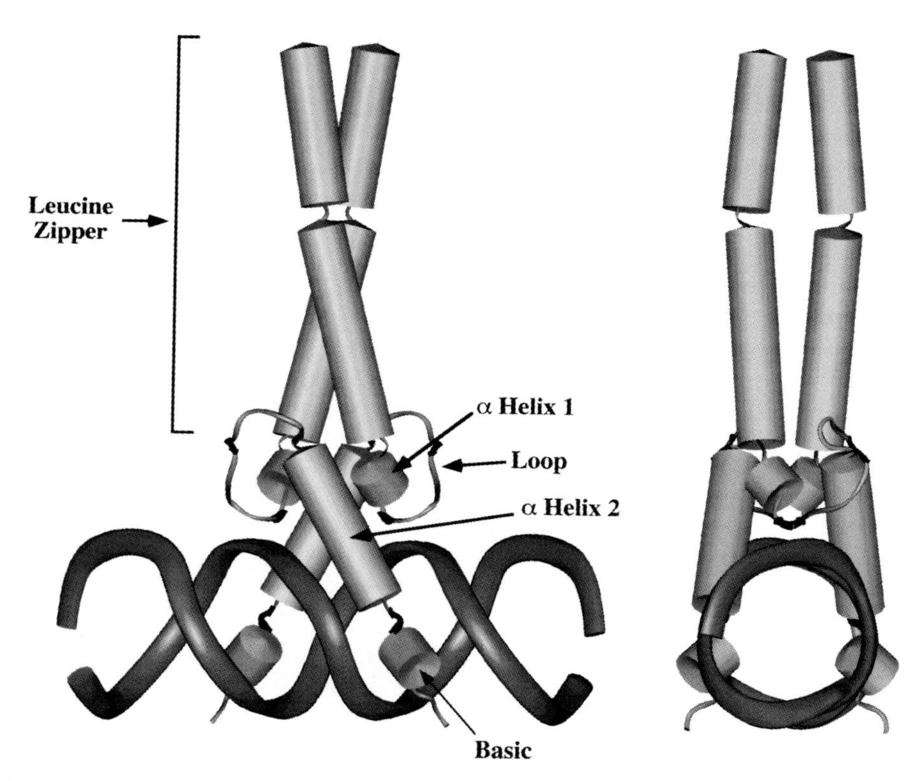

Figure 3.10. Structure of a Helix-Loop-Helix Leucine Zipper-DNA Complex.
Two views of a DNA-protein complex formed between the MAX homodimer and its cognate DNA are shown. Residues in helix 1 and the basic domains form specific contacts with nucleotides from the major groove of the DNA target sequence. Note the extensive helical leucine zippers that allow the formation of the DNA-binding domain through homodimerization of two MAX monomers. PDB Access No. 1AN2.

Zinc Fingers in Various Shapes and Sizes

Several types of DNA-binding domains are based on 'zinc fingers' motifs. Zinc fingers are formed by a defined arrangement of cysteine- and histidine residues which centrally coordinate a single zinc atom (reviewed in Berg *et al.*, 1990; Schwabe and Klug, 1994; Klug and Schwabe, 1995; Berg and Shi, 1996). Not all zinc fingers are, however, involved in DNA binding and there is evidence that some serve as protein-protein interaction motifs (e.g. Galcheva-Gargova *et al.*, 1996), RNA-binding domains (e.g. Curtis *et al.*, 1997), or simply act as spacers between adjacent DNA-binding domains (e.g. Nolte *et al.*, 1998). More than ten different types of zinc fingers with different functions have been distinguished (Schwabe and Klug, 1994).

In DNA-binding proteins, one of the most commonly found zinc finger motifs is the 'Cys_2His_2' (or 'C_2H_2') motif, which may occur in up to 1% of all the genes encoded by the human genome (Hoovers *et al.*, 1992). Apart from the highly conserved and precisely spaced cysteine and histidine residues, these motifs also contain several characteristic hydrophobic and aromatic residues that generate a stabilizing hydrophobic core within each zinc finger domain. The three-dimensional structure of C_2H_2 zinc fingers reveals a typical arrangement of two antiparallel β-strands near the N-terminal end, followed by a C-terminal α-helix (Figure 3.11.; Pavletich and Pabo, 1991). When bound to DNA, the N-terminal portion of the α-helix of most zinc-fingers projects into the major groove of double-helical DNA and the protruding amino acid-side chains make a series of specific contacts with particular nucleotides (e.g. arginine residues often interact with guanine and asparagine residues with adenine; see Suzuki *et al.*, 1994). The α-helical portion of zinc fingers is therefore sometimes referred to as the 'reading head', because it 'reads' the specific nucleotide-sequences present in the target DNA.

The majority of single zinc-finger domains confer sequence-specific recognition of a up to three nucleotides (Pavletich and Pabo, 1991), but contacts with four and even five adjacent base pairs have been observed in at least some instances (Wuttke *et al.*, 1997). Since many gene-specific transcription factors need to be able to interact more extensively with their DNA target sequences, it is not unusual for the DNA-binding domains of transcription factors to contain multiple zinc fingers in a tandem arrangement. One of the

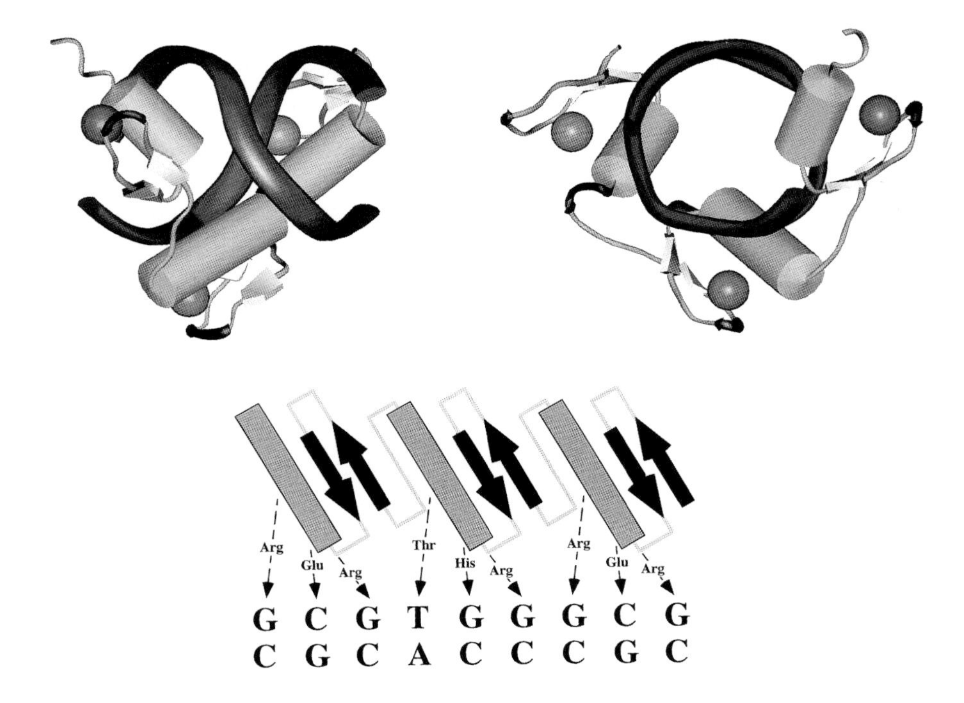

Figure 3.11. Sequence-Specific Interaction between Zinc-Fingers and DNA.
Top: High resolution structure of three tandem zinc fingers with double-stranded DNA. The complex is shown from a sideway view and along the helical axis (Pavletich and Pabo, 1991). The 'reading head' α-helices lie in the major groove of DNA and contact three different nucleotides each. *Bottom:* A schematic diagram of some of the molecular interactions between specific amino acids of the α-helices and and various nucleotides. Note the regular arrangement of two β-strands and one α-helix in each zinc finger motif. Arginine residues are used repeatedly to contact guanine residues.
PDB Access No. 1AAY.

best studied examples, the 5S rRNA-specific transcription factor TFIIIA, contains 9 distinct zinc finger motifs and it has been shown that especially the the first three zinc fingers play an important role in the sequence-specific binding of TFIIIA to its DNA target sequence (Miller *et al.*, 1985; Wuttke *et al.*, 1997; Nolte *et al.*, 1998; see also Chapter 6). A high-resolution crystallographic analysis of the TFIIIA zinc finger-DNA complex revealed clusters of zinc fingers contacting distinct DNA motifs via the major groove of the double helix in a ribbon-like configuration (fingers 1–3 and 7 to 9), whereas some of intervening zinc-fingers do not contact DNA at all, or make unusual DNA contacts via the minor groove (Nolte *et al.*, 1998). Although in many cases it is thought that individual zinc fingers are structurally and functionally autonomous, substantial protein-protein contacts between adjacent finger motifs may occur in at least some transcription factors (Wuttke *et al.*, 1998). DNA-binding domains consisting of 'multi-fingered' domains, like the one found in TFIIIA, seem to display a high degree of flexibility and it is often not possible to predict from the primary sequences of the individual zinc finger motifs whether they are involved in DNA-binding, or whether they are part of flexible linker structures. Despite such limitations, it has been possible to establish some preliminary rules that allowed in several cases a fairly accurate prediction of the DNA sequences bound by mutated zinc fingers (Desjarlais and Berg, 1992; Thukral *et al.*, 1992) or even led to the *de novo* design of an artificial zinc finger/POU DNA-binding domain conforming to a predicted DNA target sequence-specificity (Pomerantz *et al.*, 1995).

The glucocorticoid- and estrogen receptors, important gene-specific transcription factors involved in the hormonal regulation (see Chapter 10), contain a variant zinc-finger motif with two α-helices coordinating two zinc ions with cysteines (Hard *et al.*, 1990; Schwabe *et al.*, 1990).

Other Types of DNA-Binding Domains

The HTH, HLH and zinc-finger domains constitute the most widely-used DNA-binding motifs and are present in many gene-specific transcription factors. Although these three fundamental motifs are structurally distinct, they all share the common feature of predominantly using amino acid residues arranged in α-helices to contact specific nucleotide sequences. Numerous other types of

DNA-binding domains use very different principles for sequence-specific recognition of DNA sequences. An extensive description of the large variety of characterized DNA-binding domain variants is beyond the scope of this book, but a few of the most interesting 'non-standard' examples will be discussed below to illustrate alternative DNA-binding motifs.

The bacterial transcriptional repressors P22 Arc and MetJ contain a antiparallel β-sheet formation that interacts sequence-specifically with the major groove of DNA (Figure 3.12; Somers and Philips, 1992; Raumann *et al.*, 1994a,b). The sequence-specificity of this interaction is determined via the amino acid residues present in the β-sheets, but up to now it has not been possible to establish simple rules for predicting the interactions of individual amino acid side chains with particular nucleotides (He *et al.*, 1992; Brown *et al.*, 1994; Raumann *et al.*, 1994b). Another unusual, and up to now unique, form of contacting DNA sequence-specifically occurs in the important tumor suppressor protein p53. p53 is a gene-specific transcription factor that controls the expression of many genes whose products are implicated in apoptosis and growth control (e.g. Okamoto and Beach, 1994; Buckbinder *et al.*, 1995; Miyashita and Reed, 1995). p53 contacts specific nucleotide residues via the major groove of DNA through a structurally unique loop-sheet-helix motif and an additional single amino acid that is located outside this motif (Figure 3.13; Cho *et al.*, 1994). This structure is easily altered through single amino acid substitutions and thus explains why mutations in the DNA-binding domain of p53 are detected in a large percentage of primary tumors (Hollstein *et al.*, 1991; Ko and Prives, 1996).

Transcription Factors without DNA-Binding Domains

The ability of a transcription factor to bind to specific DNA sequences seems an absolute prerequisite for them to carry out their gene-specific functions. Nevertheless, there are some very potent transcription factors, including the activator protein VP16 from the herpes simplex virus, that lack a DNA-binding domain. Instead, VP16 associates with at least two transcription factors, Oct-1 and HCF, that normally occur within the host cell (Figure 3.14; Lai *et al.*, 1992; Pomerantz *et al.*, 1992; Walker *et al.*, 1994). This VP16/Oct-1/HCF complex binds to particular target sequences, the 'TAATGARAT' motif,

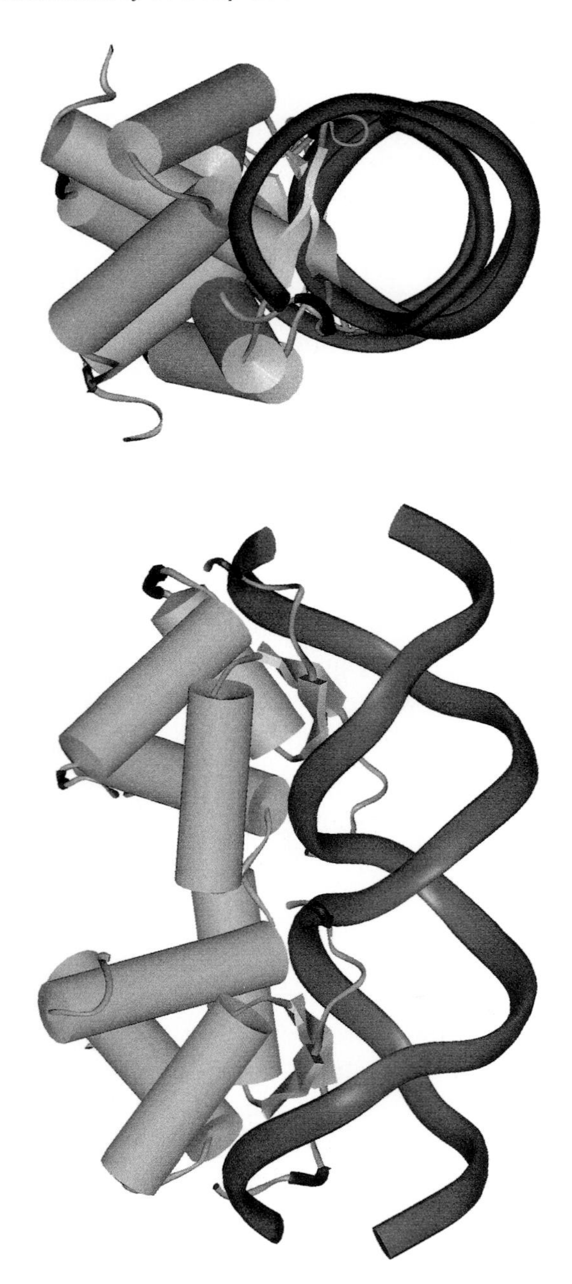

Figure 3.12. Specific DNA Recognition with Antiparalled β-Sheets.
Two views of a DNA-protein complex formed between a P22 Arc repressor DNA-binding domain dimer and its cognate DNA are shown. The DNA-binding surface interacting with nucleotides via the major groove is formed by amino acid residues arranged in an antiparallel β sheet. Note the slight bending of the double-helical DNA along its helical axis induced by the protein-DNA interactions. PDB Access No. 1PAR.

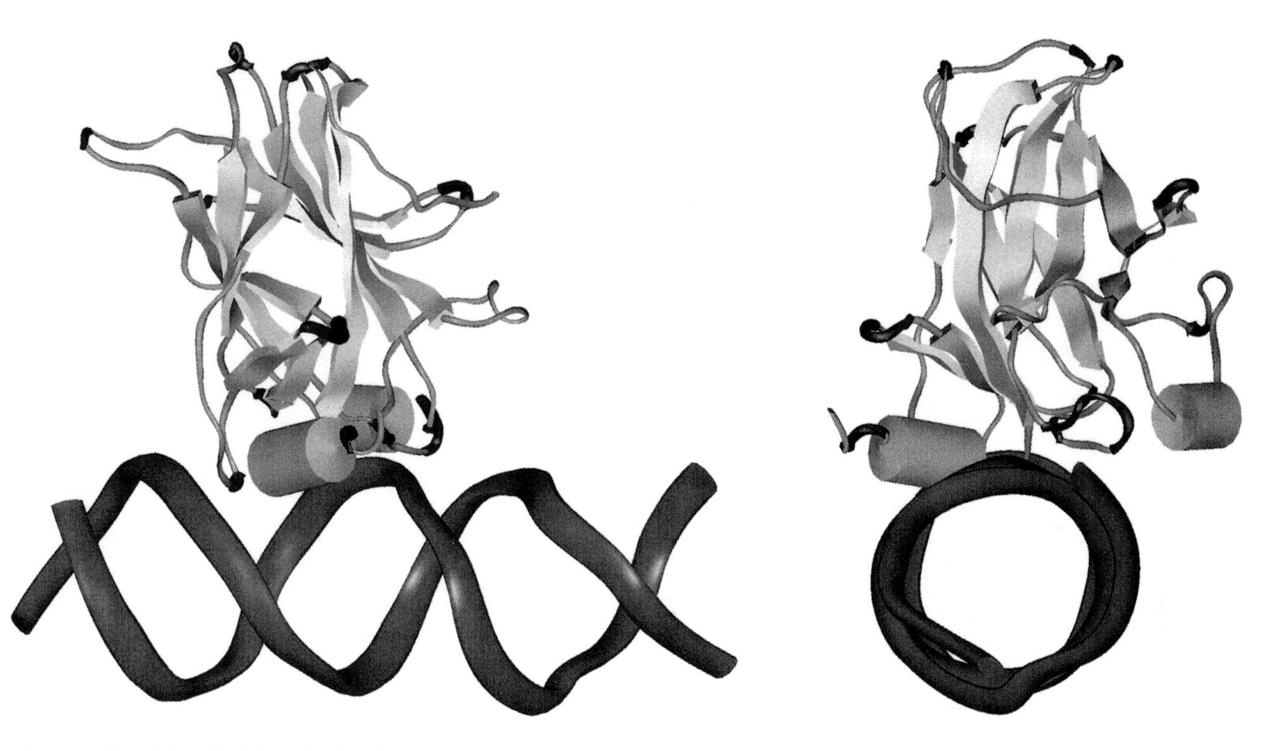

Figure 3.13. The p53 DNA-Binding Domain.
Two views of a DNA-protein complex formed between the p53 DNA-binding domain and its cognate DNA are shown. Note the complex arrangement of β-sheets required to precisely position the main elements of the DNA interactions surface, consisting essentially of an α helix and a loop. Tumor-causing mutations can easily disrupt this complex structure and abolish the DNA-binding ability of p53. PDB Access No. 1TSR.

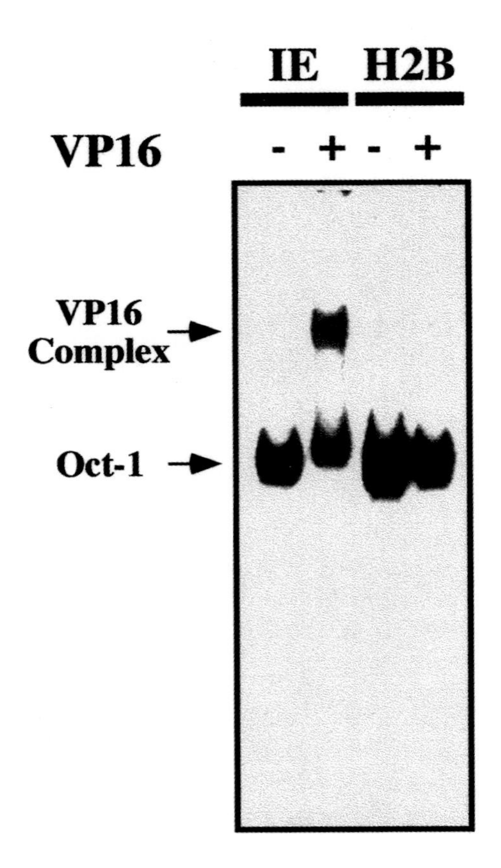

Figure 3.14. An Oct-1/VP16-Containing Complex Specifically Recognizes the 'TAATGARAT' Motif Present in the Herpes Early Promoter.
An electrophoretic gel mobility shift assay (EMSA) of Oct-1/VP16-containing complexes on two promoters is shown. Oct-1 recognizes target motifs present in the promoters of both the VP16 'Immediate Early' and histone H2B genes and gives rise to a specifically shifted band (labelled with an arrow). VP16 can, however, only interact to give a 'supershifted' complex with Oct-1 on the VP16 'Immediate Early' promoter, but not with Oct-1 bound to H2B due to sequence differences in the Oct-1 recognition motifs present in the two promoters. From Walker *et al.* (1994).

present in the promoters of the viral immediate-early genes and substantially stimulates their transcription. The VP16 transactivation domain is one of the major model systems for learning more about the mechanisms of activator function and we will encounter it again in the section below. Another viral transcriptional activator, the adenovirus E1a protein, interacts promiscuously with a diverse set of different cellular gene-specific transcription factors and thus allows the transcriptional reprograming of a large number of genes without the need for any universally-shared promoter motifs between them (Liu and Green, 1994).

3.4. Activation Domains

Once gene-specific transcription factors are bound to their DNA target sequences in the vicinity of the promoter(s), other parts of these proteins — the activation domains — spring into action. Activation domains are functionally defined as amino acid sequences that, when suitably connected to a DNA-binding domain, stimulate the rate of transcription from a specific promoter (e.g. Brent and Ptashne, 1985). Many gene-specific transcription factors contain several distinct activation domains that can function either independently of each other, or cooperatively with each other. In one of the best-studied examples, the human transcriptional activator Sp1, up to four distinct activation domains have been distinguished by deletion and point mutagenesis studies (Figure 3.15; Kadonaga *et al.*, 1987; Kadonaga *et al.*, 1988; Courey and Tjian, 1988). Two of them, A and B, map to the N-terminal region and contain glutamine-rich regions and adjacent serine/threonine-rich domains. When linked to a heterologous DNA-binding domain, each of these two domains strongly activates transcription, indicating that they contain all the structural information necessary for stimulating the rate of transcription from specific promoters (e.g. Gill *et al.*, 1994; Emami *et al.*, 1995). The other two activation domain (C and D) flank the DNA-binding domain of Sp1 and only activate transcription rather poorly by themselves, but substantially increase the effectiveness of domains A and B (Courey and Tjian, 1988).

As we can already see from this particular example, many activation domains clearly stand out on the primary amino acid level due to their unusual amino

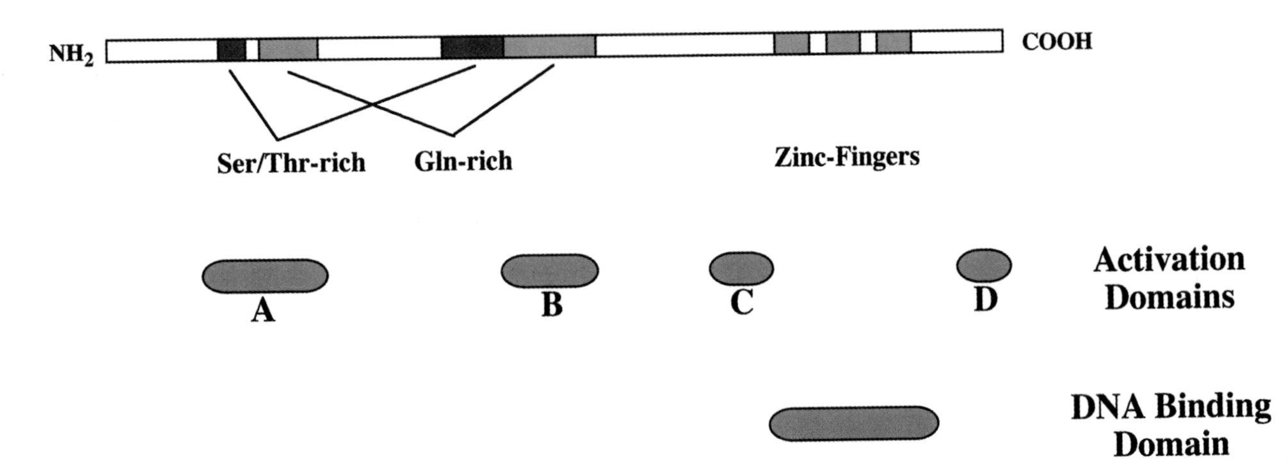

Figure 3.15. Organization of Functional Domains in the Human Gene-Specific Transcription Factor Sp1.
The primary amino acid sequence of Sp1 shows the presence of two glutamine and serine/threonine-rich regions. The glutamine-rich regions coincide with the main activation domains ('A' and 'B') located near the N-terminus. Two minor activation domains ('C' and 'D') cooperate synergistically with A and B. The DNA-binding domain consists of three distinct zinc-finger regions and allows Sp1 to bind sequence-specifically to G/C rich target sequences located in the promoters of many human genes (see also Figure 3.1.).

acid composition and often include regions rich in acidic residues (aspartic and glutamic acid), prolines, glutamines or isovalines (reviewed by Mitchel & Tjian, 1989; Johnson *et al.*, 1993; Triezenberg, 1995). It therefore looked as if these amino acids were involved in forming distinct protein motifs that interacted with different types of targets in the basal transcriptional machinery. Subsequent studies revealed, however, that the most critical residues required for activation functions were often not these abundantly-present residues, but a small number of bulky hydrophobic amino acid side chains interspersed in between them (e.g. Ruden, 1992; Regier *et al.*, 1993; Gill *et al.*, 1994; Lin *et al.*, 1994; Drysdale *et al.*, 1998; but see Gerber *et al.*, 1994, for an exception). It is therefore clear that the classification of different types of activation domains according to amino acid composition is superficial and possibly misleading (Triezenberg, 1995), but many researchers continue to refer (in the absence of a better nomenclature!) to 'acidic', 'glutamine-rich', 'proline-rich' and 'isoleucine-rich' domains. It is important to understand that these names reflect mainly the historic developments of the field and do not retain much of the original meaning anymore. In fact, one of the more surprising aspects of activation domains seems to be precisely the fact that they are seemingly without defined tertiary structure, and probably only adopt some resemblance of structural order when they are complexed with other proteins that are the targets for their regulatory interactions (see below). Recent experimental data also suggest that different activation domains have very different effects on stimulating the rate of transcript initiation and elongation (Chapter 5), implying that transcriptional activation is not a single phenomenon, but the combination of multiple effects acting on distinct $RNAP_{II}$ functions.

VP16 Transactivator as a Model System for the Functional Dissection of Transcriptional Activation Domains

VP16 is a potent activator of the Herpes simplex virus responsible for initiating high levels of transcription of the 'immediate early' genes during its infection cycle. Its 'classic' acidic-type activation domain has been subjected to extensive mutagenesis to define the functionally active amino acid residues. As we will see below, VP16 interacts apparently specifically with basal transcription factors and may therefore stimulate transcription at least partially by enhancing the rate of recruitment of the basal transcriptional machinery to the promoter start sites. In addition, it is likely that VP16 stimulates the rate of transcript initiation through interactions with components of the TFIID complex. Finally, VP16 also potently increases RNAP processivity and thus stimulates the rate of elongation (see Chapter 5 for more information). Overall, VP16-activated transcription is therefore almost certainly the consequence of numerous 'small' activation steps at different stages of the transcription cycle. At the moment it is not clear whether such behaviour is characteristic of all transcriptional activators or if viral activators are exceptional in this respect. It can be imagined that viruses had to evolve highly efficient activator proteins to efficiently hijack the control of the cellular transcriptional machinery and to compete effectively for limited amounts of basal factors during the early stages of the infection cycle. It is quite possible that many cellular transcriptional activators play a more restricted and subtle role, because their function is not to dominate transcriptional control mechanisms, but to work in concert with numerous other transcription factors. This thought needs to be kept in mind before generalizing any findings obtained from experiments involving VP16 and other viral activators. From a practical point of view, however, the high levels of activation achievable in defined *in vitro* systems have made VP16 one of the favourite model systems for attempts to understand the phenomenon of transcriptional activation on the molecular level.

Structure and Function of the VP16 Activation Domain

The acidic activation domain (AAD) of VP16 has been specifically mapped to the carboxy-terminal 78 amino acids (residues 413–490 in the intact protein; Triezenberg *et al.*, 1988; Cousens *et al.*, 1989). This minimal activation domain can be molecularly 'transplanted' to a DNA-binding motif to generate artificial (but fully functional) gene-specific transcriptional activator with altered promoter-binding specificities (Sadowski *et al.*, 1988). Visual inspection of the minimal activation domain of VP16 reveals an abundance of acidic residues (aspartic and glutamic acid; Figure 3.16; Cress and Triezenberg, 1991).

In all cases analyzed to date, including the activation domains of the transcription factors GAL4, GCN4, VP16, NF-κB, p65 and glucocorticoid receptor, it was found with various biophysical methods (including fluorescence

420 430 440 450 460 470 480 490

TAPITDVSLGDELRLDGEEVDMTPADALDDFDLEMLGDVESPSPGMTHDPVSYGALDVDDFEFEQMFTDAMGIDDFGG

Figure 3.16. Primary Sequence of the Acidic Activation Domain of Herpes Virus VP16.
The C-terminal part of the VP16 protein is shown with acidic amino acid residues highlighted in bold. From Cress and Triezenberg, 1991.

analysis, circular dichroism and nuclear magentic resonance studies) that activation domains generally form completely unstructured random coil polypeptide chains (O'Hare and Williams, 1992; Shen *et al.*, 1996). Under certain conditions, such as presence of hydrophobic solvents and low pH, the unstructured AADs are able assume defined secondary structures such as α-helices and β-sheets (Donaldson and Capone, 1992; Van Hoy *et al.*, 1993). It is likely that this ability of activation domains to undergo such a structural transition from a disorderly to a orderly structure plays a major role in their function.

So what are the potential targets controlled by interactions with VP16? Unusually perhaps, one of the main obstacles in the identification of the physiologically relevant targets within the basal transcriptional machinery is due to the fact that VP16 binds to numerous proteins with reasonable high affinity. Under *in vitro* conditions the VP16 AAD binds specifically to the TATA-binding protein (TBP), and mutagenesis studies have generally revealed a positive correlation between the ability of the mutants to bind TBP and the degree of transcription activation obtained in *in vitro* transcription assays (e.g. Stringer *et al.*, 1990; Ingles *et al.*, 1991; Nishikawa *et al.*, 1997). Specific protein interactions between the VP16 AAD and other basal factors, such as TFIIB and TFIIH, have also been observed (Lin *et al.*, 1991; Goodrich *et al.*, 1993; Roberts *et al.*, 1994; Xiao *et al.*, 1994). VP16 may thus be able to stimulate transcription (at least partially) by speeding up the rate of pre-initiation complex assembly by enhancing the rate of recruitment of specific basal transcription factors and/or structurally stabilizing various intermediate basal factor complexes (e.g. Choy *et al.*, 1993; Hahn, 1993).

Although the *in vitro* data concerning the various interactions between VP16 and basal factors is compelling, and may be functionally relevant under *in vivo* conditions, it does not constitute the whole story. *In vitro* transcription systems containing highly purified basal factors, $RNAP_{II}$ and TBP are incapable of supporting VP16-activated transcription, thus demonstrating that some important components are missing. One of this components is hypothesized to be the human TBP-associated factor, $hTAF_{II}31$ (or the *Drosophila* homolog, $dTAF_{II}40$, which was used for much of the original work). This TAF interacts specifically with the VP16 AAD (Goodrich *et al.* 1993) and is required for mediating VP16-mediated activated transcription in *in vitro* transcription assays in the presence of artificially reconstituted TFIID complexes (Chen *et al.*, 1994; see also Chapter 4). The N-terminal 140 amino acids of $hTAF_{II}31$ are sufficient for association with the VP16 activation domain spanning residues 452–490. The characterization of this interaction has led to one of the most detailed models of the structural basis describing the molecular interactions between an activation domain and its target (Uesugi *et al.*, 1997). High resolution NMR studies have shown that the VP16 activation domain undergoes a localized structural change in the presence of an $hTAF_{II}31$ fragment that is compatible with the formation of an α-helix. There is good evidence for direct contacts between the VP16 residues D^{472}, F^{479}, L^{483} and $hTAF_{II}31$, as deduced from strong chemical shifts in the NMR spectra of isotopically-labelled VP16 upon contact with $hTAF_{II}31$ (Figure 3.17). Furthermore, plotting of the position of these residues on the surface of the proposed α-helical structure shows that they are all pointing in the same direction, thus strongly supporting the concept of an interaction helix that fits into a surface groove of $hTAF_{II}31$ (Figure 3.18). The position of the structurally crucial amino acids correlates well with the region of VP16 that has been shown to be relevant for TAF -binding and activation under *in vitro* conditions (Figure 3.19).

The key features of a model incorporating these observations can be summarized as follows (based on Uesugi *et al.*, 1997): The acidic residues present in acidic activation domains probably facilitate electrostatic ('long distance') interactions between VP16 and $hTAF_{II}31$. This relatively non-specific interaction brings the two protein molecules into close proximity and allows specific hydrophobic contacts (involving among others the key residues identified by the NMR studies) to be made. Formation of these critical contacts requires a structural transition of the activation domain from an unfolded state into an α-helical conformation. The VP16/$hTAF_{II}31$ interaction is relatively weak, but the highly cooperative nature of the transition between the folded and unfolded state of the activation domain could be used to send some form of transcriptional activation signal for further processing into the TFIID complex (see Chapter 4). The VP16 AAD also undergoes restructuring into a more highly ordered conformation in the presence of TBP and, to a lesser extent, TFIIB (Shen *et al.*, 1996), indicating that a similar restructuring event of the AAD also occurs during the interaction with basal factors.

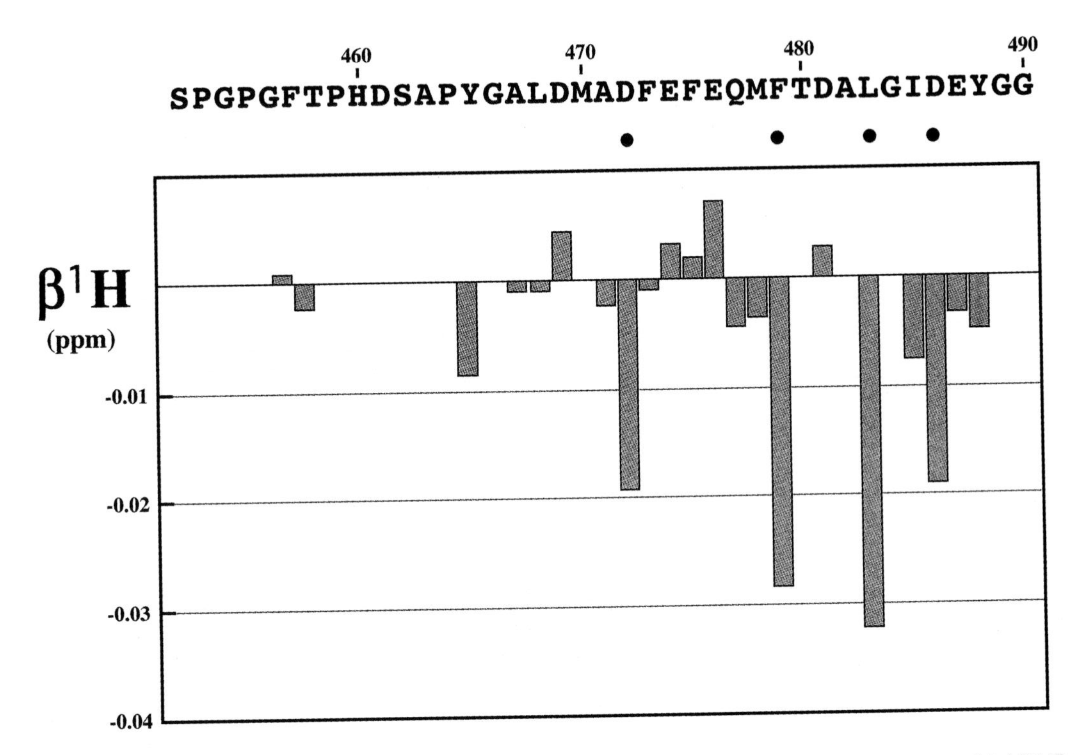

Figure 3.17. β **¹Hydrogen Chemical Shifts Induced in the VP16 Activation Domain During Interaction with $hTAF_{II}31$.** The amino acid sequence of the VP16 activation domain is shown in single amino acid letter code. Underneath the corresponding β ¹H chemical shifts that occur in the presence of the N-terminal 140 amino acids of $hTAF_{II}31$ are shown. Note the prominent negative shifts that are observed for the amino acid residues labelled by a black dot (including asp^{472}, phe^{479}, leu^{483} and asp^{486}), indicating substantial structural rearrangement of portions of the VP16 peptide backbone including these particular residues during the binding with the $hTAF_{II}31$ fragment. After Uesugi *et al.*, 1997.

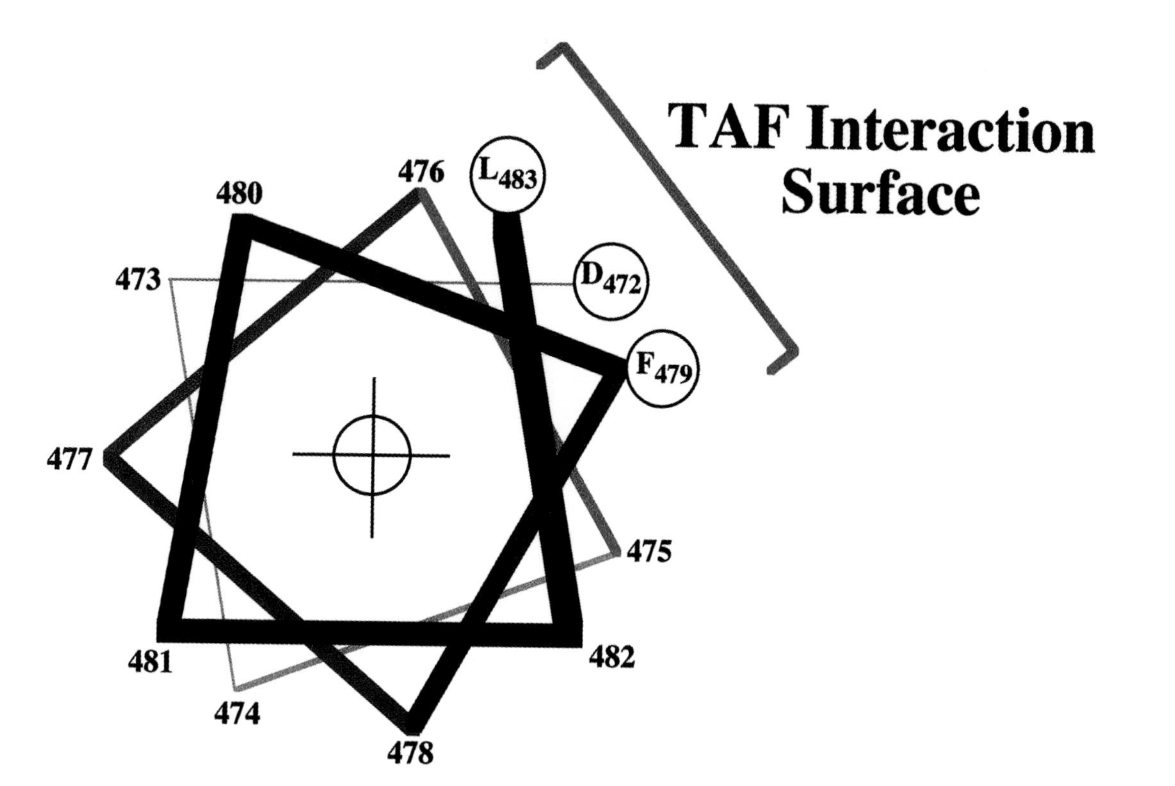

Figure 3.18. The Residues Involved In the Interaction with $hTAF_{II}31$ are Located on the Same Side of an α-Helix. Plotting of the position of the three amino acids displaying substantial chemical shifts during their interaction with $hTAF_{II}31$ shows that they are facing into a similar direction, if this portion of the activation domain undergoes a structural transition from random coil to α-helix. All three residues would therefore be available for specific contacts with the proposed interaction surface of $hTAF_{II}31$. After Uesugi *et al.*, 1997.

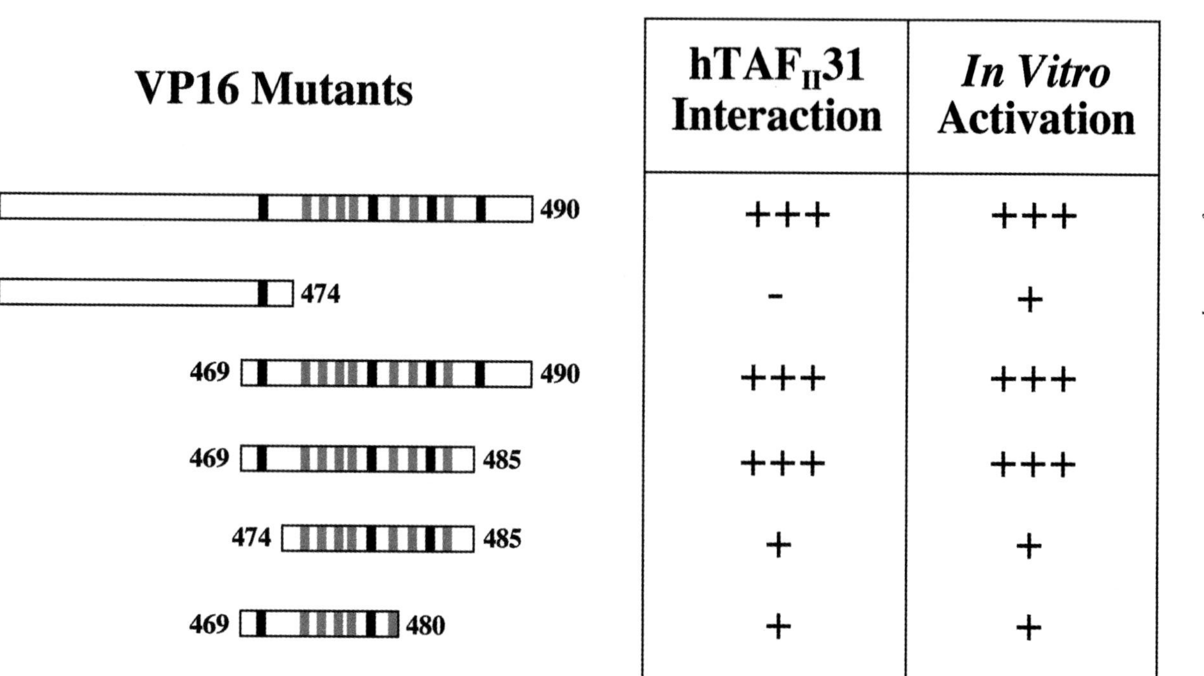

Figure 3.19. The Interaction of VP16 Mutants with hTAF$_{II}$31 and their Ability to Stimulate Transcription Correlate with Each Other.

A number of VP16 mutants with specific deletions in the activation domain (left) were tested for their ability to interact with hTAF$_{II}$31 and to activate transcription in an *in vitro* transcription assay. The good correlation between these two assays supports the proposed functional involvement of hTAF$_{II}$31 in mediating activation by VP16. After Uesugi *et al.*, 1997.

All the available data supports the view that the flexible nature of the VP16 activation domain (and perhaps of most, if not all, other activation domains) is a key functional feature explaining many of the observed properties of activators. Presumably the unfolded state of activation domains increases their chance of encountering their targets within the general transcriptional machinery and also increases the range of proteins suitable as productive interaction partners. As mentioned earlier, it is very difficult to assess at this stage in time whether the full target range of VP16 includes a variety of basal transcription factors and TAFs, or whether the data supporting such an interpretation is predominantly based on artefactual *in vitro* observations. It is reasonable, however, to assume that activation domains have been designed during evolution to be able to interact with a wide range of other transcription factors. Most eukaryotic promoters contain numerous gene-specific transcriptional activators and repressors bound to DNA sequences within the promoter region. It is likely that there is some form of competion for a limited set of interactions with components of the general transcriptional machinery that will result in activation due to kinetic or allosteric changes affecting the recruitment and/or function of the basal transcription factors. Maybe we also have to consider the function of activation domains as being specifically designed to look for good fits with their primary targets, but being able to interact with secondary targets if their primary target is not available or already occupied by a competing activation domain.

The model also solves another problem: the interactions between activation domains and the basal transcriptional machinery have to be specific, but should only last transiently. Any strong regulatory protein-protein contacts are intrinsically problematic because they would make gene control mechanisms rather sluggish in response to changing circumstances. The structural remodelling of an activation domain upon contact with its target as a mechanism may therefore have systematically evolved to maximize the specificity of the molecular recognition, while simultanously minimizing the strength of interaction.

Transcriptional Synergism: More-Than-Additive Effects

Very few, if any, gene-specific transcription factors work in isolation from each other. Although it is still difficult at this stage to quote an absolute number, it seems likely that the transcription of most eukaryotic genes is controlled by several dozen gene-specific transcription factors. Many of these factors cooperate functionally with each other, resulting in 'more-than-additive' effects (synergism) through a number of different mechanisms.

The simplest synergistic effect is based on cooperative binding of identical of different gene-specific transcription factors to adjacent DNA motifs within the regulatory region of a promoter. Many transcription factors do not only interact with DNA sequences through their DNA-binding motifs, but also make extensive protein-protein contacts with each other that greatly stabilize their specific association within a functional promoter complex (e.g. Giniger and Ptashne, 1988; Lin *et al.*, 1988; Carey *et al.*, 1990; Janson and Petterson, 1990; Tanaka, 1996; Kim and Maniatis, 1997). Apart from an increase in DNA-binding stability, the cooperative binding of many gene-specific transcription factors allows them to position themselves optimally for contacting and recruiting components of the general transcriptional machinery (Carey, 1998). Various other forms of synergism are thought to be due to additive interactions of different transcriptional activators with separate components of the basal transcriptional machinery (Carey *et al.*, 1990; Lin *et al.*, 1990; Chi *et al.*, 1995) or with different types of coactivators (see Chapter 4; Sauer *et al.*, 1995). According to this view, most gene-specific activators can only stimulate the basal transcriptional machinery through a single pathway and only to a limited extent. The presence of multiple activators that are able to stimulate several distinct rate-limiting steps simultanously in a non-competititve manner would therefore result in a substantially increased stimulation of pre-initiation complex (PIC) activity. We have already seen that the potent viral activator VP16 is probably acting at a number of different steps during PIC assembly, in addition to its specific interaction with a TFIID complex component. At a later stage it will become obvious that certain gene-specific transcriptional activators stimulate either the initiation of new transcripts or the elongation capacity of $RNAP_{II}$ in a predominant manner (Chapter 5). A combination of an activator that stimulates the initiation rate, and another activator that stimulates $RNAP_{II}$ elongation, will therefore in most cases result in very high degrees of synergism due to the complementary nature of the stimulatory functions. There are indeed several known examples, where particular combinations of gene-specific transcription factors have been shown to work optimally with each other. The HIV-transactivator protein Tat synergizes well with the human transcription factor Sp1, but not with VP16 (Figure 3.20; Southgate and Green 1991; Kamine and Chinnadurai, 1992).

Finally, especially under *in vivo* conditions or in *in vitro* systems containing reconstituted chromatin templates, the effect of chromatin structure on transcriptional synergism is very evident (e.g. Chang and Gralla, 1994). Using different artifical chromatin templates with a variable number of binding sites for a VP16 AAD-containing transcriptional activator, Chang and Gralla observed a substantial synergistic effect between templates containing a single or multiple binding sites (Figure 3.21). The most likely explanation for this threshold is that the presence of a single binding site for a transcriptional activator may not be sufficient to keep the transcription intitation site free of nucleosomes, which interfere with transcription. This interpretation is compatible with many other studies on chromatin remodelling and nucleosome displacement that indicate that this is a stepwise process that can greatly benefit from the functional cooperation of different types of gene-specific transcription factors (see Chapter 7 for more information on this topic).

3.5. Functional Modifications of Gene-Specific Transcription Factors

Post-Translational Phosphorylation Controls the Activity of Some Gene-Specific Transcription Factors

In common with other nuclear proteins, many gene-specific and basal transcription factors are phosphorylated at multiple sites under various *in vivo* growth conditions (reviewed in Treismann, 1996). Several cytoplasmic signalling pathways eventually lead to the activation of extensive kinase cascades that specifically alter the phosphorylation patterns of distinct serine, threonine (and more rarely, tyrosine) residues present in a vast number of cytoplasmic and nuclear proteins. In addition to such signalling pathways responding to external stimuli, there are the cyclin-dependent kinase families that act at various stages of the cell cycle to control the different molecular events leading up to mitosis. Changing the functional properties of specific

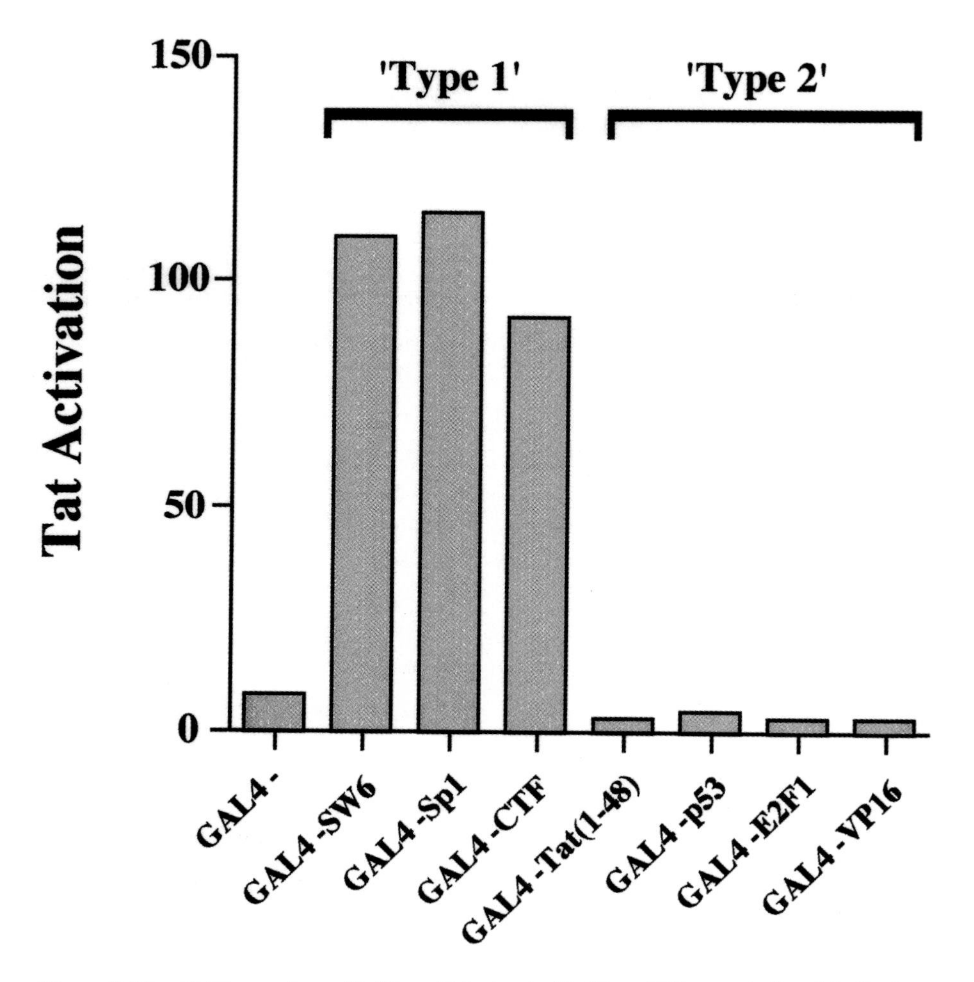

Figure 3.20. Tat Synergises with 'Type I', but not with 'Type 2' Transcriptional Activators. HeLa cells were transfected with a suitable reporter construct, an expression construct for the HIV-viral transcriptional activator Tat and a series of additional expression constructs containing a variety of different activation domains fused to the GAL4 DNA-binding domain. DNA-binding domains from the gene-specific activators SW6, Spl and CTF synergize with Tat-mediated activation and give rise to a substantial level of specific transcription from the reporter construct. Tat does not synergize with itself or with the activation domains derived from the gene-specific activators p53, E2F1 or VP16. After Blau *et al.*, 1996.

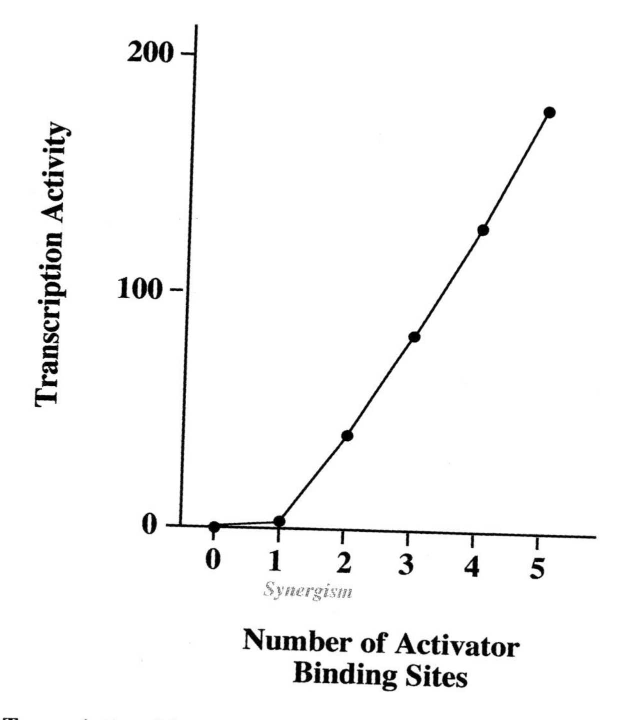

Figure 3.21. Transcriptional Synergism on Nucleosomal Templates.
A variety of templates differing in the number of binding sites for a gene-specific transcription activator were tested for specific transcription under *in vitro* conditions. Note that the templates containing a single activator binding site transcribe only at a rather low rate. Addition of additional sites increases the rate dramatically and indicate a synergistic effect.
After Chang and Gralla, 1994.

transcription factors through post-translational modifications, especially phosphorylation, allows cells to modulate and coordinate various gene expression patterns in a transient and fully-reversible manner. The overall effect of such kinase-mediated phosphorylation of gene-specific transcription factors can be either stimulatory or inhibitory on the transcription of specific genes. Whereas the majority transcriptional activators do not absolutely depend on such post-translational modifications for activity (at least under *in vitro* conditions), there are a few well-known cases where phosphorylation of particular amino acid residues within the activation- or DNA-binding domains is of major functional importance.

The cAMP-regulated transcription factor CREB is one of the most clear-cut examples of transcriptional regulation through changes in the post-translational modification state (Yamamoto *et al.*, 1988). In eukaryotic cells the cAMP/CREB complex plays a major role as a signalling molecule during the induction of hormonally controlled genes. In response to increased cAMP levels, a specific serine residue (S^{133}) present in CREB is phosphorylated (Gonzalez and Montminy, 1989) by protein kinase A (PKA), and only this phosphorylated version of CREB associates with a specific set of coactivators (p300 and CREB-binding protein [CBP]; Chrivia *et al.*, 1993; Parker *et al.*, 1996; see also Chapter 4). Phosphorylation of CREB constitutes a reversible way of stimulating transcription of target genes in response to cAMP induction and up to 50% of cellular CREB molecules contain phosphorylated S^{133} in response to strong stimuli (Hagiwara *et al.*, 1992). The 'KIX' domain of the coactivator CBP (see Chapter 4), that is involved in binding to the phosphorylated CREB activation domain, is highly conserved in higher eukaryotes and various studies have revealed a close contact between S^{133} and the KIX domain (Parker *et al.*, 1996; Rhadakrishnan *et al.*, 1997; see also Chapter 4 for more details).

Other well-documented examples of functional changes due to differential phosphorylation include the POU-containing transcription factors. The POU DNA-binding domains present in Oct-1 and Pit-1 are subject to post-translational modification during various stages of the cell-cycle and in response to homeostatic regulators (Kapiloff *et al.*, 1991; Roberts *et al.*, 1991; Segil *et al.*, 1991; Caelles *et al.*, 1995). During the mitotic stage of the cell cycle the hyperphosphorylation of specific residues within the Oct-1 POU domain prevents Oct-1 from binding to target sites present near many cell cycle-regulated genes, such as the human histone 2B promoter (Figure 3.22; Segil *et al.,* 1991). The S phase-specific transcription of the histone H2B gene is highly dependent on the evolutionarily conserved Oct-1 binding motif under both *in vivo* and *in vitro* conditions. Since the abundance and rate of *de novo* synthesis of Oct-1 protein does not vary at the different stages of the cell-cycle, the specific inhibition of its DNA-binding activity by phosphorylation

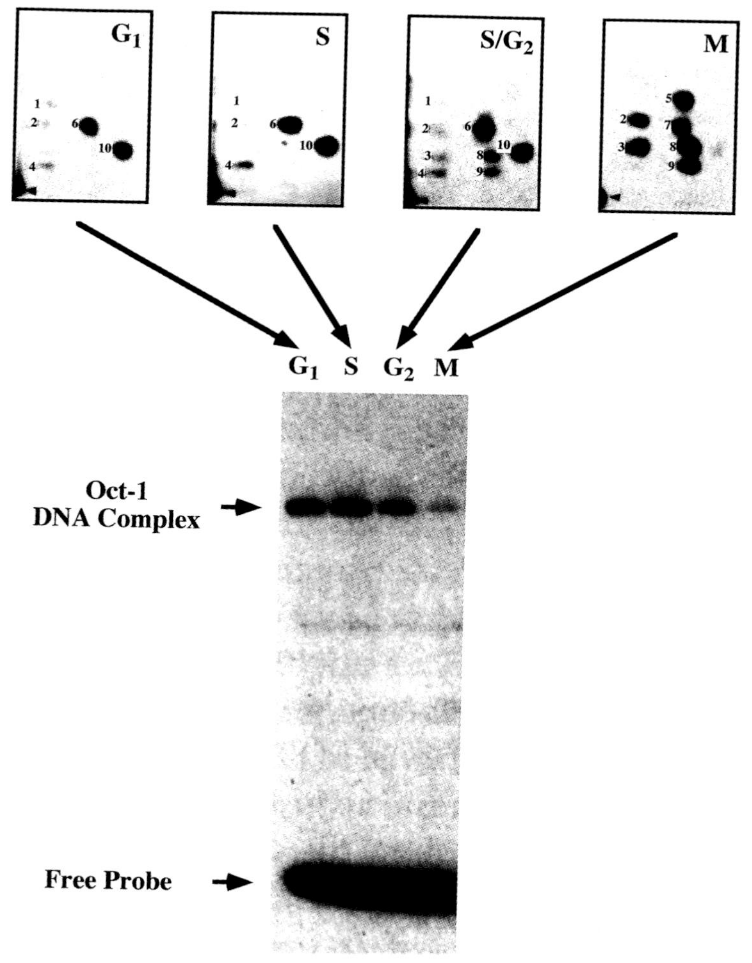

Figure 3.22. Cell Cycle Stage-Specific DNA-Binding of Differentially Phosphorylated Oct-1 Isoforms.

Top: Tryptic digests of *in vivo* ^{32}P-labelled Oct-1 extracted from cells at different stages of the cell cycle reveal differential phosphorylation patterns of various sites on Oct-1 (from Roberts *et al.*, 1991). *Bottom:* An oligonucleotide containing the Oct-1 binding site from the human histone 2B promoter was incubated with extracts prepared from cells at various stages of the cell cycle. The electrophoretic mobility shift assay shown indicates that DNA-binding of the distinctly phosphorylated Oct-1 isoform is inhibited during mitosis. After Segil *et al.*, 1991.

temporarily shuts down high levels of transcription from a large number of target genes that depend on activation by Oct-1 (Roberts *et al.*, 1991). The inhibition of DNA-binding activity of transcription factors, such as Oct-1, may be a necessary step to remove transcription factors from chromosomes prior to condensation (e.g. Nurse, 1990). Other observations, however, including the fact that phosphorylation of the Pit-1 POU domain specifically inhibits the binding of Pit-1 to particular DNA motifs, but does not affect binding to others (Kapiloff *et al.*, 1991), suggest that this is probably not the main reason. It is far more likely that the cell cycle-specific changes in the post-translational modification states of gene-specific transcription factors reflect a specific functional adaptation to the changing transcription patterns and prepare the transcriptional machinery for post-mitotic gene expression programs.

Acetylation Increases the DNA Sequence-Specific Binding of p53

Whereas phosphorylation of transcription factors is a wide-spread and well-documented phenomenon, there is also convincing evidence for the functional role of much rarer forms of post-translational modifications of transcription factors. The gene-specific transcription factor p53 contains an unusually-structured DNA-binding domain (see above), which is usually masked in most cellular p53 molecules by a regulatory domain located near the C-terminus (Hupp *et al.*, 1992; Meek, 1994; Gu and Roeder, 1997). This C-terminal tail is thought to interact with the p53 DNA-binding domain and thus prevents it from binding with high affinity to its target sites (Hupp *et al.*, 1992; Gu and Roeder, 1997). Acetylation of a series of lysine residues within the inhibitory C-terminal regulatory domain (Figure 3.23) seems to neutralize this interaction and convert p53 from a latent to an activated transcription factor with a substantially enhanced capacity for specific DNA-binding. The acetylation of this region of p53 is carried out by the p300 acetyltransferase, an important coactivator for several types of transcription factor which we will encounter again in Chapter 4. It is currently not understood how the acetyltransferase activity of p300 is controlled, but the effect of many specific deacetylase inhibitors on general cell growth suggest that the post-translational acetylation status of p53 could be of substantial regulatory significance in normal and cancer cells.

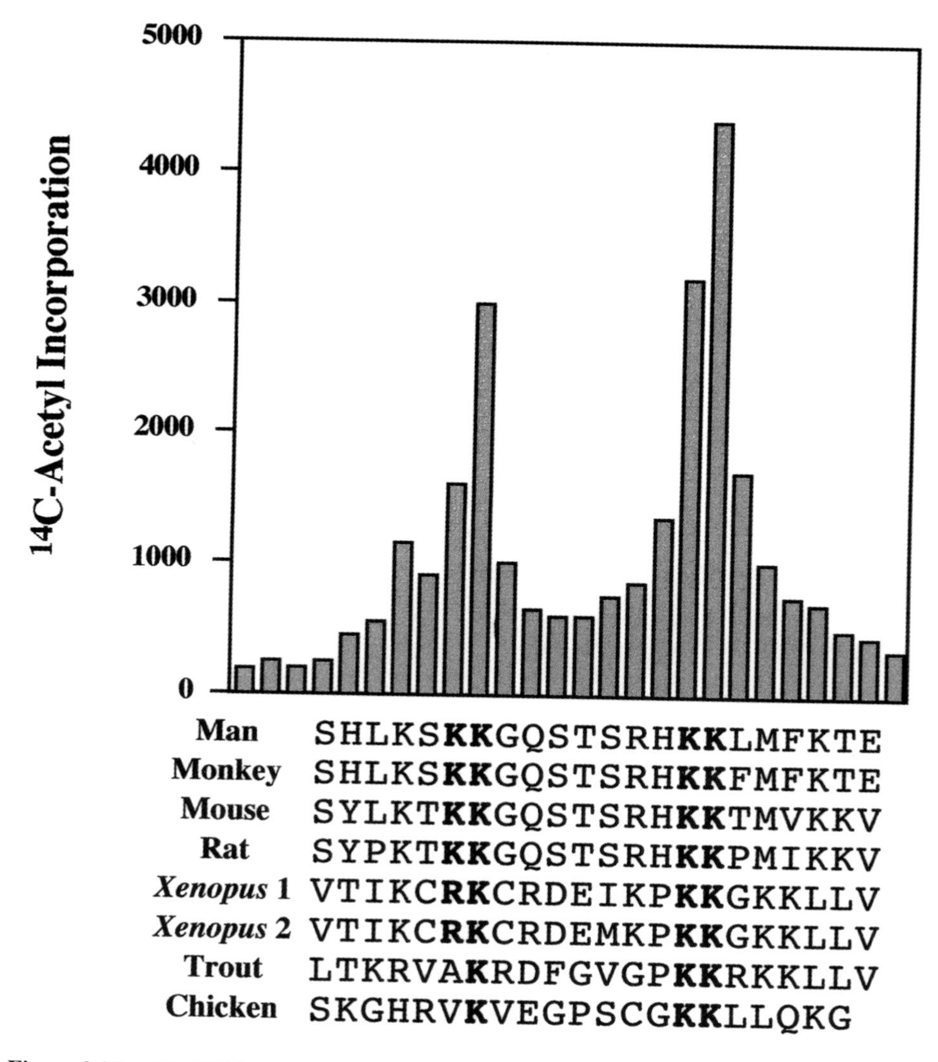

Figure 3.23. The P300 Acetylation Target Residues of p53 Occur in the Highly Conserved C-Terminal Regulatory Domain.

The amount of acetylation occuring at different residues of human p53 is shown in the histogram at the top of the figure. Two clusters of lysine residues ('K') are predominantly labelled. These lysine residues (or the functionally similar arginine 'R') are highly conserved in higher vertebrates and essentially also present in bird and fish p53 proteins. After Gu and Roeder, 1997.

O-Linked Glycosylation of Gene-Specific Transcription Factors

Despite the fact that glycosylation is often thought to be restricted to membrane proteins, there several well-documented instances of gene-specific transcription factors displaying specific O-linked glycosylation patterns (Jackson and Tjian, 1988; 1989). Endogenous human Sp1 contains on average around eight distinct O-linked N-acetylglucosamine (GlcNAc) monosaccharide residues, and binding of the lectin wheat germ agglutinin to these residues interferes with the activation (but not DNA binding-) function of Sp1 (Figure 3.24; Jackson and Tjian, 1988). Although we do not currently understand the full functional significance of post-translational glycosylation of gene-specific transcription factors, it is likely that the O-linked GlcNAc residues play a regulatory role and may actually make an important contribution to the structure and/or function of the transcription activation domains. In Sp1 the majority of glycosylation sites are located within the N-terminal half of the protein, which coincides with the known position of several glutamine-rich activation domains (Figure 3.15).

Although glycosylation does not affect the DNA-binding affinity and specificity of Sp1, there is good evidence that this is not a universal phenomenon. The presence of O-linked sugar groups in the basic C-terminus of p53 activates the DNA-binding domain in a manner similar to the one described above for the acetylation modification (Shaw *et al.*, 1996). Glycosylation of the inhibitory domain is thought to prevent the specific intramolecular interaction of the C-terminus with the DNA-binding domain and thus allows it to become available for DNA binding.

3.6. Gene-Specific Transcription Factors Shape Overall Promoter Topology

The Reason for DNA Looping

We have already seen in Chapter 2 that a substantial amount of evidence suggests that certain basal transcription factors, such as TFIID and TFIIF, rearrange the path of double-stranded DNA through the pre-initiation complex (PIC). The overall cellular strategy behind promoter looping is almost certainly due to the spatial constraints imposed by the linear nature of DNA molecules.

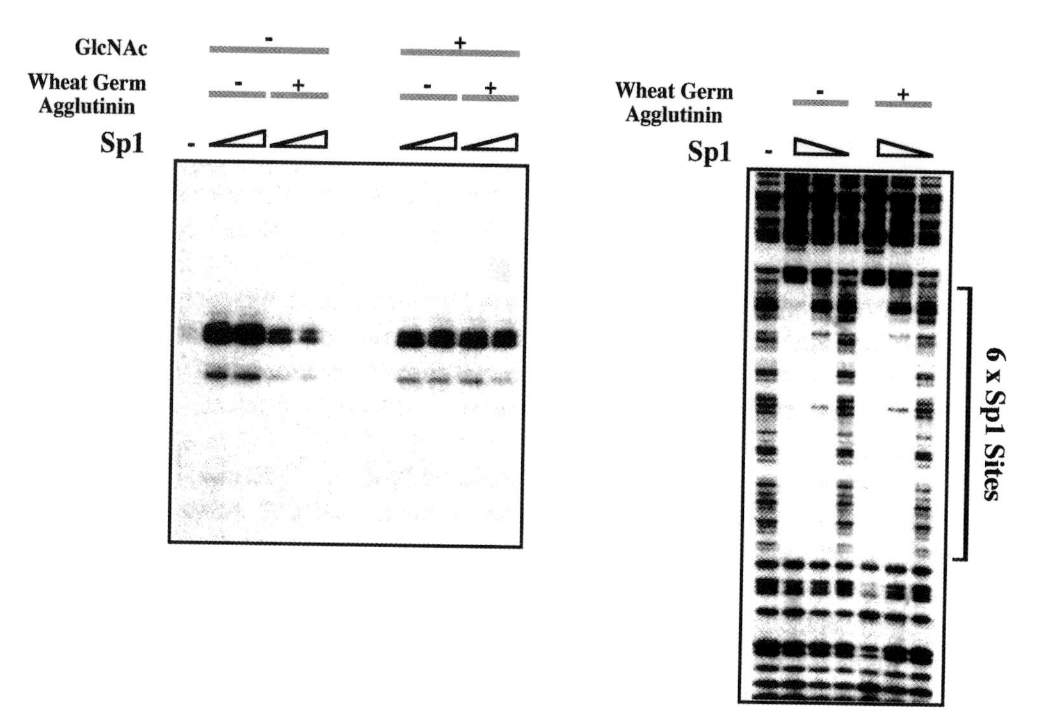

Figure 3.24. Wheat Germ Agglutinin Inhibits Activation by Sp1 Without Interfering with Specific DNA-Binding. Wheat germ agglutinin (WGA) binds specifically to the O-linked GlcNAc residues present in endogenous Sp1. *Left Panel:* Addition of WGA reduces Sp1-activated transcription in a defined *in vitro* system approximately 3- to 4-fold. The presence of free GlcNAc in the reaction mixture effectively prevents the inhibition by competitively binding to WGA and preventing it from binding to Sp1. *Right Panel:* The specific DNA-binding of Sp1 to six separate target sites present in the SV40 early promoter is not affected by the presence of WGA. After Jackson and Tjian, 1988.

Many eukaryotic genes require the assembly of a complex basal transcriptional machinery around the transcription start site (Chapter 2), which then needs to be regulated by numerous gene-specific transcription factors. In order for these activators and repressors to interact with the various components of the PIC, looping of the intervening DNA sequence becomes an absolute necessity for bringing the interacting proteins into close spatial vicinity. It is likely, that the significance of the three-dimensional promoter structure will become increasingly acknowledged in the forseeable future and supported at some stage with high-resolution structural data of the three-dimensional configuration of an entire promoter-transcription factor complex (the 'enhanceosome'; Falvo *et al.*, 1995; Carey, 1998). At the moment we can, however, only infer the important role of promoter topology from various individual bacterial and eukaryotic examples, where it has been studied in sufficient detail.

Many gene-specific activators cause substantial DNA-bending as a direct consequence of the interaction of the DNA-binding domain with its target sequence. The bending of double-stranded DNA is energetically unfavourable but can be accomplished through the insertion of hydrophobic amino acids (such as the 'leucine-lever' or 'phenylalanine wedge' Kim *et al.*, 1993; Schumacher *et al.*, 1994) between particular nucleotides to widen the minor groove, which allows tighter and more specific contacts at the DNA/protein interfaces. Such events usually introduce strong kinks (45–120°) into the DNA and have important architectural effects on the structure of promoter/ transcription factor complexes. In other cases, looping of promoter DNA is the consequence of subsequent protein-protein interactions/formation and higher order transcription complexes involving proteins located on DNA target sites that are spread widely over the promoter. The formation of DNA loops based on such protein-protein interaction is obviously favoured by the DNA-bending induced by other transcription factors and in many cases the overall promoter topology is ultimately determined by many different types of protein-protein and protein-DNA contacts.

The functional relevance of DNA-looping and establishment of a distinct three-dimensional promoter architecture has been extensively demonstrated in several bacterial systems (e.g. Hoover *et al.,* 1990; Perez-Martin and Espinosa, 1993; Ansari *et al.,* 1995; Dethiollaz *et al.*, 1996; Wyman *et al.*, 1997). One of the best-known example occurs in the *lac* operon, where it has been shown that DNA-looping, induced by the presence of two sequence-specific DNA-binding domains on the opposite ends of the *lac* repressor tetramer, causes the formation of various combinatorial looping patterns, involving the three distinct operator sites present in various positions near the transcription start site of the *lac* operon (reviewed in Matthews and Nichols, 1998). The spacing and sequences of the ancilliary operator sites (O_2 and O_3), in relation to the main operator O_1 is of prime importance, and influences the stability of the resulting 'repressor loop' more than ten-fold (Kramer *et al.*, 1987; Eismann and Muller-Hill, 1990). Interestingly, the formation of the *lac*-repressor loop between O_1 and O_3 is also strongly stabilized through a 90° kink induced by the CAP activator protein bound between the two sites (see Figure 4.1. in Chapter 4).

In eukaryotic systems the evidence for specific promoter looping is less extensive, but this is probably mostly due to the fact that the question has not been comprehensively addressed in a sufficient number of model systems. There is some intriguing evidence suggesting that several gene-specific transcription factors, such as HMG-I/Y, LEF-1 and YY1, exert their main regulatory functions on various cellular promoters through DNA looping (Natesan and Gilman, 1993; Falvo *et al.*, 1995; Giese *et al.*, 1995). Other transcription factors, such as Sp1, have been shown to specifically oligomerize into higher order complexes and cause DNA-looping between distant DNA target sites under *in vitro* conditions (Figure 3.25; Mastrangelo *et al.*, 1991). *In vivo*, Sp1 molecules bound to sites in the thymidine kinase promoter located 1.7 kb away from the transcript initiation site can interact synergistically with more proximally-bound Sp1 factors (Courey *et al.*, 1989). Such observations suggest that the gene-specific transcription factors that activate transcription from large distances almost certainly depend on such quaternary associations to exert their stimulatory effect on the proximally-located preinitiation complex.

3.7. Conclusions

Gene-specific transcription factors interact with the basal transcriptional machinery on individual promoters, resulting in the controlled and coordinated expression of genes during many important biological processes. Transcriptional

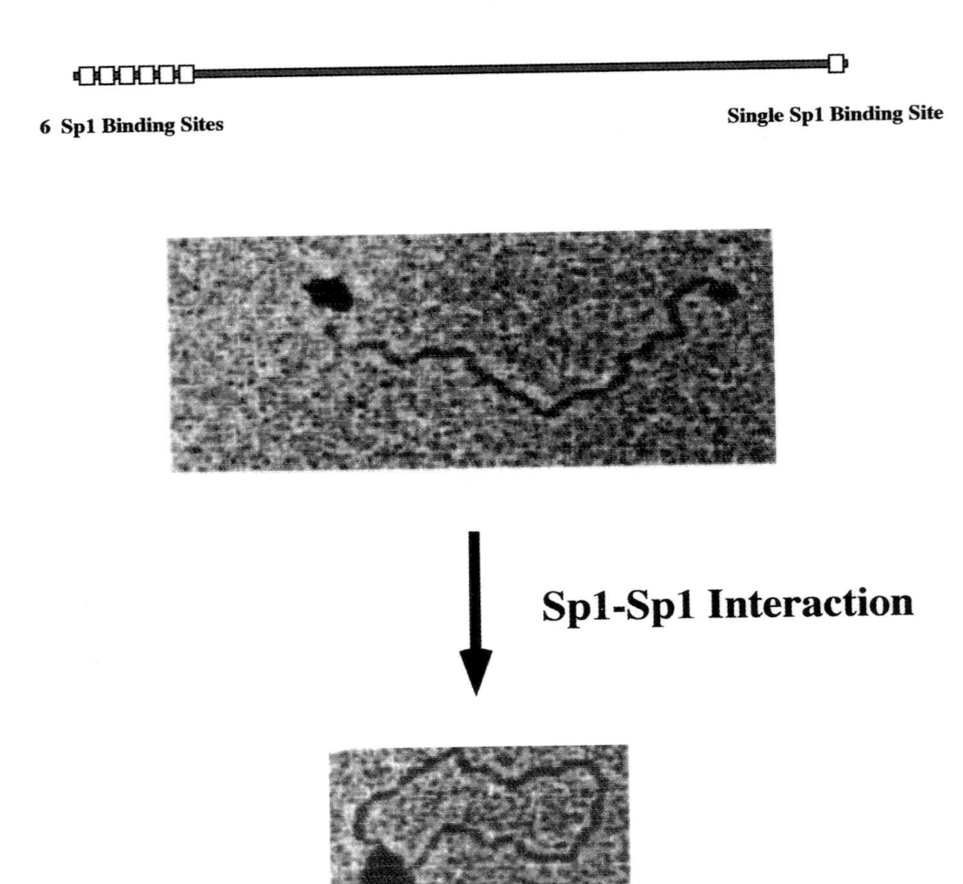

Figure 3.25. Evidence for DNA Looping Through Protein-Protein Contacts Between Distantly-Located Sp1 Molecules.
A linear DNA molecule containing Sp1 sites at opposite ends can undergo loop formation through specific contacts between Sp1 molecules bound to them.

activation (rather than repression) is the major mode of regulating gene expression in eukaryotes.

Most gene-specific transcription factors contain well-defined protein domains that carry out distinct functions. Frequently occuring DNA-binding domains are based on variations of the helix-turn-helix, helix-loop-helix, and zinc finger motifs. The domains present in eukaryotic transcription factors that are responsible for stimulating the rate of transcription often contain a high proportion of particular amino acids, such as acidic residues, prolines and glutamines. Recent studies have indicated that strategically-placed hydrophobic residues play an important role in mediating specific contacts between activation domains and components of the basal transcriptional machinery. Although we are still far from understanding the structure and function of activation domains at the molecular level, there are many preliminary indications that they are unstructured in isolation and only take up specific conformations in the presence of suitable molecular interaction partners.

Many gene-specific transcription factors are post-translationally modified and in some cases a clear correlation between specific modifications and changes in transcription factor functions has been established.

References

WWW Sites of Interest

http://transfac.gbf-braunschweig.de/TRANSFAC/

A database systematically listing structural and functional information about more than 2300 gene-specific transcription factors and their promoter target sequences.

Research Literature

Ansari, A.Z., Bradner, J.E., and O'Halloran, T.V. (1995). DNA-bend modulation in a repressor-to-activator switching mechanism. *Nature* 374, 371–375.

Assa-Munt, N., Mortishire-Smith, R.J., Aurora, R., Herr, W., and Wright, P.E. (1993). The solution structure of the Oct-1 POU-specific domain reveals a striking similarity to the bacteriophage λ repressor DNA binding domain. *Cell* 73, 193–205.

Aurora, R., and Herr, W. (1992). Segments of the POU domain influence one another's DNA binding specificity. *Mol. Cell. Biol.* 12, 455–467.

Ayer, D.E., Kretzner, L., and Eisenmann, R.N. (1993). Mad: a heterodimeric partner for Max that antagonizes Myc transcriptional activity. *Cell* 72, 211–222.

Bandara, L.R., Girling, R., La Thangue, N.G. (1997). Apoptosis induced in mammalian cells by small peptides that functionally antagonize the Rb-regulated E2F transcription factor. *Nature Biotechnol.* 15, 896–901.

Banerji, J., Rusconi, S., and Schaffner, W. (1981). Expression of a β-globin gene is enhanced by remote SV40 sequences. *Cell* 33, 729–740.

Bauerle, P.A., and Baltimore, D. (1996). NFκB: ten years after. *Cell* 87, 13–20.

Beals, C.R., Clipstone, N.A., Ho, S.N., and Crabtree, G.R. (1997). Nuclear localization of NF-ATc by calcineurin-dependent, cyclosporin-sensitive intramolecular interaction. *Genes Dev.* 11, 824–834.

Benezra. R., Davis, R.L., Locksun, D., Turner, D.L., and Weintraub, H. (1990). The protein Id: a negative regulator of helix-loop-helix DNA binding proteins. *Cell* 61, 49–59.

Berg, J.M. (1990). Zinc finger domains: hypothesis and current knowledge. *Annu. Rev. Biophys. Biophys. Chem.* 19, 405–421.

Berg, J.M., and Shi, Y. (1996). The galvanization of biology: a growing appreciation for the roles of zinc. *Science* 271, 1081–1085.

Blackwood, E., and Eisenmann, R.N. (1991). Max: a helix-loop-helix zipper protein that forms a sequence-specific DNA-binding complex with Myc. *Science* 251, 1211–1217.

Borelli, E., Hen, R., and Chambon, P. (1984). Adenovirus-2 E1a products repress enhancer-induced stimulation of transcription. *Nature* 312, 608–612.

Brent, R., and Ptashne, M. (1985). A eukaryotic transcriptional activator bearing the DNA specificity of a prokaryotic repressor. *Cell* 43, 729–736.

Brown, B.M., Milla, M.E., Smith, T.L., and Sauer, R.T. (1994). Scanning mutagenesis of Arc repressor as a functional probe of operator recognition. *Nature Struct. Biol.* 1, 164–168.

Buckbinder, L., Talbott, R., Velasco-Miguel, S., Takenaka, I., Faha, B., Seizinger, B.R., and Kley, N. (1995). Induction of the growth inhibitor IGF-binding protein 3 by p53. *Nature* 377, 646–649.

Caelles, C., Hennemann, H., and Karin, M. (1995). M phase-specific phosphorylation of the POU transcription factor GHF-1 by a cell cycle-regulated protein kinase inhibits DNA binding. *Mol. Cell. Biol.* 15, 6694–6701.

Carey, M. (1998). The enhanceosome and transcriptional synergy. *Cell* 92, 5–8.

Carey, M., Lin, Y.S., Green, M.R., and Ptashne, M. (1990). A mechanism for synergistic activation of a mammalian gene by GAL4 derivatives. *Nature* 345, 361–364.

Ceska, T.A., Lamers, M., Monaci, P., Nicosia, A., Cortese, R., and Suck, D. (1993). The X-ray structure of an atypical homeodomain present in the rat liver transcription factor LFB1/HNF1 and implications for DNA-binding. *EMBO J.* 12, 1805–1810.

Chang, C., and Gralla, J.D. (1994). A critical role for chromatin in mounting a synergistic transcriptional response to GAL4-VP16. *Mol. Cell. Biol.* 14, 5175–5181.

Chen, J.-L., Attardi, L.D., Verrijzer, C.P., Yokomori, K., and Tjian, R. (1994). Assembly of recombinant TFIID reveals differential requirements for distinct transcriptional activators. *Cell* 79, 93–105.

Chi, T., Liebermann, P., Ellwood, K. and Carey, M. (1995). A general mechanism for transcriptional synergy of eukaryotic activators. *Nature* 377, 254–257.

Cho, Y., Gorina, S., Jeffrey, P.D., and Pavletich, N.P. (1994). Crystal structure of a p53 tumor suppressor-DNA complex: understanding tumorigenic mutations. *Science* 265, 346–355.

Choy, B., and Green, M.R. (1993). Eukaryotic activators function during multiple steps of preinitiation complex assembly. *Nature* 366, 531–536.

Chrivia, J.C., Kwok, R.P., Lamb, N., Hagiwara, M., Montminy, M., and Goodman, R.H. (1993). Phosphorylated CREB binds specifically to nuclear protein CBP. *Nature* 365, 855–859.

Clark, L., Pollock, R.M., and Hay, R.T. (1988). Identification and purification of EBP1: a HeLa cell protein that binds to a region overlapping the 'core' of the SV40 enhancer. *Genes Dev.* 2, 991–1002.

Courey, A.J., and Tjian, R. (1988). Analysis of Sp1 *in vivo* reveals multiple transcriptional domains, including a novel glutamine-rich activation motif. *Cell* 55, 887–898.

Courey, A.J., Holtzman, D.A., Jackson, S.P., and Tjian, R. (1989). Synergistic activation by the glutamine-rich domains of human transcription factor Sp1. *Cell* 59, 827–836.

Cousens, D.J., Greaves, R., Goding, C.R., and O'Hare, P. (1989). The C-terminal 79 amino acids of the herpes simplex virus regulatory protein, Vmw65, efficiently activate transcription in yeast and mammalian cells in chimeric DNA-binding proteins. *EMBO J.* 8, 2337–2342.

Cress, A., and Triezenberg, S.J. (1991). Nucleotide and deduced amino acid sequence of the gene encoding virion protein 16 of herpes simplex virus type 2. *Gene* 103, 235–238.

Curtis, D., Treiber, D.K., Tao, F., Zamore, P.D., Williamson, J.R., and Lehmann, R. (1997). A CCHC metal-binding domain in Nanos is essential for translational regulation. *EMBO J.* 16, 834–843.

Dekker, N., Cox, M., Boelens, R., Verrijzer, C.P., and Kaptein, R. (1993). Solution structure of the POU-specific DNA-binding domain of Oct-1. *Nature* 362, 852–855.

Desjarlais, J.R., and Berg, J.M. (1992). Redesigning the DNA-binding specificity of a zinc finger protein: a data base-guided approach. *Proteins Struct. Funct. Genet.* 12, 101–104.

Desplan, C., Theis, J., and O'Farrell, P.H. (1988). The sequence-specificity of homeodomain-DNA interaction. *Cell* 54, 1081–1090.

Dessain, S., Gross, C.T., Kuziora, M., and McGinnis, W. (1992). Antp-type homeodomains have distinct DNA binding specificities that correlate with their target specificities in the embryo. *EMBO J.* 11, 991–1002.

Dethilollaz, S., Eichenberger, P., and Geiselmann, J. (1996). Influence of DNA geometry on transcriptional activation in *Escherichia coli*. *EMBO J.* 15, 5449–5458.

Donaldson, L., and Capone, J.P. (1992). Purification and characterization of the carboxy-terminal transactivation domain of Vmw65 from herpes simplex virus type 1. *J. Biol. Chem.* 267, 1411–1414.

Drysdale, C.M., Jackson, B.M., McVeigh, R., Klebanow, E.R., Bai, Y., Kokubo, T., Swanson, M., Nakatani, Y., Weil, P.A., and Hinnebusch, A.G. (1998). The Gcn4p activation domain interacts specifically *in vitro* with RNA polymerase II holoenzyme, TFIID, and the Adap-Gcn5p coactivator complex. *Mol. Cell. Biol.* 18, 1711–1724.

Dynan, W.S., and Tjian, R. (1983). The promoter-specific transcription factor Sp1 binds to upstream sequences in the SV40 early promoter. *Cell* 35, 79–87.

Eismann, E.R., and Muller-Hill, B. (1990). *lac* repressor forms stable loops *in vitro* with supercoiled wild-type *lac* DNA containing all three natural *lac* operators. *J. Mol. Biol.* 213, 763–775.

Ellenberger, T.E., Brandl, C.J., Struhl, K., and Harrison, S.C. (1992). The GCN4 basic region leucine zipper binds DNA as a dimer of uninterupted α helices: crystal structure of the protein-DNA complex. *Cell* 71, 1223–1237.

Ellenberger, T.E., Fass, D., Arnaud, M., and Harrison, S.C. (1994). Crystal structure of transcription factor E47: E-box recognition by a basic region helix-loop-helix dimer. *Genes Dev.* 8, 970–980.

Emami, K.H., Navarre, W.W., and Smale, S.T. (1995). Core promoter specificities of the Sp1 and VP16 transcriptional activation domains. *Mol. Cell. Biol.* 15, 5906–5916.

Falvo, J.V., Thanos, D., and Maniatis, T. (1995). Reversal of intrinsic DNA bends in the IFNβ gene enhancer by transcription factors and the architectural protein HMG I(Y). *Cell* 83, 1101–1111.

Ferre-D'Amare, A.R., Prendergast, G.C., Ziff, E.B., and Burley, S.K. (1993). Recognition by Max of its cognate DNA through a dimeric b/HLH/Z domain. *Nature* 363, 38–45.

Ferre-D'Amare, A.R., Pognonec, G.C., Roeder, R.G. and Burley, S.K. (1994). Structure and function of the b/HLH/Z domain of USF. *EMBO J.* 13, 180–189.

Galcheva-Gargova, Z., Konstantinov, K.N., Wu, I-H., Klier, G., Barrett, T., and Davies, R. (1996). Binding of a zinc finger protein ZPR1 to the epidermal growth factor receptor. *Science* 272, 1797–1802.

Gerber, H.-P., Seipel, K., Georgiev, O., Hofferer, M., Hug, M., Rusconi, S., and Schaffner, W. (1994). Transcriptional activation modulated by homopolymeric glutamine and proline stretches. *Science* 263, 808–810.

Gerster, T., Matthias, P., Thali, M., Jiricny, J., and Schaffner, W. (1987). Cell type specificity elements of the immunoglobulin heavy chain enhancer. *EMBO J.* 6, 1323–1330.

Giese, K., Kingsley, C., Kirshner, J.R., and Grosschedl, R. (1995). Assembly and function of a TCR alpha enhancer complex is dependent on LEF1-induced DNA bending and multiple protein-protein interactions. *Genes Dev.* 9, 995–1008.

Gill, G., Pacal, E., Tseng, Z.H., and Tjian, R. (1994). A glutamine-rich hydrophobic patch in transcription factor Sp1 contacts the dTAF$_{II}$110 component of the *Drosophila* TFIID complex and mediates transcriptional activation. *Proc. Natl. Acad. Sci. USA* 91, 192–196.

Gillies, S.D., Morrison, S.L., Oi, V.T., and Tonegawa, S. (1983). A tissue-specific transcription enhancer element is located in the major intron of a rearranged immunoglobulin heavy chain gene. *Cell* 33, 717–728.

Giniger, E., and Ptashne, M. (1988). Cooperative DNA binding of the yeast transcription factor GAL4. *Proc. Natl. Acad. Sci. USA* 85, 382–386.

Glass, C.K., Rose, D.W., and Rosenfeld, M.G. (1997). Nuclear receptor coactivators. *Curr. Opin. Cell Biol.* 9, 222–232.

Gonzalez, G.A., and Montminy, M.R. (1989). Cyclic AMP stimulates somatostatin gene transcription by phosphorylation of CREB at serine-133. *Cell* 59, 675–680.

Goodrich, J.A., Hoey, T., Thut, C.J., Admon, A., and Tjian, R. (1993). *Drosophila* $TAF_{II}40$ interacts with both a VP16 activation domain and the basal transcription factor TFIIB. *Cell* 75, 519–530.

Goodbourn, S., Burstein, H., and Maniatis, T. (1986). The human β-interferon gene enhancer is under negative control. *Cell* 45, 601–610.

Gorman, C.M., Rigby, P.W.J. and Lane, D.P. (1985). Negative regulation of viral enhancers in undifferentiated embryonic stem cells. *Cell* 42, 519–526.

Gstaiger, M., Georgiev, O., van Leeuwen, H., van der Vliet, P., and Schaffner, W. (1996). The B cell coactivator Bob1 shows DNA sequence-dependent complex formation with Oct-1/Oct-2 factors, leading to differential promoter activation. *EMBO J.* 15, 2781–2790.

Gu, W., and Roeder, R. (1997). Activation of p53 sequence-specific binding by acetylation of the p53 C-terminal domain. *Cell* 90, 595–606.

Hagiwara, M., Alberts, A., Brindle, P., Meinkoth, J., Feramisco, J., Deng, T., Karin, M., Shenolikar, S., and Montminy, M. (1992). Transcriptional attenuation following cAMP induction requires PP-1-mediated dephosphorylation of CREB. *Cell* 70, 105–113.

Hahn, S. (1993). Transcription: efficiency in activation. *Nature* 363, 672–673.

Haigh, A., Greaves, R., and O'Hare, P. (1990). Interference in the assembly of a virus-host transcription complex by peptide competition. *Nature* 344, 257–259.

Hard, T., Kellenbach, E., Boelens, R., Maler, B.A., Dahlman, K., Freedman, L.P., Carlstedt-Duke, J., Yamamoto, K.R., Gustafsson, J.A., and Kaptein, R. (1990). Solution structure of the glucocorticoid receptor DNA-binding domain. *Science* 249, 157–160.

Harrison, C.J., Bohm, A.A., and Nelson, H.C.M. (1994). Crystal structure of the DNA-binding domain of the heat shock transcription factor. *Science* 263, 224–227.

He, Y.-Y., McNally, T., Manfield, I., Navratil, O., Old, I.G., Philips, S.E., Saint, G.I., and Stockley, P.G. (1992). Probing Met repressor-operator recognition in solution. *Nature* 359, 431–433.

Herr, W., and Cleary, M.A. (1995). The POU domain: versatility in transcriptional regulation by a flexible two-in-one DNA binding domain. *Genes Dev.* 9, 1679–1693.

Hollstein, M., Sidransky, D., Vogelstein, B., and Harris, C.C. (1991). p53 mutations in human cancers. *Science* 253, 49–53.

Hoover, T.R., Santero, E., Porter, S., and Kustu, S. (1990). The integration host factor stimulates interaction of RNA polymerase with *NIFA*, the transcriptional activator for nitrogen fixation operons. *Cell* 63, 11–22.

Hoovers, J.M.N., Mannens, M., Joh, R., Bliek, J., van Heyningen, V., Porteous, D.J., Leschot, N.J., Westerveld, A., and Little, P.F. (1992). High resolution localization of 69 potential human zinc finger protein genes: a number are clustered. *Genomics* 12, 254–263.

Hupp, T.R., Meek, D.W., Midgley, C.A., and Lane, D.P. (1992). Regulation of specific DNA binding function of p53. *Cell* 71, 875–886.

Ingles, C.J., Shales, M., Cress, W.D., Triezenberg, S.J., and Greenblatt, J. (1991). Reduced binding of TFIID to transcriptionally compromised mutants of VP16. *Nature* 351, 588–590.

Jackson, S.P., and Tjian, R. (1988). O-glycosylation of eukaryotic transcription factors: implications for mechanisms of transcriptional regulation. *Cell* 55, 125–133.

Jackson, S.P., and Tjian, R. (1989). Purification and analysis of RNA polymerase II transcription factors by using wheat germ agglutinin affinity chromatography. *Proc. Natl. Acad. Sci. USA* 86, 1781–1785.

Jacobson, E.M., Li, P., Leon-del-Rio, A., Rosenfeld, M.G., and Aggarwal, A.K. (1997). Structure of Pit-1 POU domain bound to DNA as a dimer: unexpected arrangement and flexibility. *Genes Dev.* 11, 198–212.

Johnson, P.F., Sterneck, E., and Williams, S.C. (1993). Activation domains of transcriptional regulatory proteins. *J. Nutr. Biochem.* 4, 386–398.

Jones, N.C., Rigby, P.W.J., and Ziff, E.B. (1988). *Trans*-acting protein factors and the regulation of eukaryotic transcription: lessons from studies on DNA tumor viruses. *Genes Dev.* 2, 267–281.

Jordan, S.R., and Pabo, C.O. (1988). Structure of the λ complex at 2.5Å resolution: details of the repressor-operator interactions. *Science* 242, 893–899.

Kadonaga, J.T., and Tjian, R. (1986). Affinity purification of sequence-specific DNA binding proteins. *Proc. Natl. Acad. Sci. USA* 83, 5889–5893.

Kadonaga, J.T., Carner, K.R., Masiarz, F.R., and Tjian, R. (1987). Isolation of cDNA encoding transcription factor Sp1 and functional analysis of the DNA-binding domain. *Cell* 51, 1079–1090.

Kadonaga, J.T., Courey, A.J., Ladika, J. and Tjian, R. (1988). Distinct regions of Sp1 modulate DNA binding and transcriptional activation. *Science* 242, 1566–1570.

Kamine, J., and Chinnadurai, G. (1992). Synergistic activation of the human immunodeficiency virus type 1 promoter by the viral TAT protein and cellular transcription factor Sp1. *J. Virol.* 66, 3932–3936.

Kapiloff, M.S., Farkash, Y., Wegner, M., and Rosenfeld, M.G. (1991). Variable effects of phosphorylation of Pit-1 dictated by the DNA response elements. *Science* 253, 786–789.

Kaptein, R., Zuiderseg, E.R., Scheek, R.M., Boelens, R., and van Gunsteren, W.F. (1985). A protein structure from nuclear magnetic resonance data. *lac* repressor headpiece. *J. Mol. Biol.* 182, 179–182.

Kim, T.K., and Maniatis, T. (1997). The mechanism of transcriptional synergy of an *in vitro* assembled interferon-β enhanceosome. *Mol. Cell* 1, 119–129.

Kim, Y. , Geiger, J.H., Hahn, S., and Sigler, P.B. (1993). Crystal structure of a yeast TBP/TATA-box complex. *Nature* 365, 512–520.

Kim, J.L., Nikolov, D.B., and Burley, S.K. (1993). Co-crystal structure of TBP recognizing the minor groove of a TATA element. *Nature* 365, 520–527.

Kissinger, C.R., Liu, B.S., Kornberg, T.B., and Pabo, C.O. (1990). Crystal structure of an *engrailed* homeodomain-DNA complex at 2.8Å resolution: a framework for understanding homeodomain-DNA interactions. *Cell* 63, 579–590.

Klemm, J.D., Rould, M.A., Aurora, R., Herr, W., amd Pabo, C.O. (1994). Crystal structure of the Oct-1 POU domain bound to the octamer site: DNA recognition with tethered DNA-binding modules. *Cell* 77, 21–32.

Klug, A., and Schwabe, J.W.R. (1995). Zinc fingers. *FASEB J.* 9, 597–604.

Ko, L.J., and Prives, C. (1996). p53: puzzle and paradigm. *Genes Dev.* 10, 1054–1072.

Kramer, H., Niemoller, M., Amouyal, M., Revet, B., von Wilcken-Bergmann, B., and Muller-Hill, B. (1987). *lac* repressor forms loops with linear DNA carrying two suitably spaced *lac* operators. *EMBO J.* 6, 1481–1491.

Lai, J.S., Cleary, M.A., and Herr, W.,(1992). A single amino acid exchange transfers VP16-induced positive control from the Oct-1 to the Oct-2 homeo domain. *Genes Dev.* 6, 2058–2065.

Lee, W., Mitchell, P., and Tjian, R. (1987). Purified transcription factor AP-1 interacts with TPA-inducible enhancer elements. *Cell* 49, 741–752.

Leonardo, M., Pierce, J., and Baltimore, D. (1987). Protein binding sites in immunoglobulin gene enhancers determine transcriptional activity and inducibility. *Science* 236, 1573–1577.

Leiting, B., De Francesco, R., Tomei, L., Cortese, R., Otting, G., and Wuthrich, K. (1993). The three-dimensional NMR-solution structure of the polypeptide fragment 195-286 of the LFB1/HNF1 transcription factor from rat liver comprises a non-classical homeodomain. *EMBO J.* 12, 1797–1803.

Lin, Y.S., Carey, M.F., Ptashne, M., and Green, M.R. (1988). GAL4 derivatives function alone and synergistically with mammalian activators *in vitro. Cell* 54, 659–664.

Lin, Y.S., Carey, M., Ptashne, M. and Green, M.R. (1990). How different eukaryotic transcriptional activators can cooperate promiscuously. *Nature* 345, 359–361.

Lin, Y.S., Ha, I., Maldonado, E., Reinberg, D., and Green, M.R. (1991). Binding of a general transcription factor TFIIB to an acidic activating region. *Nature* 353, 569–571.

Lin, J., Chen, J., Elenbaas, B., and Levine, A.J. (1994). Several hydrophobic amino acids in the p53 amino-terminal domain are required for transcriptional activation, binding to mdm-2 and the adenovirus E1B 55-kd protein. *Genes Dev.* 8, 1235–1246.

Liu, F., and Green, M.R. (1994). Promoter targeting by adenovirus E1a through interaction with different cellular DNA-binding domains. *Nature* 368, 520–525.

Ma, P.C., Rould, M.A., Weintraub, H., and Pabo, C.O. (1994). Crystal structure of the MyoD bHLH domain complex: perspectives on DNA recognition and implications for transcriptional activation. *Cell* 77, 451–459.

Macchi, M., Bornert, J.-M., Davidson, I., Kanno, M., Rosales, R., Vigneron, M., Xiao, J.-H., Fromental, C. and Chambon, P. (1989). The SV40 TC-II(κB) enhanson binds ubiquitous and cell type specifically induced nuclear proteins from lymphoid and non-lymphoid cell lines. *EMBO J.* 8, 4215–4227.

Mangelsdorf, D.J., Thummel, C., Beato, M., Herrlich, P., Schutz, G., Umesono, K., Blumberg, B., Kastner, P., Mark, M., Chambon, P., and Evans, R.M. (1995). The nuclear receptor superfamily: the second decade. *Cell* 83, 835–839.

Mastrangelo, I.A., Courey, A.J., Wall, J.S., Jackson, S.P., and Hough, P.V.C. (1991). DNA looping and Sp1 multimer links: a mechanism for transcriptional synergism and enhancement. *Proc. Natl. Acad. Sci. USA* 88, 5670–5674.

Matthews, K.S., and Nichols, J.C. (1998). Lactose repressor protein: functional properties and structure. *Progr. Nucl. Acids Res.* 58, 127–164.

McKay, D.B., and Steitz, T.A. (1981). Structure of catabolite gene activator protein at 2.9Å resolution suggest binding to left-handed B-DNA. *Nature* 290, 744–749.

Meek, D. (1994). Post-translational modification of p53. *Semin. Cancer Biol.* 5, 203–210.

Mercurio, F., and Karin, M. (1989). Transcription factors AP-3 and AP-2 interact with the SV40 enhancer in a mutually exclusive manner. *EMBO J.* 8, 1455–1460.

Miyashita, T., and Reed, J.C. (1995). Tumor suppressor p53 is a direct transcriptional activator of the human *bax* gene. *Cell* 80, 293–299.

Miller, J., McLachlan, A.D., and Klug, A. (1985). Repetitive zinc-binding domains in the protein transcription factor IIIA from *Xenopus* oocytes. *EMBO J.* 4, 1609–1614.

Mitchell, P.J., and Tjian, R. (1989). Transcriptional regulation in mammalian cells by sequence-specific DNA binding proteins. *Science* 245, 371–378.

Mitchell, P.J., Wang, C., and Tjian, R. (1987). Positive and negative regulation of transcription *in vitro*: enhancer-binding protein AP-2 is inhibited by SV40 T antigen. *Cell* 50, 847–861.

Murre, C., McCaw, P.S., and Baltimore, D. (1989). A new DNA-binding and dimerization motif in immunogloulin enhancer binding, daughterless, MyoD and myc proteins. *Cell* 56, 777–783.

Natesan, S., and Gilman, M.Z. (1993). DNA bending and orientation-dependent function of YY1 in the *c-fos* promoter. *Genes Dev.* 7, 2497–2509.

Nishikawa, J., Kokubo, T., Horikoshi, M., Roeder, R.G., and Nakatani, Y. (1997). *Drosophila* $TAF_{II}230$ and the transcriptional activator VP16 bind competitively to the TATA box-binding domain of the TATA box-binding protein. *Proc. Natl. Acad. Sci. USA* 94, 85–90.

Nolte, R.T., Colin, R.M., Harrison, S.C., and Brown, R.S. (1998). Differing roles for zinc fingers in DNA recognition: structure of a six-finger transcription factor IIIA complex. *Proc. Natl. Acad. Sci. USA* 95, 2938–3941.

Nurse, P. (1990). Universal control mechanism regulating onset of M-phase. *Nature* 344, 503–508.

O'Hare, P., and Williams, G. (1992). Structural studies of the acidic transactivation domain of the Vmw65 protein of herpes simplex virus using ^{1}H NMR. *Biochemistry* 31, 4150–4156.

Okamoto, K., and Beach, D. (1994). Cyclin G is a transcriptional target of the p53 tumor supressor protein. *EMBO J.*, 13, 4816–4822.

Pabo, C.O., and Lewis, M. (1982). The operator-binding domain of lambda repressor: structure and DNA recognition. *Nature* 298, 443–447.

Pabo, C.O., and Sauer, R.T. (1992). Transcription factors: structural families and principles of DNA recognition. *Annu. Rev. Biochem.* 61, 1053–1095.

Parker, D., Ferreri, K., Nakjima, T., LaMorte, V.J., Evans, R., Koerber, S.C., Hoeger, C., and Montminy, M. (1996). Phosphorylation of CREB at Ser-133 induces complex formation with CPB via a direct mechanism. *Mol. Cell. Biol.* 16, 694–703.

Pavletich, N.P., and Pabo, C.O. (1991). Zinc finger-DNA recognition: crystal structure of a Zif268-DNA complex at 2.1Å. *Science* 252, 809–817.

Perez-Martin, J., and Espinosa, M. (1993). Protein-induced bending as a transcriptional switch. *Science* 260, 805–807.

Pesce, S., and Benezra, R. (1993). The loop region of the helix-loop-helix protein Id1 is critical for its dominant negative activity. *Mol. Cell. Biol.* 13, 7874–7880.

Pierce, J.W., Read, M.A., Ding, H., Luscinskas, F.W., and Collins, T. (1996). Salicylates inhibit I-kappa B-alpha phosphorylation, endothelial-leukocyte adhesion molecule expression and neutrophil transmigration. *J. Immunol.* 156, 3961–3969.

Pognonec, P., Roeder, R.G., and Burley, S.K. (1994). Structure and function of the b/HLH/Z domain of USF. *EMBO J.* 13, 180–189.

Pomerantz, J.L., Kristie, T.M., and Pabo, C.O. (1992). Recognition of the surface of a homeo domain protein. *Genes Dev.* 6, 2047–2057.

Pomerantz, J.L., Sharp, P.A., and Pabo, C.O. (1995). Structure-based design of transcription factors. *Science* 267, 93–96.

Prendergast, G.C., Lawe, D., and Ziff, E.B. (1991). Association of Myn, the murine homolog of max, with c-Myc stimulates methylation-sensitive DNA binding and ras cotransformation. *Cell* 65, 395–407.

Raumann, B.E., Rould, M.A., Pabo, C.O., and Sauer, R.T. (1994a). DNA recognition by β-sheeets in the Arc repressor-operator crystal structure. *Nature* 367, 754–757.

Raumann, B.E., Brown, B.M., and Sauer, R.T. (1994b). Major groove DNA recognition by β-sheets: the ribbon-helix-helix family of gene regulatory proteins. *Curr. Opin. Struct. Biol.* 4, 36–43.

Regier, J.L., Shen, F., and Triezenberg, S.J. (1993). Pattern of aromatic and hydrophobic amino acids critical for one of two subdomains of the VP16 transcriptional activator. *Proc. Natl. Acad. Sci. USA* 90, 883–887.

Rhadakrishnan, I., Perez-Alvarado, G.C., Parker, D., Dyson, H.J., Montminy, M.C., and Wright, P.E. (1997). Solution structure of the KIX domain of CBP bound to the transactivation domain of CREB: a model for activator:coactivator interactions. *Cell* 91, 741–752.

Roberts, S.G.E., and Green, M.R. (1994). Activator-induced conformational change in general transcription factor TFIIB. *Nature* 371, 717–720.

Roberts, S.B., Segil, N., and Heintz, H. (1991). Differential phosphorylation of the transcription factor Oct1 during the cell cycle. *Science* 253, 1022–1026.

Ruden, D.M. (1992). Activating regions of yeast transcription factors must have both acidic and hydrophobic amino acids. *Chromosoma* 101, 342–348.

Sadowski, I., Ma, J., Triezenberg, S., and Ptashne, M. (1988). GAL4-VP16 is an unusually potent transcriptional activator. *Nature* 335, 563–564.

Schevitz, R.W., Otwinowski, Z., Joachimiak, A., Lawson, C.I., and Sigler, P.B. (1985). The three-dimensional structure of trp repressor. *Nature* 317, 782–786.

Schneider, T.D. (1996). Reading of DNA sequence logos: prediction of major groove binding by information theory. *Meth. Enzymol.* 274, 445–455.

Schumacher, M.A., Choi, K.Y., Zalkin, H., and Brennan, R.G. (1994). Crystal structure of the *LacI* family member, *PurR*, bound to DNA: minor groove binding by α helices. *Science* 266, 763–770.

Schultz, S.C., Shields, G.C., and Steitz, T.A. (1991). Crystal structure of the CAP-DNA complex: the DNA is bent by 90°. *Science* 253, 1001–1007.

Schwabe, J.W.R., and Klug, A. (1994). Zinc mining for protein domains. *Nature Struct. Biol.* 1, 345–349.

Schwabe, J.W.R., Neuhans, D., and Rhodes, D. (1990). Solution structure of the DNA-binding domain of the estrogen receptor. *Nature* 348, 458–461.

Segil, N., Roberts, S.B., and Heitz, N. (1991). Mitotic phosphorylation of the Oct-1 homeodomain and regulation of Oct-1 DNA binding activity. *Science* 254, 1814–1816.

Serfling, E., Jasin, M., and Schaffner, W. (1985). Enhancers and eukaryotic gene transcription. *Trends Genet.* 1, 224–230.

Shaw, P., Freeman, J., Bovey, R., and Iggo, R. (1996). Regulation of specific DNA binding by p53: evidence for a role for O-linked glycosylation and charged residues at the carboxy-terminus. *Oncogene* 12, 921–930.

Shibasaki, F., Price, E.R., Milan, D., and McKeon, F. (1996). Role of kinases and the phosphatase calcineurin in the nuclear shuttling of transcription factor NF-AT4. *Nature* 382, 370–373.

Somers, W.S., and Philips, S.E. (1992). Crystal structure of the Met repressor-operator complex at 2.8Å resolution reveals DNA recognition by β-strands. *Nature* 359, 387–393.

studies could not exclude the possibility of additional proteins being involved, photocrosslinking studies provided strong evidence favouring direct physical contacts between CAP and the C-terminal domain of α (Chen *et al.*, 1994a). A cleavable, ^{125}I-labeled crosslinking agent, when attached to a precisely defined amino acid residue located within the CAP activation domain, transfers the radioactive label specifically to the α subunit. The reaction depends strictly on the simultanous presence of CAP, *E. coli* RNAP and *lac* promoter DNA, thus proving that the CAP-α interaction occurs on a promoter-bound transcription initiation complex rather than free in solution (Figure 4.2). None of the other RNAP subunits are specifically labelled under these experimental conditions, and proteolytic mapping of the labeled position within the crosslinked α-subunit reveal that the target site for CAP-α contacts is, in good agreement with the

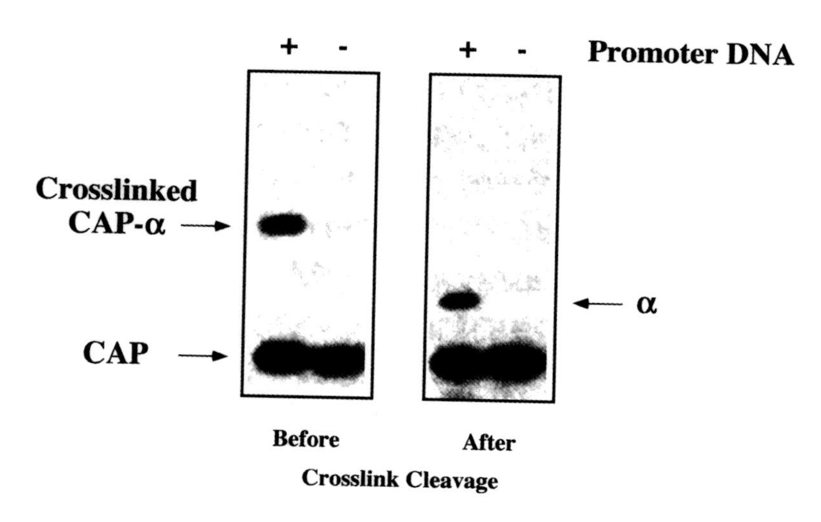

Figure 4.2. The *E. coli* Activator Protein CAP Specifically Contacts the α Subunit of RNA Polymerase.

CAP is specifically labelled with a radioactive, cleavable UV-activated crosslinking agent. Incubation with RNAP results in the formation of a cross-linked CAP-α complex in presence of suitable promoter DNA, but not without it (left gel). After cleavage of the crosslinking agent the radioactive label remains exclusively attached to α. Note that there is no evidence for similar interactions with any of the other RNAP subunits (β′, β and σ^{70}), which proves the high degree of selectivity of the CAP-α interaction. Data from Chen *et al.* (1994a).

previously obtained genetic data, located in the C-terminal domain of the α subunit. We can therefore conclude that the CAP activation region and the C-terminal domain of the α subunit of *E. coli* RNAP are in close physical proximity within the bacterial transcription initiation complex, and probably interact with each other through direct protein-protein contacts. The structure of the C-terminal domain of the *E. coli* RNAP subunit α is known, and the region involved in the CAP interaction event corresponds to a single helical region (Figure 4.3; Jeon *et al.*, 1995)

Direct Activator/RNAP Contacts Account For Many, But Not All, Bacterial Activator Functions

The precedent for direct interactions between transcriptional activators and specific RNAP subunits set by CAP/α has been amply confirmed during the last few years for a number of other bacterial transcriptional activators. The transcriptional activator λcI protein contacts the C-terminal domain of the *E. coli* σ^{70} RNAP subunit (Li *et al.*, 1994; Kudell and Hochschild, 1994), and another bacteriophage transcription factor, the N4SSB protein, interacts with the β' subunit (Miller *et al.*, 1997). Although no transcriptional activator has yet been shown to communicate with the β subunit, it looks likely that such examples could be found in the forseeable future.

The picture that emerges from these examples implies that simple, direct protein-protein contacts between DNA-bound activators and *E. coli* RNAP are sufficient to account for many aspects of stimulated transcription in bacteria. This theory has been put to the test, and essentially confirmed, by recent experimental work involving the creation of arbitrary and artificial protein-interaction surfaces (Dove *et al.*, 1997; Dove and Hochschild, 1998). In these experiments the strength of protein-interactions between the modified RNAP and activator constructs was directly proportional to the degree of transcriptional activation obtained, suggesting that direct recruitment of RNAP to a promoter via DNA-bound activator proteins can account for many aspects of transcriptional activator function in bacteria. It is clear, however, that this simple model does not completely account for the action of all types of bacterial activators on all promoters. There are multiple kinetically limiting steps preceeding the transition from the initiation to the elongation mode in bacterial RNAPs and it is likely that many of them can be speeded up by allosteric

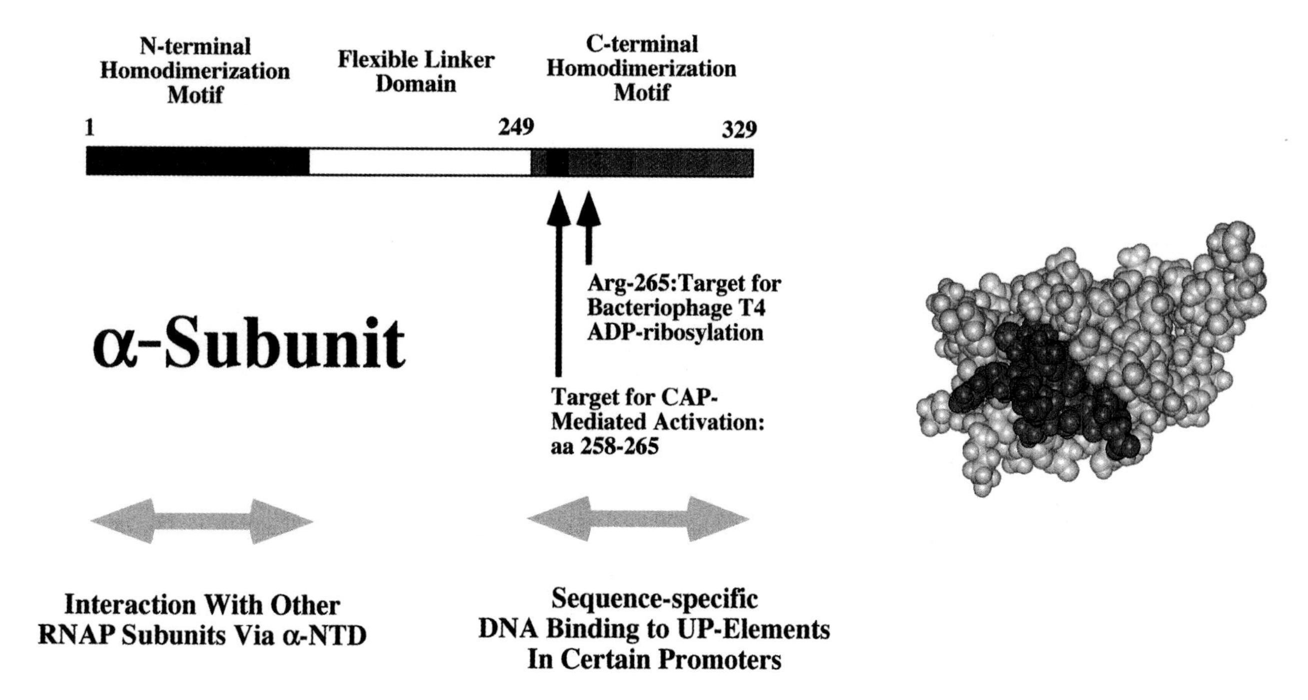

Figure 4.3. Functional Domains of the *E. coli* RNAP α Subunit.
Left: The α subunit consists of an N-terminal and C-terminal domain that are separated by a flexible linker. Various functionally-defined regions in the individual domains are identified. *Right*: The high resolution structure of the C-terminal domain of the *E. coli* RNA α subunit is shown. All the amino acid residues that have been implicated in supporting activated transcription by CAP map to a narrow, contiguous region shown in black. PDB Access No. 1COO.

changes. Transcriptional activators may be able to induce, or stabilize, such energetically favourable conformations by 'pushing the right buttons' through specific (rather than arbitrary!) contacts with defined areas of RNAP. Such allosterically-induced changes may, for example, increase the rate of promoter clearance by RNAP (Menendez *et al.*, 1987), or facilitate the formation of open complexes (e.g. Wyman *et al.*, 1997). It is also likely that particular transcriptional activators, such as CAP, use the RNAP recruitment mode if they are located close to the transcription start site, but utilize another mechanism, such as DNA-bending, to activate RNAP transcription if their binding site is located too far away to allow direct protein:protein contacts with the promoter-bound RNAP (Figure 4.4; Dethiollaz *et al.*, 1996).

These selected examples illustrate that, even in a 'simple' model system such as *E. coli*, it already becomes impossible to account for transcriptional activation by just one single mechanism. Below we will encounter numerous eukaryotic examples where it becomes even more obvious that there are probably dozens of (often redundant) pathways for transcriptional activators to stimulate RNA polymerase activity. This is an important message to keep constantly in mind as we work our way, step-by-step, through several of the best-documented cases of activator functions in these highly complex systems.

4.2. Direct Contacts between Eukaryotic Activators and Eukaryotic RNA Polymerases?

The bacterial systems described above are probably not particularly good model systems for helping us to understand in detail the function of gene-specific activators in eukaryotic systems. Eukaryotic RNAPs are surrounded by basal factors, which make direct contacts between RNAP and gene-specific transcription factors rather difficult. Also, in bacteria there are usually only one or two gene-specific transcription factors that regulate the expression of an individual gene (e.g. *lac* repressor and CAP protein control transcription of the *lac* operon). On the other hand the regulation of eukaryotic promoters by several dozen gene-specific transcription factors is not unusual (see e.g. Yuh *et al.*, 1998, for a recent detailed analysis of a complex promoter), which would make it difficult to envisage direct contacts between all these activators/

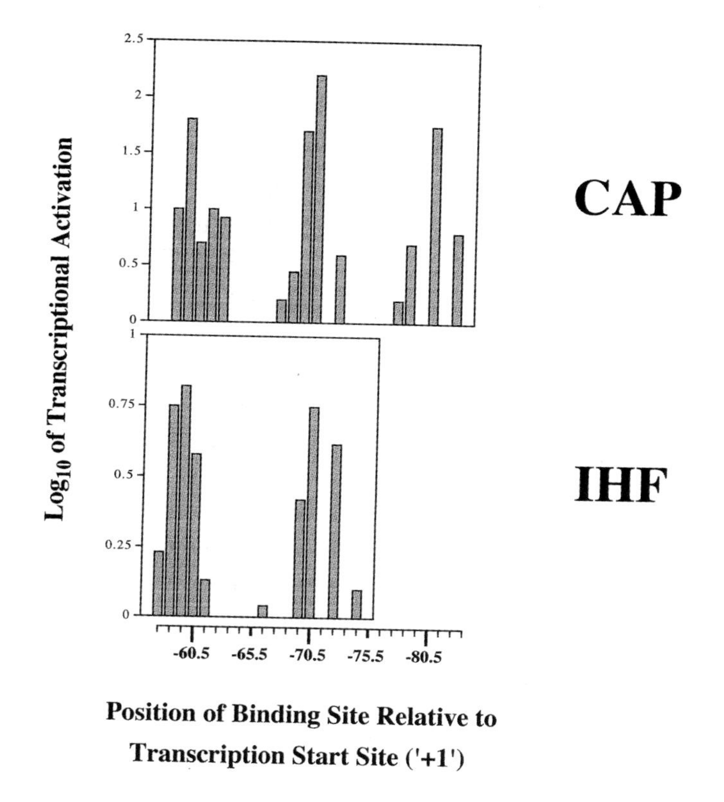

Figure 4.4. DNA Curvature Makes an Important Contribution to Transcriptional Activation in *E. coli.*

In two separate experiments the binding sites of the bacterial transcriptional activator CAP and the DNA-bending protein IHF were placed at variable distances from the transcription start site of the *malT* promoter of *E. coli*. The resulting constructs were assayed for transcriptional activation *in vivo*. In the case of CAP (top graph), three positional maxima can be observed when the binding site is placed at either positions –60.5, –71.5 or –81.5. The ten basepair periodicity of these optimal positions corresponds to the pitch of the DNA double-helix and suggests that CAP needs to be precisely positioned on the DNA template relative to the RNA polymerase bound near the start site at position +1. Interestingly, a very similar effect is seen if binding site for the DNA-bending factor IHF are placed at variable disatnces within the *malT* promoter (bottom graph). Since IHF does not contact RNA polymerase directly, the result suggests that DNA bending alone, although in a rather rigid geometry relative to RNA polymerase, is sufficient to result in substantial levels of activation.

Data replotted from Dethiollaz *et al.* (1996).

repressors and a single RNAP. Many of our current molecular models suggest that the vast majority of gene-specific transcription factors interact with various coactivators and basal factors, rather than RNAP itself, to convey their regulatory message.

There is, however, at least one interesting experimental observation suggesting that in certain (exceptional?) situations eukaryotic transcriptional activators may stimulate transcription by RNAPs through direct physical contact between the activation domains and a particular RNAP subunit. A transactivator, protein X, encoded by the human hepatitis B virus interacts directly with the human RBP5 RNAP subunit *in vitro* and *in vivo* in a manner that is dependent on the presence of a functional activation domain (Cheong *et al.*, 1995; Lin *et al.*, 1997). RPB5 is one of the subunits present in all three nuclear RNAPs (see Chapter 1), and protein X is known to stimulate transcription of genes transcribed by $RNAP_I$, $RNAP_{II}$ and $RNAP_{III}$. It is therefore quite possible that an unusual transcription factor with such a wide-ranging activation spectrum, such as protein X, has specifically evolved a way of stimulating transcription though contacting an invariable part of eukaryotic RNAPs. It remains to be seen whether this model withstands the test of time and whether other examples of such direct eukaryotic activator/RNAP contacts will be found in future investigations.

4.3. The Coactivator Concept in Eukaryotic Transcription

Discovery of TFIID-Based Coactivators

Many investigators studying eukaryotic transcription processes during the 1980s focused predominantly on the molecular characterization of various promoter elements ('enhancers'), and on the gene-specific transcription factors binding to them. This work led to the identification of many important gene-specific transcriptional activators that we now know play an important role in controlling the rate of transcription of cellular genes (Chapter 3). A major shift of interest towards characterizing the components of the basal transcriptional machinery itself became only distinctly discernible twoards the end of that decade, especially when it became increasingly clear that further advances in understanding regulated gene expression depended crucially on defining the way gene-specific transcription factors interact with the general transcriptional apparatus. Previous studies had already pointed towards TFIID as a likely target for direct interactions with transcriptional activators (Sawadogo and Roeder, 1985; Horikoshi *et al.*, 1988; Nakajima *et al.*, 1988). These observations helped to focus the efforts of several laboratories on the biochemical characterization of TFIID, and the cloning of the genes encoding it. Purification of sufficient amounts of TFIID from the small quantities present in nuclear extracts presented a significant biochemical challenge, but was eventually achieved in the yeast system by many different groups (Buratowski *et al.*, 1988; Cavallini *et al.*, 1989; Hahn *et al.*, 1989; Horikoshi *et al.*, 1989; Schmidt *et al.*, 1989). The purified yeast 'TFIID' was a single, relatively small polypeptide, and its amino acid sequence was sufficiently highly conserved so that the 'TFIID' homologs from higher eukaryotic species could be isolated by sequence homology during the early 1990s (e.g. Hoey *et al.*, 1990; Peterson *et al.*, 1990). Recombinant versions of 'TFIID' from all these species supported transcription in fractionated *in vitro* transcription systems depleted of endogenous TFIID activity. There was, however, a very troublesome aspect. Recombinant 'TFIID' worked fine for *in vitro* reconstituted basal transcription, but addition of gene-specific transcriptional activators (such as VP16 or Sp1) failed to make any difference, i.e. there was no additional stimulation of the rate of transcription (Figure 4.5; Pugh and Tjian, 1990; Dynlacht *et al.*, 1991; Tanese *et al.*, 1991). At this stage other discrepancies (such as the smaller footprint of recombinant 'TFIID' compared to endogenous TFIID) became increasingly obvious, which lead to the proposal that there were additional factors present in endogenous TFIID that were necessary for supporting activated levels of transcription. These 'coactivators' would not be required for basal levels of transcription, but were hypothesized to provide an interface through which transcriptional activators could communicate their 'activation signal' to the basal transcriptional machinery (Pugh and Tjian, 1990). The realization that the hitherto cloned 'TFIID' did not represent the complete TFIID molecule necessitated an adjustment of the previously-used nomenclature. The term 'TFIID' is nowadays reserved for the full TFIID complex, and the cloned protein historically referred to as 'TFIID' was renamed the 'TATA-binding protein' (TBP). It is especially important to be aware of this nomenclature change when reading research papers from the early 1990s, where there are many instances of publications

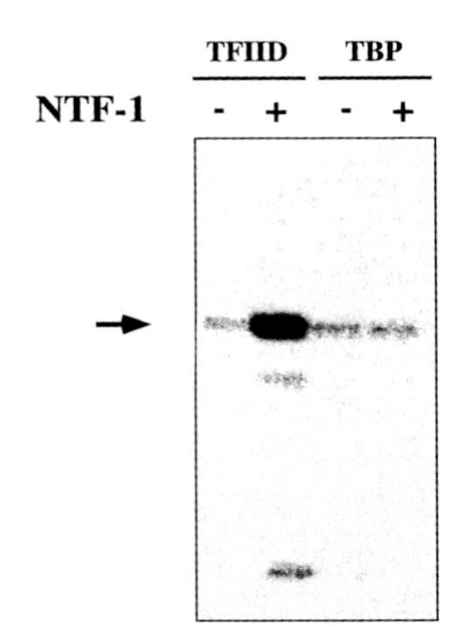

Figure 4.5. TBP Supports Basal Levels of Transcription but Fails to Support Activation by Gene-Specific Transcription Factors.
The gene-specific transcriptional activator NTF-1 was incubated with suitable DNA template in a reconstituted *in vitro* transcription system lacking the TFIID fraction. Addition of TFIID supported activated levels of transcription, whereas substitution of TFIID with purified recombinant TBP only allowed basal levels of transcription in presence or absence of the gene-specific transcription factor. Such experiments show that TBP does not fully substitute for TFIID during activated transcription and suggested the necessity for TBP-associated coactivators. The specific full-length transcript is labelled with an arrow. Data from Dynlacht *et al.* (1991).

using the term 'TFIID' for describing results concerning the structure and function of TBP.

The TFIID Complex Contains TBP and TBP-Associated Factors (TAFs)

Immunoprecipitation experiments with antibodies directed against TBP did indeed reveal the presence of numerous other copurifying proteins. Biochemical assays demonstrated that this collection of TBP-associated factors (TAF_{II}s; the roman numerals indicate that they are specific for the $RNAP_{II}$ system) did indeed contain coactivator activities (Dynlacht *et al.*, 1991; Tanese *et al.*, 1991; Zhou *et al.*, 1992; Chiang *et al.*, 1993; Chiang and Roeder, 1995). Ironically, later studies also revealed that even yeast TBP is associated with TAFs *in vivo* (Poon and Weil, 1993; Reese *et al.*, 1994). The yeast TAFs seem to be much less avidly bound to TBP, a fact that originally allowed the biochemical isolation of TBP as a single polypeptide from this organism. Many studies carried out since the original development of the coactivator concept support the view that several distinct types of coactivators within TFIID cooperate functionally with particular activators.

4.4. TFIID as a Specific Coactivator Complex

Structure of the TFIID complex

The biochemical characterization of the protein composition of TFIID complexes from different eukaryotic sources revealed that they all contain, in addition to TBP, between 8–12 TAF_{II}s (Figure 4.6; Dynlacht *et al.*, 1991; Tanese *et al.*, 1991; Zhou et al., 1992; Poon and Weil, 1993; Reese *et al.*, 1994; Poon *et al.*, 1995; reviewed in Burley and Roeder, 1996; Verrijzer and Tjian, 1996). Some of the TAFs play a major role in recognizing sequence-specific promoter elements ($dTAF_{II}60$, $dTAF_{II}150$; see Chapter 2) and thus essentially fulfill a 'basal' transcription function, but many of the other TAFs seem to be exclusively utilized as transcriptional coactivators. TAFs show a high degree of sequence conservation across the entire eukaryotic range (Weinzierl *et al.*, 1993a,b; Reese *et al.*, 1994; Poon *et al.*, 1995), but are absent in the two other evolutionary domains, archaea and bacteria (Bult *et al.*, 1996; Qureshi *et al.*, 1997).

Certain TAFs Act as Transcriptional Coactivators

The discovery that addition of purified TAFs to recombinant TBP allowed a variety of activators to stimulate transcription in a fractionated *in vitro* transcription system immediately suggested that the TAFs were the sought-after coactivators (Dynlacht *et al.*, 1991; Tanese *et al.*, 1991). The interaction between Sp1 and $dTAF_{II}110$, that correlated under *in vitro* and *in vivo* conditions with the presence of functional activation domains, provided the first piece of molecular evidence for direct contacts between an activator and coactivator (Figure 4.7; Goodrich *et al.*, 1993; Hoey *et al.*, 1993; Gill *et al.*, 1994; Jacq *et al.*, 1994; Chiang and Roeder, 1995). Subsequent studies with the different types of activation domains helped to established the concept that different activators contact distinct TAFs within the TFIID complex (Goodrich *et al.*,

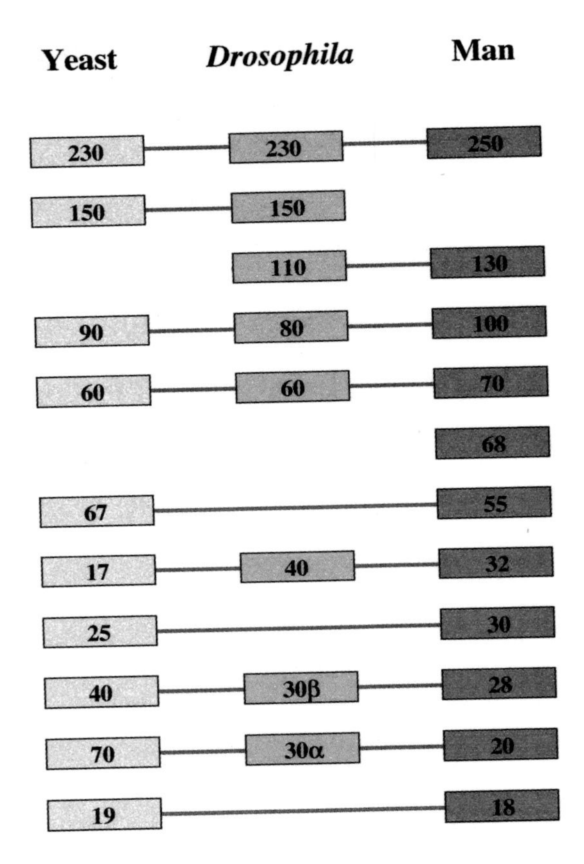

Figure 4.6. TAF Composition and Conservation Across the Eukaryotic Range.
The known TAFs present in the TFIID complexes from three different species (yeast, *Drosophila*, and man) are shown schematically, with grey bars linking TAFs that display sequence homology to each other. Modified from Tansey and Herr (1997).

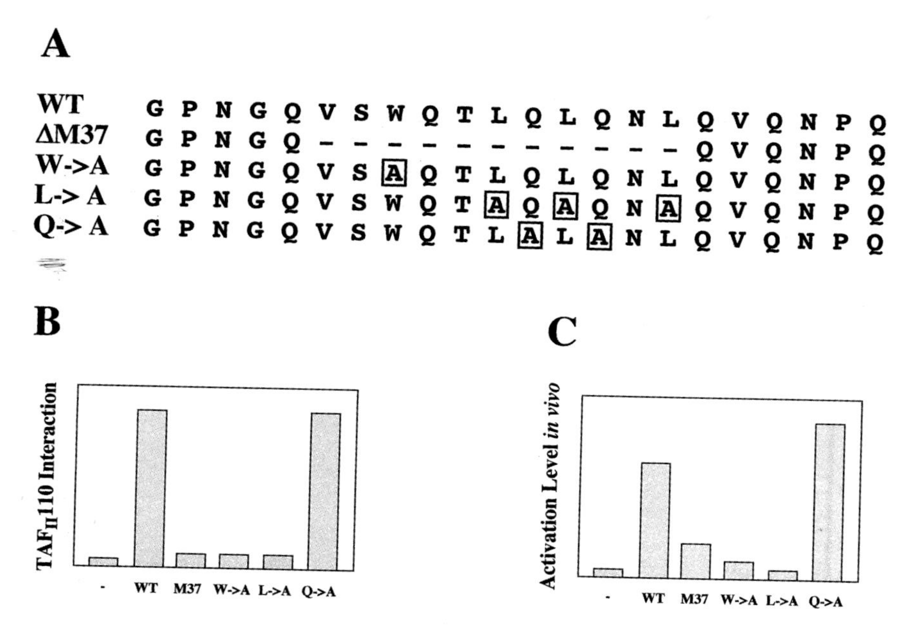

Figure 4.7. The Ability of Sp1 Mutants to Interact with TAF$_{II}$110 *In Vitro* Correlates with Their Ability to Stimulate Transcription *In Vivo*.

Four different mutations in an Sp1 glutamine-rich activation domain (*Panel A*) where tested for their ability to interact with TAF$_{II}$110 in a yeast two-hybrid protein-protein interaction assay and their ability to activate transcription in transfected cell lines. Substitutions of two central glutamine ('Q') residues with alanine ('A') had little effect on the affinity of the dTAF$_{II}$110/Sp1 interaction (*Panel B*), or the ability of the protein to support high levels of transcription *in vivo* (*Panel C*). Deletion of eleven amino acid residues (in ΔM37), or substitution of hydrophobic amino acids (leucine ['L'] and tryptophan ['W']) with alanine, abolishes the interaction with dTAF$_{II}$110 and severely affects the ability of the mutants to stimulate transcription in transfected cells. The data demonstrates the central role of dTAF$_{II}$110 in transcriptional activation by Sp1. Data from Gill *et al.* (1994).

1993; Chen *et al.*, 1994b; Ferreri *et al.*, 1994). The most clear-cut proof of this concept was provided by Chen *et al.* (1994b), who assembled a variety of partial and complete TFIID complexes from recombinant TBP and TAFs *in vitro*. Functional testing of various partial TFIID complexes with different transcriptional activators showed that certain partial complexes consisting of TBP and a limited subset of TAFs were fully capable of supporting activated transcription levels with particular transcriptional activators, but not with others (Figure 4.8). Activation by Sp1 was strictly dependent on the presence of

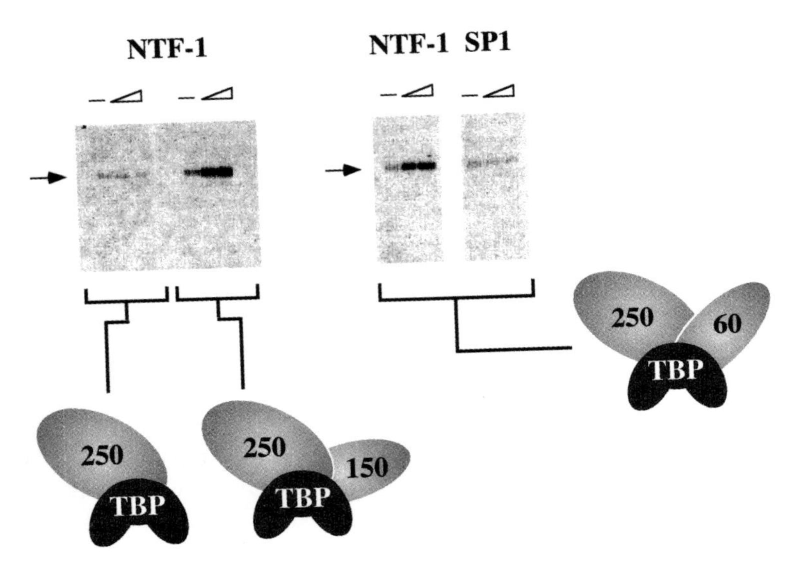

Figure 4.8. Partially Reconstituted TFIID Complexes Reveal a High Degree of Specificity between TAF/Activator Interactions.
The gene-specific transcriptional activator NTF-1 can stimulate activated levels of transcription through interactions with different TAFs. The combination of $dTAF_{II}250$/TBP only supports basal levels of transcription. Addition of $TAF_{II}150$ to this minimal complex allows activated levels of transcription to occur in presence of NTF-1. An alternatively-configured complex, consisting of $TAF_{II}250$/TBP and $TAF_{II}60$ also supports NTF-1 activated transcription, but fails to allow Sp1 to stimulate transcription. This experiment shows that different gene-specific transcription factors (NTF-1, Sp1) use different TAFs as coactivators and that certain gene-specific transcription factors can activate transcription through multiple independent TAF interactions ($TAF_{II}60$ or $TAF_{II}150$). Data from Chen *et al.* (1994b).

$TAF_{II}110$ in the recombinant TFIID complexes, whereas another gene-specific transcription factor (NTF-1) was able to stimulate transcription via two types of TFIID complexes with very different TAF compositions. This result confirmed the previously observed specificity of the Sp1-$TAF_{II}110$ interaction (Figure 4.7), and illustrated that other activators, such as NTF-1, can make specific and functionally productive contacts with at least two different TAF_{II}s. Below we will see how multiple TAF targets for a transcriptional activator could have important consequences for the synergistic function of transcription factors *in vivo*.

Synergistic Effects based on TAF-Activator Interactions Controlling Embryonic Development?

Many developmental decisions during embryogenesis depend on the establishment of sharp boundaries between adjacent cells as a consequence of diffuse 'positional' information present as directed gradients. During *Drosophila* development the initial anterior-posterior positional information (specifying the establishment of segment-specific differences along the anterior-posterior axis) is laid down as a continuous gradient of the transcription factor *bicoid* (Figure 4.9). *bicoid* protein is present at its highest concentration at the anterior end of the embryo, and at its lowest concentration at the posterior end during the first few hours of embryonic development. The continuous gradient between these two extreme positions specifies the expression boundaries of another set of genes, the 'gap' genes. The gap genes are only regionally expressed, and therefore refine the approximate positional information laid down by the *bicoid* gradient into a more highly localized spatial pattern. Genetic studies with *bicoid* mutants demonstrated that the expression borders of the gap gene '*hunchback*' is directly controlled by the regional variations of *bicoid* concentrations. If the level of *bicoid* falls below a certain threshold along the anterior-posterior axis, the transcription level of *hunchback* mRNA falls sharply in that particular region of the developing *Drosophila* embryo (Figure 4.9; Driever and Nusslein-Volhard, 1988; Simpson-Brose *et al.*, 1994; reviewed in Jackle and Sauer, 1993). This recurrent theme of progressive spatial refinement of transcription patterns, and the establishment of distinct gene expression thresholds, occurs in many types of developmental mechanisms involved in

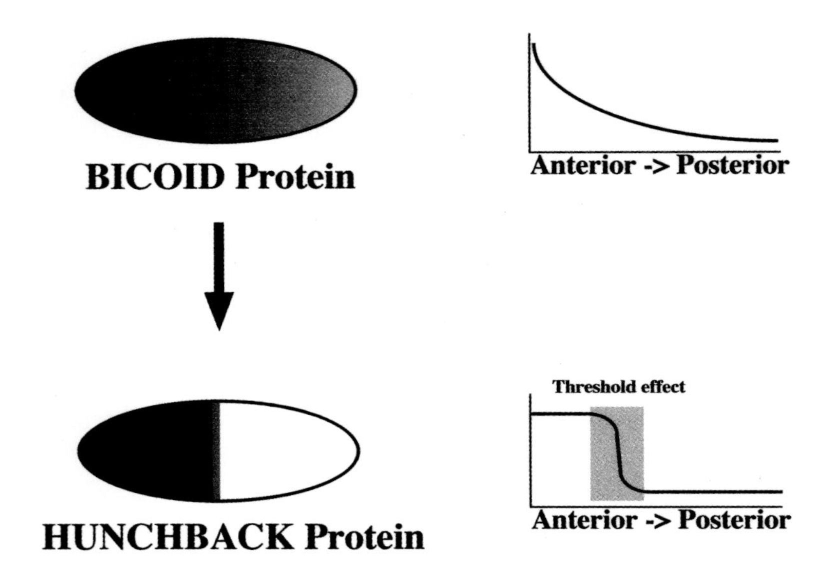

Figure 4.9. Spatial Expression Patterns of the *bicoid* and *hunchback* Proteins in Developing *Drosophila* Embryos.
Immunolocalization studies have revealed that the *bicoid* protein is distributed within the early stages of developing *Drosophila* embro as a smooth gradient with maximum expression at the anterior end. This *bicoid* gradient specifies the spatially restricted expression domain of another transcription factor, *hunchback*, at a later stage. The expression levels of both proteins are also indicated on a graphs on the right hand side, showing the quantity of protein present relative to the position along the embryonic anterior-posterior axis. Note how the smooth *bicoid* gradient gives rise to a very distinct posterior border of *hunchback* protein expression. See text for more details about possible molecular mechanisms that might be responsible for this effect.

pattern formation, but up to now it has been difficult to understand the molecular events that give rise to them. The molecular analysis of the binding characteristics of the *bicoid* transcription factor to the *hunchback* promoter, in combination with the identification of the TAFs that interact with *bicoid*, have led to the formulation of a working model that explains, at least to a certain extent, how such threshold phenomena could in principle originate on the molecular level through the interactions between gene-specific transcription factors and the coactivators of the general transcriptional machinery (Sauer *et al.*, 1995a,b).

The *hunchback* promoter contains several *bicoid* binding sites that vary from each other in affinity (Figure 4.10). Certain high affinity sites will be bound within cells, even if the *bicoid* concentration is relatively low (especially in the posterior half of the embryo), but will only result in relatively low levels of activation of *hunchback* transcription. If the concentration of *bicoid* increases (towards the middle part of the embryo), some additional sites will bind the transcription factor, resulting in a substantial increase of *hunchback* expression.

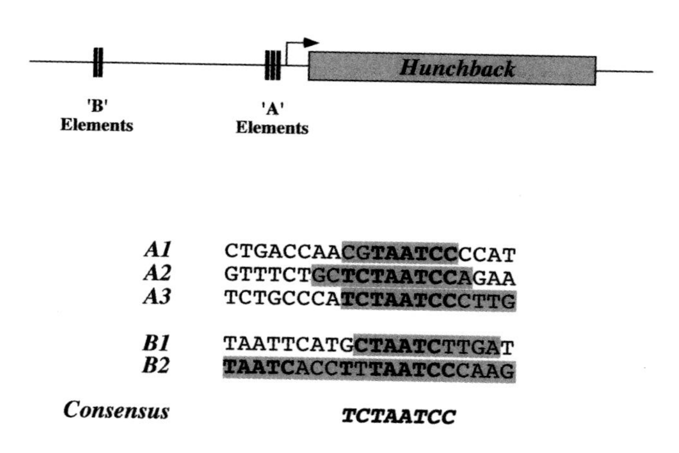

Figure 4.10. Multiple Binding Sites for the Transcription Factor *bicoid* in the *hunchback* Promoter Region.
Top: Two distinct clusters ('A' and 'B' elements; shown as vertical bold lines) of *bicoid* binding sites are located at different distances relatively to the coding region of the *hunchback* gene (shown as a grey box). *Bottom*: Sequence of the *bicoid* binding sites found in the 'A' and 'B' elements. The DNA regions contacted by *bicoid* protein are shown shaded in grey. Nucleotides that conform to the consensus sequence are indicated in bold. Note the presence of two consensus sequences and the extended footprinting pattern of *bicoid* on the B2 site.

This effect can be mimicked under defined *in vitro* conditions on promoters containing a variable number of *bicoid* binding sites in the presence of saturating amounts of *bicoid* (Figure 4.11). One molecule of *bicoid* bound upstream of the transcription start site results in a 3-fold level of activation, but two *bicoid* molecules give rise to a 10-fold stimulation. This increase from 3-fold to 10-

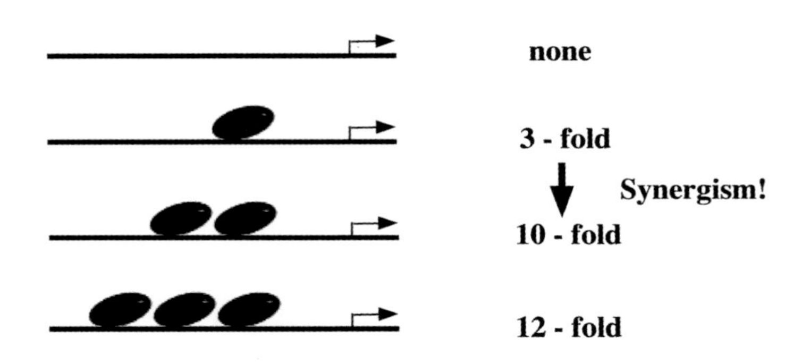

Figure 4.11. Synergistic Action of *bicoid* under *In Vitro* Assay Conditions.
A *hunchback* promoter with variable numbers of *bicoid* binding sites was tested for *in vitro* transcriptional activity in the presence of saturating amounts of purified *bicoid* protein. Note the big 'jump' in activity between templates containing one and two *bicoid* binding sites, indicating a synergistic effect. The addition of a third *bicoid* binding site only improves the rate of transcription marginally.

fold is synergistic because it is higher than expected from purely additive effects. Addition of a third *bicoid* binding site to the promoter results in an even higher, but essentially only additive effect on *hunchback* transcription. This experiment suggests that there are precisely two different target sites through which *bicoid* molecules can stimulate the transcriptional machinery, which results in a high level of transcription if both of them are simultanously targeted. Analysis of the functional domain organization of *bicoid* revealed indeed two distinct and independent activation domains that interact with different TAF_{II}s in the TFIID complex. A 'glutamine-rich' activation domain was found to bind to $TAF_{II}110$, and an 'alanine-rich' domain specifically contacts $TAF_{II}60$ (Sauer *et al.*, 1995b). These results suggest that the posterior expression boundary is partly established as a consequence of the synergistic differences between one or two *bicoid* molecules bound to the *hunchback* promoter, and stimulating transcription through either only one or two distinct TAFs, respectively. Binding of another *bicoid* molecule to one of the low affinity sites on the *hunchback* promoters makes only a slight difference because the other two *bicoid* molecules are already contacting the two available target TAFs. The establishment of a *hunchback* expression domain with a distinct posterior border is further facilitated through positive feedback action of the *hunchback* transcription factor on its own promoter. This means that relatively low expression of *hunchback* protein at an initial stage is sufficient to establish a positive feedback loop that helps to establish and maintain an 'all-or-nothing' expression state of the gene. This is especially important due to the fact that *bicoid* protein is only present in early embryos and therefore would not be present in sufficient quantities in late embryos to sustain high levels of transcription from the *hunchback* promoter. The *hunchback* protein is a transcriptional activator in its own right and required for the high level transcription of homeotic genes, such as the thorax-specifying gene product *Ultrabithorax*, at a later stage of embryogenesis.

TAFs Contain Intrinsic Enzymatic Activites

The discovery of specific contacts between activators and $TAF_{II}s$ raises several interesting questions about the functional consequences of such interactions. How can an interaction between two specific proteins that are not in direct physical contact with $RNAP_{II}$ lead to an increased level of transcription?

The simplest explanation for the ability of TFIID to support activated transcription level is that the interaction between transcription factor activation domains and $TAF_{II}s$ enhances the recruitment of TFIID to the promoter, or subsequently stabilizes the DNA-bound TFIID complex. This hypothesis is supported by results indicating that Sp1 facilitates the binding of TFIID (but not of TBP alone) to the initiator promoter element (Kaufmann and Smale, 1994), and suggests a general role of $TAF_{II}s$ in the enhanced recruitment of basal factors in response to activators (Choy and Green, 1993). According to this view most transcriptional activators would mainly contribute to the establishment of a stable transcription initiation complex in a rather straightforward manner, which might bear some functional resemblance to the recruitment of bacterial RNAP by the *E. coli* CAP protein (see above).

A more sophisticated theory, which is not incompatible with the recruitment hypothesis described above, postulates that the binding of an activator to a coactivator induces allosteric changes in the coactivator. Such an event could lead to structural changes in the three-dimensional conformation of the TFIID complex itself, or alter other protein contacts within the basal initiation complex. Another intriguing observation relevant in this context is the discovery of at least two separate enzymatic activities present in the largest TFIID subunit, $TAF_{II}250$. $TAF_{II}250$ is a crucial polypeptide subunit that plays an important architectural role during the assembly and maintenance of the TFIID complex (Weinzierl *et al.*, 1993a; Wang and Tjian, 1994), and affects the DNA-binding properties of TBP (Kokubo *et al.*, 1993). The kinase activity of $TAF_{II}250$ has been shown to phosphorylate the large subunit of TFIIF *in vitro* (Ruppert and Tjian, 1995; Dikstein *et al.*, 1996a; O'Brien and Tjian, 1998). The degree of phosphorylation of TFIIF is closely associated with its activity (Kitajima *et al.*, 1994), and it is therefore feasible that the kinase domain of $TAF_{II}250$ may play an important role in controlling the quantity of active TFIIF present in cells at a given stage. The other enzymatic activity present in $TAF_{II}250$, a histone acetyltransferase (abbreviated as 'HAT'), is almost certainly involved in the acetylation of highly conserved lysine residues present in the N-terminal domain of histones (Mizzen *et al.*, 1996). The hyperacetylation of histones is known to facilitate an enhanced access of transcription factors to chromatin-covered DNA templates and the formation of 'open' chromatin structures that facilitate readthrough by RNAPs (see Chapter 7; reviewed in Brownell and Allis, 1996). The presence of a HAT in $TAF_{II}250$ therefore suggests that TFIID may use this activity to increase access for other transcription factors and $RNAP_{II}$ by establishing an active chromatin configuration in the immediate vicinity of the transcription start site. At the moment we do not know, however, how this HAT activity is controlled, and it will be interesting to see whether allosteric changes in the structure of the TFIID complex are required to trigger it.

4.5. Tissue-specific TFIID Complexes

The idea of having gene-specific activators acting on an invariant basal transcriptional machinery was one of the important concepts that has served well for establishing an initial intellectual framework for scientists working on various aspects of eukaryotic transcriptional regulation. During the last few years, however, data has started to accumulate pointing at the fact that there are cell-type and tissue-specific TFIID variants that may carry out specialized functions during cellular differention. In single-cell eukaryotic organisms, such as the yeast *S. cerevisiae,* we can say with absolute certainty that there is only a single set of TBP and TAFs (based on the recently completed genomic sequence). Some higher eukaryotic organisms, however, contain variants of TBP and TAFs that are not universally expressed in every cell type.

Cell Type-Specific TAF_{II}s

As a general $RNAP_{II}$ basal factor, TFIID was expected to occur in an essentially invariant form in different tissues and cell types. For much of the early work on establishing the composition and function of TFIID, nuclear extracts from the human HeLa cell line served as the starting material (e.g. Tanese *et al.*, 1991). These cells could be conveniently grown on the large scale required for such studies, but the resulting TFIID preparation represented only the type of TFIID complex found in this particular cell type. Similarly, although *Drosophila* TFIID was isolated from highly differentiated embryos containing many different cell- and tissue types, the purified TFIID represented only the most abundant form, because some of the possible tissue-specific variants would have been mixed up during the homogenization stage with the other less-abundant forms of TFIID (Dynlacht *et al.*, 1991). For a long time it therefore was not possible to say with absolute certainty whether the discovered TAFs were invariably present in all types of TFIID complexes, or whether tissue-specific TFIID complexes with different types of TAFs existed. Several biochemical studies provided the first hints for the presence of biochemically distinct TFIID fractions with different functional properties (Timmers and Sharp, 1991; Brou *et al.*, 1993; Chiang *et al.*, 1993; Jacq *et al.*, 1994), but it was difficult to exclude the possibility that such heterogeneity might have been the consequence of the instability of the TFIID complex during the fractionation procedures used in these experiments. Other reports of tissue-specific coactivator functions came from studies of mammalian lymphocyte cell lines (Luo and Roeder, 1995; Strubin *et al.*, 1995), but these were not found to be associated with TFIID (see below for more information about other types of coactivator activities present in eukaryotic cells).

A systematic search for differences in TFIID complexes in different cell-lines by Dikstein *et al.* (1996) finally revealed conclusive evidence for the presence of a specific human TFIID subunit, $hTAF_{II}105$, that was exclusively present in the TFIID complex isolated from a highly differentiated B ('Daudi') cell line. The primary sequence of $hTAF_{II}105$ reveals a C-terminal domain that is very similar to $dTAF_{II}110$ and its human homolog, $hTAF_{II}130$, i.e. TFIID subunits that we have already encountered as coactivators for the transcriptional activators Sp1 and *bicoid* (see above; Hoey *et al.*, 1993; Sauer *et al.*, 1995a,b). Substantial sequence differences in the N-terminal domain between $dTAF_{II}110$/$hTAF_{II}130$ and $hTAF_{II}105$ suggest that $hTAF_{II}105$ has evolved a specialized role as a coactivator for a subset of genes that are specifically expressed in B-cells. This interpretation is supported by the facts that the N-terminal domains of $dTAF_{II}110$/$hTAF_{II}130$ are responsible for interaction with transcriptional activators (c.f. Dikstein *et al.*, 1996) and that overexpression of $hTAF_{II}105$ specifically interferes with the expression certain genes in B cells, probably through a competition mechanism resulting in the formation of non-productive complexes ('squelching'; Gill and Ptashne, 1988).

It is likely that $hTAF_{II}105$ represents only the first of many examples of cell type- and tissue specific variants of TFIID. The discovery of specifc TAF_{II}s that are only expressed in particular cells already blurs the clear distinction that has been (artificially) created between gene-specific transcription factors and the basal transcriptional machinery. In the following section we will see that this distinction may become even less tenable in the light of the recent discovery of the existence of tissue-specific variants of the fundamental basal factor, TBP.

Cell Type-Specific TBP Variants

The first example of a variant form of TBP was a protein, TBP-related factor ('TRF') that was initially characterized in *Drosophila* during the early 1990s. The unusual and puzzling phenotype of certain TRF mutations pointed to a gene with a function in the central nervous system and male gonad development (Crowley *et al.*, 1993). TRF displays a high degree of sequence homology to TBP (especially on the DNA-binding surface), and is able to bind to consensus TATA-boxes *in vitro*. Observations, that it apparently failed to interact with other basal factors (TFIIA, TFIIB) and could not support basal transcription made a role of TRF as a basal TBP-like transcription factor appear somewhat unlikely (see Buratowski [1997] for an interesting historical background of the 'multiple TBP' concept).

This view changed recently with a more detailed re-investigation of the biochemical properties of TRF (Hansen *et al.*, 1997). An improved method for preparing recombinant TRF revealed that TRF *does* interact with TFIIA and TFIIB in specific complexes in the presence of TATA box-containing DNA. It was also found that TRF can functionally substitute for TBP in reconstituted *in vitro* transcription reactions with other purified basal factors. Interestingly, start sites of different promoters are transcribed at different basal rates, depending on whether TBP or TRF is used in the transcription reactions, which indicates that TBP and TRF display a certain degree of inherent promoter selectivity (Figure 4.12). Sensitivity of TRF-directed transcription to α-amanitin shows that TRF mediates $RNAP_{II}$ transcription in a manner directly comparable to TBP.

The most exciting discovery was, however, that TRF is associated *in vivo* with a number of TAF-like polypeptides ('nTAFs') distinct from the TAFs found in the TFIID complex. The restricted expression of TRF in the developing central nervous system and gonads during *Drosophila* embryogenesis (Figure 4.13) immediately suggests that the TRF/nTAF protein complexes play a tissue-specific role in controlling gene transcription. This view is strongly supported by observations of specific TRF-binding sites on polytene chromosomes. Polytene chromosomes are highly unusual and specialized structures occuring in insect salivary glands that are generated by several rounds of DNA replication without physical separation of the sister chromatid strands. Staining of such polytene chromosomes with antibodies against specific transcription factors reveal the binding sites of such proteins along the chromosome in a way that would be impossible to detect in normal cellular chromosomes. When polytene chromosomes are incubated with labelled antibodies specific for TRF a distinct subset of genetic loci are detected that coincide precisely with the known positions of genes involved in nervous system function, male fertility and possibly tRNA transcription (Figure 4.14). A human

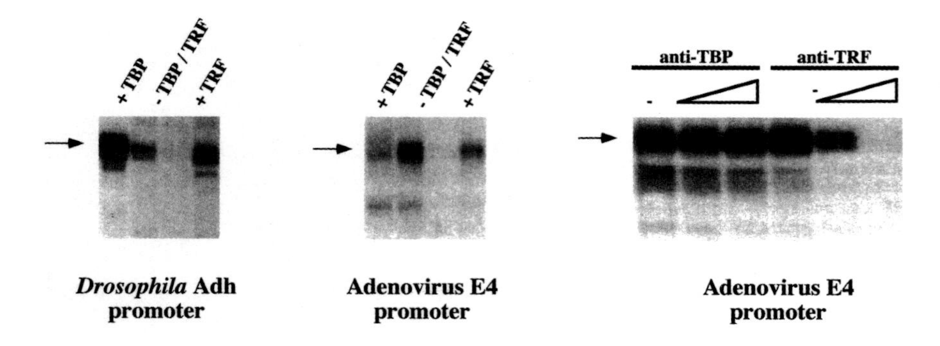

Figure 4.12. Different Promoter Selectivities of TRF and TBP.
Two different promoters were assayed for basal transcriptional activity in a fractionated *in vitro* transcription system (*left and middle panels*). In the absence of either TBP or TRF no transcription occurs. The addition of either purified TBP or TRF restores transcription to different extents. The Adh promoter is transcribed at a higher rate with TRF, whereas the adenovirus E4 promoter is more active with TBP. The specificity of the TRF-mediated basal transcription on the adenovirus E4 promoter is illustrated on the right hand panel. Addition of increasing amounts of anti-TBP antibodies does not interfere with TRF-mediated transcription, but anti-TRF antibodies specifically abolish it. Data from Hansen *et al.* (1997).

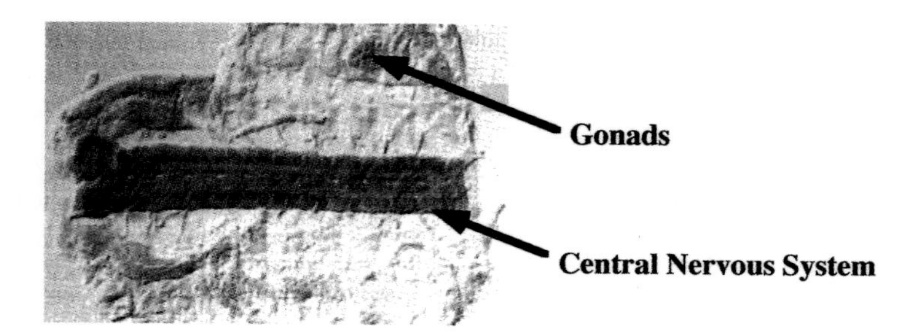

Figure 4.13. Tissue-Specific Expression of TRF in Developing *Drosophila* Embryos.
The expression of TRF was determined by staining *Drosophila* embryos with antibodies specifically directed against TRF proteins. TRF is expressed at high levels in the developing central nervous system and gonads, but not in other parts of the embryo. From Hansen *et al.*, (1997).

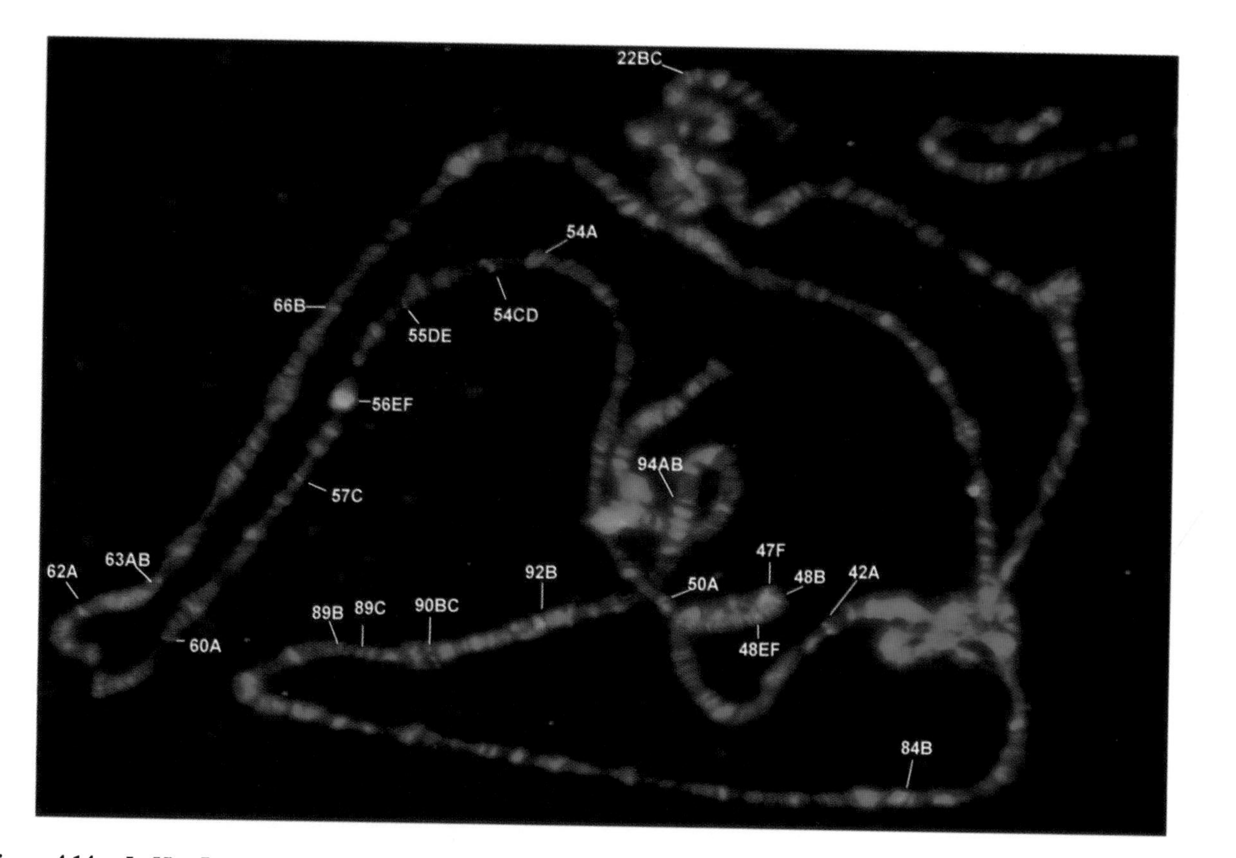

Figure 4.14. ***In Vivo*** **Immunolocalization of TRF of *Drosophila* Polytene Chromosomes.**
The polytene chromosomes are stained green and TRF is labelled red. A number of specific loci contain substantial levels of TRF, indicating that TRF preferentially participates in the transcription of a subset of genes. Image courtesy of John T.Lis, Cornell University.

homolog of TRF has already been identified (cf. Hansen *et al.* [1997]) and it seems likely that further variants of TBP, possibly expressed in different tissues or at certain developmental stages, will be discovered in the near future.

4.6. Specialized Coactivator/Activator Systems

Although many gene-specific transcription factors have been shown to functionally interact with the various components of the TFIID complex, an increasing amount of evidence suggests that eukaryotic cells contain many additional coactivator activities that are physically and functionally distinct from TAFs. Some of these seem to serve as coactivators for a relatively wide range of gene-specific activators that are part of common signalling pathways (p300/CBP), whereas others may have a much more narrowly defined specificity range and are thought to interact with only a small number of transcriptional activators (OCA-B). Presently we also do not understand whether these additional coactivators carry out activities that are not performed by the TFIID complex itself, or whether they synergize with TFIID. Below we will look at two of the best-understood examples, the p300/CBP and OCA-B coactivators, which fulfill specific roles in transcription control circuits regulating various cellular signalling functions and lymphocyte differentiation, respectively. An additional group of coactivators that plays an important role in cooperating with nuclear hormone receptors and are involved in histone acetylation will be discussed in more detail in Chapter 7.

p300/CBP: Coactivators Involved in Cell Signalling with Chromatin Remodelling Functions

The highly similar p300 and CPB nuclear proteins were originally discovered as coactivators for the adenovirus E1A protein and the gene-specific transcription factor CREB, respectively (Eckner *et al.*, 1994; Kwok *et al.*, 1994; Lundblad *et al.*, 1995; reviewed in Shikama *et al.*, 1997). Subsequently p300/CBP have also been found to interact with many other transcription factors involved in cell signalling and differentiation functions, including Jun (Arias *et al.*, 1994), Sap-1a (Janknecht and Nordheim, 1994), STAT-2 (Bhattacharya *et al.*, 1996), nuclear receptors (Chakravarti *et al.*, 1996; Kamei *et al.*, 1996), p53 (Gu *et al.*, 1997; Lill *et al.*, 1997) and helix-loop-helix activators (Eckner *et al.*, 1996; Qiu *et al.*, 1998). Microinjection of antibodies directed against p300 into muscle precursors cells specifically interferes with their differentiation program (Eckner *et al.*, 1996; Puri *et al.*, 1997), and genetic studies have shown that the p300/CBP homologs play an important role in cellular differentiation and pattern formation in invertebrate model systems (Akimaru *et al.*, 1997; Shi and Mello, 1998). A reduced level of CBP (due to haplo-insufficiency) has also been implicated in the human Rubenstein-Tabi syndrome, clinically characterized by craniofacial defects and mental retardation (Petrij *et al.*, 1995). These *in vivo* results demonstrate that p300/CBP are not general coactivators required for transcription from all $RNAP_{II}$-promoters, but play a more highly specialized role in conjunction with certain classes of gene-specific transcription factors to control cellular differentiation pathways.

Specific interactions between various gene-specific transcription factors and p300/CBP stimulate both intrinsic and associated histone acetyltransferase activities within p300/CBP and enhance the recruitment rate of $RNAP_{II}$ complexes to promoters (Kee *et al.*, 1996; Nakjima *et al.*, 1997a,b). The acetylation activity of p300/CBP is of major interest because it specifically leads to the hyperacetylation of histones, which often correlates with the formation of transcriptionally active chromatin (Figure 4.15; see also Chapter 7; Bannister and Kouzarides, 1996; Ogryzko *et al.*, 1996). Furthermore, Gu and Roeder (1997) recently discovered that p300 also mediates the specific acetylation the gene-specific transcription factors p53. The acetylation of the C-terminal domain of p53 is thought to neutralize a positively charged region, thus allowing the DNA-binding domain to adopt an active and accessible conformation (see also Chapter 2). The acetylase activities present in p300/CBP and several other coactivators, such as the previously described $TAF_{II}250$, the p300/CBP associated factor P/CAF (Yang *et al.*, 1996) and the yeast coactivator GCN5 (Kuo *et al.*, 1996) suggest that these molecules may carry out their coactivating function via similar mechanisms. Once they are activated through contacts with suitable transcription factors these coactivators could participate in creating an accessible chromatin environment, and possibly also increase the stable binding of various basal and gene-specific transcription factors to establish active transcription complexes.

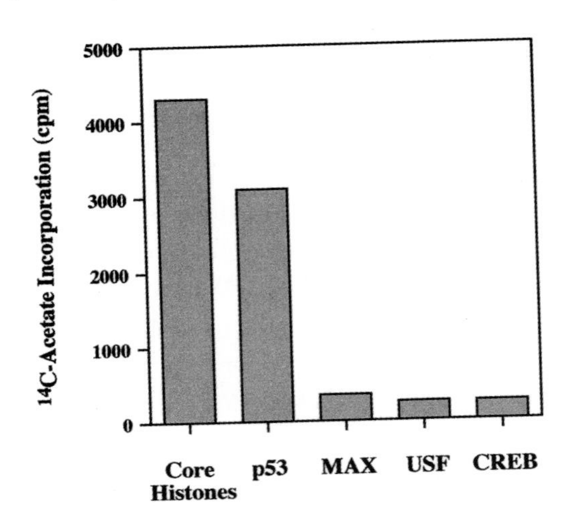

Figure 4.15. p300 Specifically Acetylates Core Histones and p53 *in vitro*.
Purified core histones, p53, and three other gene-specific transcription factors (MAX, USF and CREB) were incubated with recombinant p300 and ^{14}C-acetyl-CoA. The specific incorporation of ^{14}C-acetate is shown for the different proteins and demonstrates specific acetylation of core histones and p53. Note that MAX, USF and CREB are only acetylated at background level. After Gu and Roeder (1997).

High Resolution Structure of a Coactivator-Activator Complex

The activator-binding region of the coactivator CBP has been mapped to the so-called 'KIX' domain, which interacts directly with a subdomain of the CREB activation domain ('KID'; Figure 4.16), as well as activation domains from several other transcription factors, including Jun (Arias *et al.*, 1994); Myb (Dai *et al.*, 1996), *cubitus interruptus* (Akimaru *et al.*, 1997) and the viral protein Tax (Kwok *et al.*, 1996). Determination of the solution structure of the KID/KIX complex by NMR spectroscopy revealed a first glimpse of an activation domain bound to a the binding surface of a coactivator. We have already seen in Chapter 3 that the unstructured activation domain of the viral transactivator VP16 undergoes a specific transition to an α-helical conformation upon interaction with a human TAF_{II} (see Chapter 3; Uesugi *et al.*, 1997). The KID/KIX structure extends our knowledge of activator/coactivator interactions considerably further because it allows us to analyze the structural features of

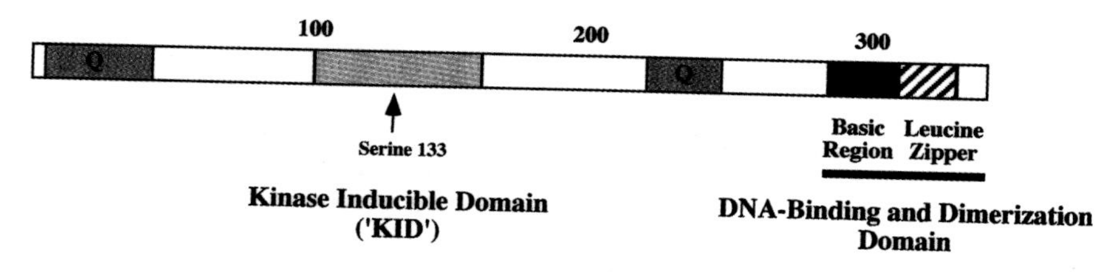

Figure 4.16. Organization of Functional Domains in the Gene-specific Transcription Factor CREB.
CREB is a member of the bZip class of transcription factors (Chapter 3) with a basic domain/leucine zipper DNA-binding domain at the C-terminal end. CREB homodimerizes via the leucine zipper. Note the position of the 'kinase-inducible domain' ('KID') containing serine^{133}, which is specifically phosphorylated to stimulate the transactivation potential of CREB. Phosphorylated KID is capable of interaction with the 'KIX' domain of the coactivator CREB-binding protein (CBP).

the coactivator surface that induce such a conformational change in more detail (Rhadakrishnan *et al.*, 1997).

Similar to the VP16 activation domain, the isolated KID region is unstructured when free in solution, but undergoes an extensive transition to an α-helical conformation after complexing to KIX (Figure 4.17). The KIX region involved in the binding of the KID activation domain forms a stable domain with a shallow groove exposed on the surface. This groove is lined with numerous hydrophobic side chains that are mainly responsible, together with several important electrostatic interactions, for the induction of the amphipathic α-helical structure of the KID activation domain during formation of the KID/KIX complex.

Although it can be argued that the KID/KIX domain only represents one particular instance of an activator/coactivator structure, and may not reveal other important structural changes that occur in the full-length proteins (rather than the isolated domains used in the study), it is likely that the information obtained is of general significance. Many of the key residues present in KID, involved in the specific interactions with the coactivator groove, are highly conserved among a number of other activation domains targeting CBP that show little, if any, other traces of sequence similarity in the surrounding regions (Figure 4.18). Similarly, a recently discovered 'hydrophobic cleft' on the human

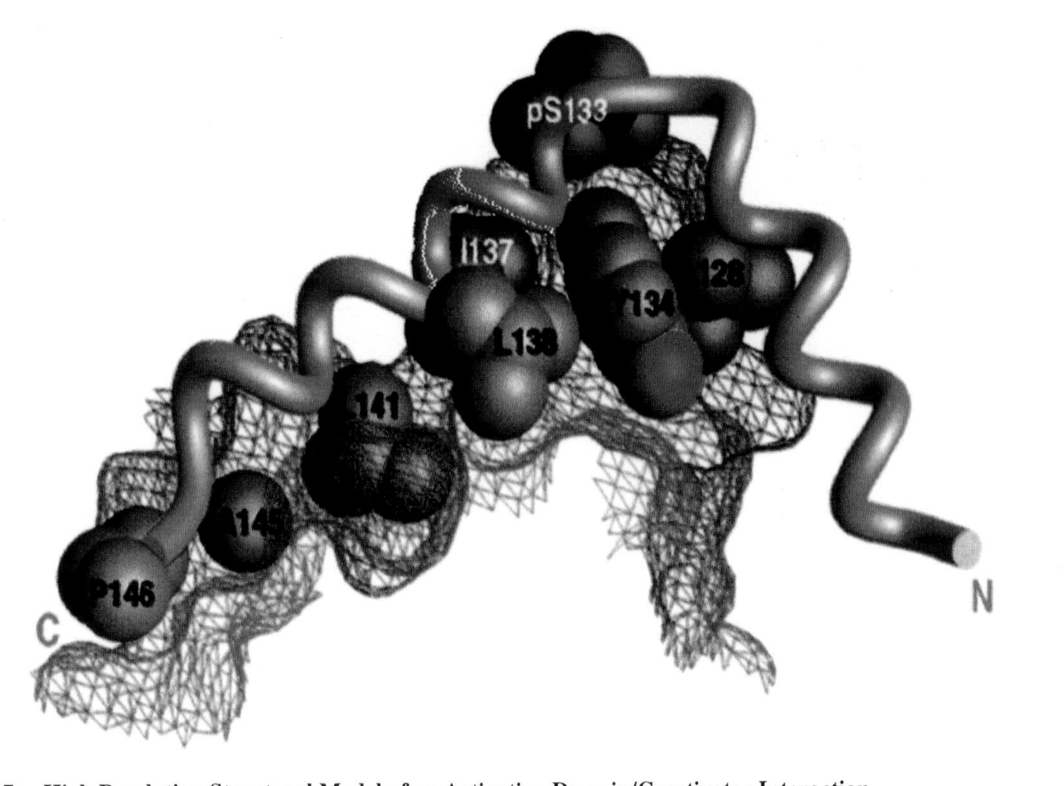

Figure 4.17. High Resolution Structural Model of an Activation Domain/Coactivator Interaction.
The surface features of the CBP-KIX domain forming a groove are shown as an orange mesh. The groove is lined with a row of hydrophobic residues (shown as green spheres). The α-helical KID activation domain is shown in pink. From Radhakrishnan *et al.* (1997).

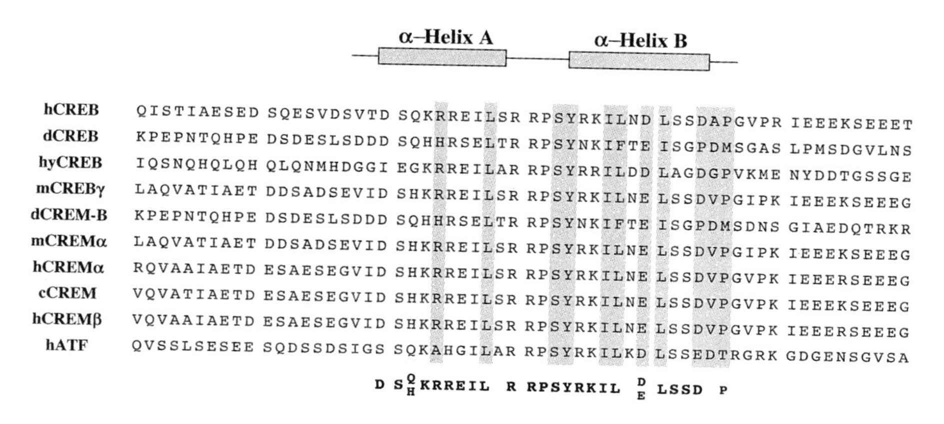

Figure 4.18. Structural Similarities in the Activation Domains Capable of Specific Interactions with p300/CBP.

All the activation domains of gene-specific transcription factors capable of interaction with p300/CBP contain a highly conserved domain capable of participating in the formation of two α-helices and may therefore interact with the KIX domain in a similar manner. Redrawn from Radhakrishnan *et al.* (1997).

thyroid hormone nuclear receptor that specifically binds to coactivators (Feng *et al.*, 1998) reinforces the general concept of the key importance of hydrophobic residues in many activator-coactivator interactions.

OCA-B: A Highly Specific Coactivator for Octamer Binding Proteins

TFIID and p300/CBP interact with a number of different types of gene-specific activators and enable them to stimulate the activity of the basal transcriptional machinery. The following example, OCA-B (also sometimes referred to as OBF or BOB), is yet another example of a coactivator activity, but its highly specialized function puts OCA-B in a class of its own at the opposite end of the specificity spectrum. Up to now OCA-B has been exclusively implicated in the coactivation function of the gene-specific transcription factor Oct-2 found mainly in B-lymphocyctes. The expression of the immunoglobulin genes in B-lymphocyctes is controlled through the binding of Oct-2 to an octamer DNA sequence motif ('ATTTGCAT') located in the vicinity of various promoters (Dreyfus *et al.*, 1987; Wirth *et al.*, 1987). Oct-2 stimulates transcription from octamer-containing promoters in non-lymphoid cell lines only rather inefficiently, which originally led to the hypothesis that it needed additional factors that are present in lymphoid cells, but absent (or inactive) in other cell types (Mitzushima-Suganuo and Roeder, 1986; Gerster *et al.*, 1990). Biochemical fractionation of B-cell extracts indeed resulted in the purification of the tissue-, promoter- and factor-specific coactivator OCA-B (Luo *et al.*, 1992). As expected from its functional properties, OCA-B is constitutively expressed in B lymphocytes, and mutations of the OCA-B gene prevent terminal B-cell differentiation in mice (Schubart *et al.*, 1996; Nielsen *et al.*, 1996).

4.7. The $RNAP_{II}$ Holoenzyme Mediator Complex

Some of the least-understood coactivator activities are present in the 'mediator' complex that forms a crucial part of the $RNAP_{II}$ holoenzyme (Chapter 2). The mediator supports activated levels of transcription by certain gene-specific activators *in vitro* in the absence of TAFs, or any of the other known coactivator activities (Figure 4.19; Kelleher *et al.*, 1990; Flanagan *et al.*, 1991; Kim *et al.*, 1994; Koleske *et al.*, 1994; Maldonado *et al.*, 1996; Myers *et al.*, 1998). A number of activation domains have been shown to associate with the $RNAP_{II}$ holoenzyme. The viral transactivator VP16 interacts directly with the mediator complex *in vitro* (Hengartner *et al.*, 1995). E1A and VP16 can also be found in an *in vivo* association with an SRB-containing human $RNAP_{II}$ holoenzyme (Gold *et al.*, 1996). Biochemical and genetic studies suggest that the yeast transcriptional activator GAL4 contacts SRB4 directly (Koh *et al.*, 1998), and GCN4 has been shown to contact several different SRB mediator components (SRB2, SRB4, and SRB7), as well as TFIID and the ADA2-ADA3 complex. This implies that at least some gene-specific transcriptional activators can use physically and functionally distinct coactivator systems to stimulate the rate of transcription from particular promoters (Drysdale *et al.*, 1998). The observed redundancy almost certainly accounts for the observation that mutations in yeast TAF_{II}-encoding genes usually have relatively little effect on the overall rate of activated transcription *in vivo* (Moqtaderi *et al.*, 1996; Walker *et al.*, 1996). Recently, a comparable observation that activated levels of transcription can still occur in a human TAF_{II}-depleted

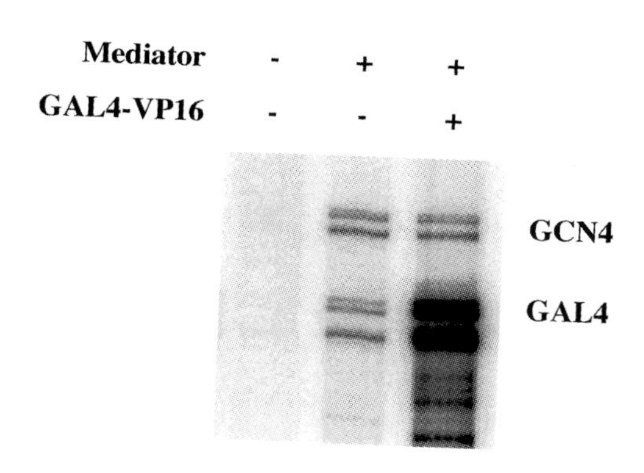

Figure 4.19. The Mediator Complex Contains Coactivator Activity.
Purified mediator was added to a reconstituted *in vitro* transcription reaction containing purified $RNAP_{II}$ and basal factors. Note that basal transcription is dependent on the presence of mediator in this system (compare the left and middle lanes). Addition of GAL4-VP16 results in a specific stimulation of transcription from the GAL4 promoter (lower set of bands), but not from the GCN4 promoter (upper set of bands). Transcription from the two promoters was monitored by using 'G-less' cassettes of different lengths. Data from Myers *et al.* (1998).

nuclear extract (possibly through a mediator-like activity) supports the concept of functionally-redundant coactivator systems present in all eukaryotic cells (Oelgeschlager *et al.*, 1998). These findings have interesting implications for theories of synergistic activation: if gene-specific transcription factors can specifically and simultanously interact with several types of coactivators that faciliate nonoverlapping functions within the basal transcriptional machinery, it is easy to understand how more-than-additive stimulation of transcription can be achieved.

Interestingly, in addition to the coactivator functions the mediator complex also contains a number of proteins that have been implicated in transcriptional repression (Figure 4.20). The RGR1 (Li et al., 1995), SIN4 and ROX3 (Song *et al.*, 1996), and several SRB proteins (SRB8, SRB9, SRB10 and SRB11; Surosky *et al.*, 1994; Kuchin *et al.*, 1995; Wahi *et al.*, 1995) have been identified in several independent genetic screens as global repressors controlling the expression of many diverse groups of genes and are *bona fide* components of the mediator complex (e.g. Myers *et al.*, 1998). The presence of these proteins

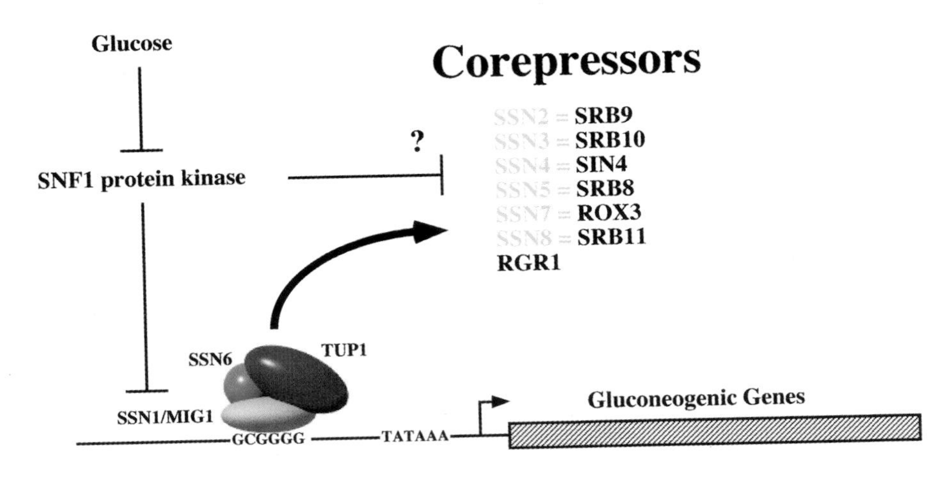

Figure 4.20. The Mediator Complex Contains Transcriptional Corepressors. Genetic screens revealed the yeast genes ('SSNx') encoding the proteins responsible for the transcriptional repression of gluconeogenic genes in the presence of high glucose levels. The sequence-specific DNA-binding ability of a the SSN1/SSN6/TUP1 repressor complex is negatively controlled via the activity of the SNF1 protein kinase, which in turn reflects the amount of glucose present. In the absence of SNF1-mediated phosphorylation, the SSN1/SSN6/TUP1 complex binds to G/C-rich promoter motifs present upstream of gluconeogenic genes and exerts its inhibitory action on the basal transcriptional machinery through a corepressor complex containining a number of other SSN genes. All these corepressors correspond to components known to be present in the mediator complex of the yeast $RNAP_{II}$ holoenzyme (SIN4, ROX3 and RGR1), or that have been isolated in SRB mutagenesis screens (SRB8, SRB9, SRB10 and SRB11) and are thus at least indirectly involved in holoenzyme function.

in the $RNAP_{II}$ holoenzyme strongly suggests that the mediator complex is involved in both positive and negative transcriptional control mechanisms. Although the experimental evidence is still rather incomplete, it is likely that other coactivator complexes, such as TFIID, may also contain similar global repressors. Such mixtures of positive and negative cofactors within single regulatory complexes presents an attractive system for integrating and distilling the numerous signals from gene-specific activators and repressors into a coherent enzymatic response by $RNAP_{II}$.

4.8. General Cofactors. The USA Fraction

The USA Fraction Contains a Variety of Coactivators and Repressors of Basal Transcription

A new group of coactivators was discovered in the 'Upstream Stimulatory Activity' ('USA') that was initially identified as a specific chromatographic fraction from mammalian cell nuclear extracts. USA complements and enhances coactivation by the TFIID complex in a general manner (Figure 4.21; Meisterernst and Roeder, 1991; Meisterernst *et al.*, 1991; Kretzschmar *et al.*, 1992, 1993, and 1994), and full transcriptional activation of at least one model system (the human HIV-1 promoter) by NF-κB or Sp1 requires the simultanous presence of both TFIID and USA (Guermah *et al.*, 1998). Even in *in vitro* transcription system reconstituted from highly purified or mostly recombinant basal transcription factors, one or more of the activities present in USA (the 'Positive Cofactors' or 'PCs') are required for activators to stimulate transcription from many test promoters to the maximally attainable level (up to 90-fold). This indicates that PCs assist in the coactivation effect directly, rather than counteract the nonspecific inhibitory effects of some contaminating proteins or nucleosomes (Ge and Roeder, 1994). Two well characterized components of USA, PC2 and PC4, are fully effective in substituting for the

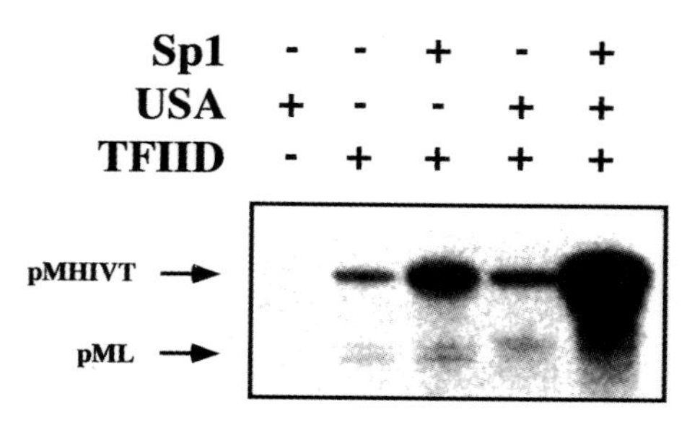

Figure 4.21. Stimulatory Effect of USA on Sp1-Mediated Activated Transcription *In Vitro*.
In this *in vitro* transcription reaction two different promoter test constructs were incubated with various combinations of fractions containing the activator Sp1, USA fraction and TFIID. The positions of the transcripts originating from a promoter containing only a TATA box (pML), and another promoter containing Sp1-binding sites (pMHIV), are indicated with labelled arrows. In the absence of TFIID neither of the two promoters is transcribed (*leftmost lane*). Addition of TFIID results in basal levels of transcription from both promoters. In presence of Sp1 the transcription from pMHIVT is specifically activated, and this level of Sp1 activation is substantially enhanced in the presence of USA (*rightmost lane*). Data from Meisterernst *et al.* (1991).

USA fraction in some *in vitro* systems and are therefore likely to represent the main active ingredients of the USA fraction (Guermah *et al.*, 1998). Apart from the coactivating PC activities, USA also contains a variety of repressors or 'Negative Cofactors' ('NCs'). Even these NCs are not necessarily bad news for activator/coactivator function, because they only repress the basal rate of transcription and do not affect activator/coactivator-mediated increases in promoter-specific transcription (Meisterernst *et al.*, 1991). By lowering the rate of 'background noise' in the system, the NCs substantially contribute to enhancing the overall rate of induction in the presence of gene-specific activators. The NCs are thought to lower the rate of basal transcription by interacting directly with TBP and thus preventing the formation of preinitiation complexes that would otherwise carry out low levels of specific transcription, even in absence of gene-specific transcriptional activators. Other NCs may be able to interact with chromatin proteins or mimic histones, and thus exert a similar effect.

The coactivator proteins in the USA are completely distinct from the other coactivators described in the previous sections. They fractionate differently from TFIID during biochemical purification procedures and enhance non-discriminantly the activity of essentially all types of activators, even if they contain very different types of activation domains (Ge and Roeder, 1994; Kaiser *et al.*, 1995). This is obviously in stark contrast to the high degree of specific interactions of certain activation domain types with selected TAFs or the specialized p300/CBP coactivator system. Although the various coactivator activities are present within a single biochemical fraction there is up to now no evidence suggesting that they are arranged in any form of stable, multisubunit complex, and some of the already characterized factors are fully active as individual polypeptides *in vitro* when they are produced in recombinant form (e.g. Meisterernst *et al.*, 1994).

4.9. Conclusions

Most bacterial activators interact with specific subunits of RNAP directly and increase the rate of transcript initiation on specific promoters mainly through a recruitment mechanism. Although this may also occur in exceptional circumstances in eukaryotic cells, the presence of basal factors necessitate a more indirect mode of communication between activators and $RNAP_{II}$. On eukaryotic promoters $RNAP_{II}$ is surrounded by several additional 'layers' of protein factors that allowed the evolution of complex control mechanisms through multiple indirect contacts with different gene-specific transcription factors. The existence of coactivators was not even considered until a decade ago, but since then they have started to dominate our theories concerning the mechanisms of activated transcription in eukaryotic systems. Coactivators come in all shapes and sizes and probably act through a variety of (functionally redundant) pathways. The proposed functions of coactivators include stimulation of the recruitment and assembly of the preinitiation transcriptional machinery, post-translational modifications of basal factors, and creation of active chromatin environments for highly transcribed genes. Although many coactivators are probably invariant from cell to cell, there is evidence for existence of tissue-specific coactivators. Other types of coactivators act in close cooperation with a limited set of gene-specific transcription factors and thus blur the distinction between basal factors and gene-specific transcription factors. The best-studied coactivators include the TAF_{II}-subunits of the TFIID complex, USA, CBP/p300 and the holoenzyme-specific mediator complex.

Research Literature

Akimaru, H., Chen, Y., Dai, P., Hou, D.-X., Nonaka, M., Smolik, S.M., Armstring, S., Goodman, R.H., and Ishii, S. (1997). *Drosophila* CBP is a coactivator of *cubitus interruptus* in *hedgehog* signalling. *Nature* 386, 735–738.

Apone, L.M., Virbasius, C.A., Reese, J.C., and Green, M.R. (1996). Yeast $TAF_{II}90$ is required for cell-cycle progression through G2/M but not for general transcription activation. *Genes Dev.* 10, 2368–2380.

Arias, J., Alberts, A.S., Brindle, P., Claret, F.X., Smeal, T., Karin, M., Feramisco, J., and Montminy, M. (1994). Activation of cAMP and mitogen responsive genes relies on a common nuclear factor. *Nature* 370, 226–229.

Bando, M., Ijuin, S., Hasegawa, S., and Horikoshi, M. (1997). The involvement of histone fold motifs in the mutual interaction between human $TAF_{II}80$ and $TAF_{II}22$. *J. Biochem.* 121, 591–597.

Bannister, A.J., and Kouzarides, T. (1996). The CBP coactivator is a histone acetyltransferase. *Nature* 384, 641–643.

Bell, A., Gaston, K., Williams, R., Chapman, K., Kolb, A., Buc, H., Minchin, S., Williams, J. and Busby, S. (1990). Mutations that alter the ability of the *Escherichia coli* cyclic AMP receptor protein to activate transcription. *Nucl. Acids Res.* 18, 7243–7250.

Berger, S.L., Pina, B., Silverman, N., Marcus, G.A., Agapite, J., Regier, J.L., Triezenberg, S.J., and Guarente, L. (1992). Genetic isolation of ADA2: a potential transcriptional adaptor required for function of certain acidic activation domains. *Cell* 70, 251–265.

Bhattacharya, S., Eckner, R., Grossman, S., Oldread, E., Arany, Z., D'Andrea, A., and Livingston, D.M. (1996). Cooperation of Stat2 and p300/CBP in signalling induced by interferon-α. *Nature* 383, 344–347.

Brownell, J.E., and Allis, C.D. (1996). Special HATs for special occasions: linking histone acetylation to chromatin assembly and gene activation. *Curr. Opin. Gen. Dev.* 6, 176–184.

Brou, C., Chaudhary, S., Davidson, I., Lutz, Y., Wu, J., Egly, J.M., Tora, L., and Chambon, P. (1993). Distinct TFIID complexes mediate the effect of different transcriptional activators. *EMBO J.* 12, 489–499.

Bult, C. J., *et al.*. (1996). Complete genome sequence of the methanogenic archaeon, *Methanococcus jannaschii*. *Science* 273, 1058–1073.

Buratowski, S. (1997). Multiple TATA-binding factors come back into style. *Cell* 91, 13–15.

Buratowski, S., Hahn, S., Sharp, P.A., and Guarente, L. (1988). Function of a yeast TATA element binding protein in a mammalian transcription system. *Nature* 334, 37–42.

Burke, T., and Kadonaga, J. (1996). *Drosophila* TFIID binds to a conserved downstream basal promoter element that is present in many TATA box-deficient promoters. *Genes Dev.* 10, 711–724.

Burley, S.K., and Roeder, R.G. (1996). Biochemistry and structural biology of transcription factor IID (TFIID). *Ann. Rev. Biochem.* 65, 769–799.

Cavallini, B., Faus, I., Matthes, H., Chipoulet, J.M., Windsor, B., Egly, J.M., and Chambon, P. (1989). Cloning of the gene encoding the yeast protein BTF1Y, which can substitute for the human TATA box-binding factor. *Proc. Natl. Acad. Sci. USA* 86, 9803–9807.

Chakravarti, D., LaMorte, V.J., Nelson, M.C., Makjima, T., Schulman, I.G., Juguilon, H., Montminy, M., and Evans, R.M. (1996). Role of CPB/P300 in nuclear receptor signalling. *Nature* 383, 99–103.

Chen, Y., Ebright, Y.W., and Ebright, R.H. (1994a). Identification of the target of a transcription activator protein by protein-protein photocrosslinking. *Science* 265, 90–92.

Chen, J.-L., Attardi, L.D., Verrijzer, C.P., Yokomori, K., and Tjian, R. (1994b). Assembly of recombinant TFIID reveals differential requirements for distinct transcriptional activators. *Cell* 79, 93–105.

Cheong, J., Yi, M., Lin, Y., and Murakami, S. (1995). Human RPB5, a subunit shared by eukaryotic RNA polymerases, binds human hepatitis B virus X protein and may play a role in X transactivation. *EMBO J.* 14, 143–150.

Chiang, C.-M., Ge, H., Wang, Z., Hoffmann, A., and Roeder, R.G. (1993). Unique TATA-binding protein-containing complexes and cofactors involved in transcription by RNA polymerase II and III. *EMBO J.* 12, 2749–2762.

Chiang, C.-M., and Roeder, R.G. (1995). Cloning of an intrinsic human TFIID subunit that interacts with multiple transcriptional activators. *Science* 267, 531–535.

Choy, B., and Green, M.R. (1993). Eukaryotic activators function during multiple steps of preinitiation complex assembly. *Nature* 366, 531–536.

Cormack, B.P., Strubin, M., Stargell, L.A., and Struhl, K. (1994). Conserved and nonconserved functions of yeast and human TATA-binding proteins. *Genes Dev.* 9, 1335–1343.

Crowley, T.E., Hoey, T. Liu, J.K., Jan, Y.N., Jan, L.Y., and Tjian, R. (1993). A new factor related to TATA-binding protein has highly restricted expression patterns in *Drosophila. Nature* 361, 557–561.

Dai, P., Akimaru, H., Tanaka, Y., Hou, D.X., Yasukawa, T., Kanei-Ishii, C., Takahashi, T., and Ishii, S. (1996). CBP as a transcriptional coactivator of c-Myb. *Genes Dev.*, 10, 528–540.

Damania, B., and Alwine, J.C. (1996). TAF-like function of SV40 large T antigen. *Genes Dev.* 10, 1369–1381.

Das, G., Hinkley, C.S., and Herr, W. (1995). Basal promoter elements as a selective determinant of transcriptional activator function. *Nature* 374, 657–660.

Dethiollaz, S., Eichenberger, P., and Geiselmann, J. (1996). Influence of DNA geometry on transcriptional activation in *Escherichia coli. EMBO J.* 15, 5449–5458.

Dikstein, R., Ruppert, S., and Tjian, R. (1996a). $TAF_{II}250$ is a bipartite protein kinase that phosphorylates the basal transcription factor RAP74. *Cell* 84, 781–790.

Dikstein, R., Zhou, S. and Tjian, R. (1996b). Human $TAF_{II}105$ is a cell type-specific TFIID subunit related to $hTAF_{II}130$. *Cell* 87, 137–146.

Dove, S.L., and Hochschild, A. (1998). Conversion of the omega subunit of *Escherichia coli* RNA polymerase into a transcriptional activator or an activation target. *Genes Dev.* 12, 745–754.

Dove, S.L., Joung, J.K., and Hochschild, A. (1997). Activation of prokaryotic transcription through arbitrary protein-protein contacts. *Nature* 386, 627–630.

Dreyfus, M., Doyen, N., and Rougeon, F. (1987). The conserved decanucleotide from the immunoglobulin heavy chain promoter induces a very high transcriptional activity in B-cells when introduced into an heterologous promoter. *EMBO J.* 6, 1685–1690.

Driever, W., and Nusslein-Volhard, C. (1988). The *bicoid* protein determines position in the *Drosophila* embryo in a concentration-dependent manner. *Cell* 54, 95–104.

Drysdale, C.M., Jackson, B.M., McVeigh, R., Klebanow, E.R., Bai, Y., Kokubo, T., Swanson, M., Nakatani, Y., Weil, P.A., and Hinnebusch, A.G. (1998). The Gcn4p activation domain interacts specifically in vitro with RNA polymerase II holoenzyme, TFIID, and the Adap-Gcn5 coactivator complex. *Mol. Cell. Biol.* 18, 1711–1724.

Dynlacht, B.D., Hoey, T., and Tjian, R. (1991). Isolation of coactivators associated with the TATA-binding protein that mediate transcriptional activation. *Cell* 66, 563–576.

Eckner, R., Ewen, M.E., Newsine, D., Gerdes, M., DeCaprio, J.A., Lawrence, J.B., and Livingston, D.M. (1994). Molecular cloning and functional analysis of the adenovirus E1A-associated 300 kD protein (p300) reveals a protein with properties of a transcriptional adapter. *Genes Dev.* 8 869–884.

Eckner, R., Yao, T.-P., Oldread, E., and Livingston, D.M. (1996). Interaction and functional collaboration of p300/CBP and bHLH proteins in muscle and B-cell differentiation. *Genes Dev.* 10, 2478–2490.

Eschenlauer, A., and Reznikoff, W. (1991). *Escherichia coli* catabolite activator protein mutants defective in positive control of *lac* operon transcription. *J. Bacteriol.* 173, 5024–5029.

Feng, W., Ribeiro, R.C.J., Wagner, R.L., Ngyuen, H., Apriletti, J.W., Fletterick, R.J., Baxter, J.D., Kushner, P.J., and West, B.L. (1998). Hormone-dependent coactivator binding to a hydrophobic cleft on nuclear receptors. *Science* 280, 1747–1749.

Ferreri, K., Gill, G., and Montminy, M. (1994). The cAMP-regulated transcription factor CREB interacts with a component of the TFIID complex. *Proc. Natl. Acad. Sci. USA* 91, 1210–1213.

Finnin, M.S., Cicero, M., Davies, C., Porter, S.J., White, S.W., and Kreuzer, K.N. (1997). The activation domain of the MotA transcription factor from bacteriophage T4. *EMBO J.* 16, 1992–2003.

Flanagan, P.M., Kelleher, R.J. III, Sayre, M.H., Tschochner, H., and Kornberg, R.D. (1991). A mediator required for activation of RNA polymerase II transcription *in vitro*. *Nature* 350, 436–438.

Ge, H., and Roeder, R.G. (1994). Purification, cloning, and characterization of a human coactivator, PC4, that mediates transcriptional activation of class II genes. *Cell* 78, 513–523.

Gerster, T., Balmaceda, C.-G., and Roeder, R.G. (1990). The cell type-specific octamer transcription factor OTF-2 has two domains required for the activation of transcription. *EMBO J.* 9, 1635–1643.

Gill, G., and Ptashne, M. (1988). Negative effect of the transcriptional activator GAL4. *Nature* 334, 721–724.

Gill, G., Pascal, E., Tseng, Z.H., and Tjian, R. (1994). A glutamine-rich hydrophobic patch in transcription factor Sp1 contacts the $dTAF_{II}110$ component of the *Drosophila* TFIID complex and mediates transcriptional activation. *Proc. Natl. Acad. Sci. USA* 91, 192–196.

Gold, M.O., Tassan, J.P., Nigg, E.A., Rice, A.P., and Herrmann, C.H. (1996). Viral transactivators E1A and VP16 interact with a large complex that is associated with CTD kinase activity and contains CDK8. *Nucl. Acids Res.* 24, 3771–3777.

Goodrich, J.A., Hoey, T., Thut, C.J., Admon, A., and Tjian, R. (1993). *Drosophila* $TAF_{II}40$ interacts with both a VP16 activation domain and the basal transcription factor TFIIB. *Cell* 75, 519–530.

Gu, W., and Roeder, R.G. (1997). Activation of p53 sequence-specific DNA binding by acetylation of the p53 C-terminal domain. *Cell* 90, 595–606.

Gu, W., Shi, X.L., and Roeder, R.G. (1997). Synergistic activation of transcription by CBP and p53. *Nature* 387, 819–823.

Guermah, M., Malik, S., and Roeder, R.G. (1998). Involvement of TFIID and USA components in transcriptional activation of the human immunodeficiency virus promoter by NFκ-B and Sp1. *Mol. Cell. Biol.* 18, 3234–3244.

Hahn, S., Buratowski, S., Sharp, P.A., and Guarente, L. (1989). Isolation of the gene encoding the yeast TATA binding protein TFIID: a gene identical to the *SPT15* suppressor of Ty element insertions. *Cell* 58, 1173–1181.

Hansen, S.K., and Tjian, R. (1995). TAFs and TFIIA mediate differential utilization of the tandem Adh promoter. *Cell* 82, 565–575.

Hansen, S.K., Takada, S., Jacobson, R.H., Lis, J.T., and Tjian, R. (1997). Transcription properties of a cell-type-specific TATA-binding protein, TRF. *Cell* 91, 71–83.

Hawley, D.K., and McClure, W.R. (1983). Compilation and analysis of *Escherichia coli* promoter DNA sequences. *Nucl. Acids Res.* 11, 2237–2255.

Hengartner, C.J., Thompson, C.M., Ahng, J., Chao, D.M., Lian, S., Koleske, A.J., Okamura, S., and Young, R.A. (1995). Association of an activator with an RNA polymerase II holoenzyme. *Genes Dev.* 9, 897–910.

Hisatake, K., Hasegawa, S., Takada, R., Nakatani, Y., Horikoshi, M., and Roeder, R.G. (1993). The p250 subunit of native TATA box-binding factor TFIID is the cell-cycle regulatory protein CCG1. *Nature* 362, 179–181.

Hoey, T., Dynlacht, B.D., Peterson, M.G., Pugh, F., and Tjian, R. (1990). Isolation and characterization of the *Drosophila* gene encoding the TATA box binding protein, TFIID. *Cell* 61, 1179–1186.

Hoey, T., Weinzierl, R.O.J., Gill, G., Chen, J.-L., Dynlacht, B.D., and Tjian, R. (1993). Molecular cloning and functional analysis of *Drosophila* TAF110 reveal properties expected of coactivators. *Cell* 72, 247–260.

Hoffmann, A., Chiang, C.M., Oelgeschlager, T., Xie, X., Burley, S.K., Nakatani, Y., and Roeder, R.G. (1996). A histone octamer-like structure within TFIID. *Nature* 380, 356–359.

Horikoshi, M., Hai, T., Lin, Y.-S., Green, M.R., and Roeder, R.G. (1988). Transcription factor ATF interacts with the TATA factor to facilitate establishment of a preinitiation complex. *Cell* 54, 1033–1042.

Horikoshi, M., Wang, C.K., Fujii, H., Cromlich, J.A., Weil, B.A., and Roeder, R.G. (1989). Cloning and structure of a yeast gene encoding a general transcription initiation factor TFIID that binds the TATA box. *Nature* 341, 299–303.

Ishihama, A. (1993). Protein-protein communication within the transcription apparatus. *J. Bacteriol.* 175, 2483–2489.

Jackle, H., and Sauer, F. (1993). Transcriptional cascades in *Drosophila*. *Curr. Opin. Cell Biol.* 5, 505–512.

Jacq, X., Brou, C., Lutz, Y., Davidson, I., Chambon, P. and Tora, L. (1994). Human $TAF_{II}30$ is present in a distinct TFIID complex and is required for transcriptional activation by the estrogen receptor. *Cell* 79, 107–117.

Janknecht, R., and Nordheim, A. (1994). Regulation of the *c-fos* promoter by the ternary complex factor Sap-1a and its coactivator CBP. *Oncogene* 12, 1961–1969.

Jeon, Y.H., Negishi, T., Shirakawa, M., Yamazaki, T., Fujita, N., Ishihama, A., and Kyogoku, Y. (1995). Solution structure of the activator contact domain of the RNA polymerase α subunit. *Science* 270, 1495–1497.

Kaiser, K., Stelzer, G. and Meisterernst, M. (1995). The coactivator p15 (PC4) initiates transcriptional activation during TFIIA-TFIID-promoter complex formation. *EMBO J.* 14, 3520–3527.

Kamei, Y., Xu, L., Heinzel, T., Torchia, J., Kurokawa, R., Gloss, B., Lin, S.-C., Heyman, R.A., Rose, D.W., Glass, C.K., and Rosenfeld, M.G. (1996). A CBP integrator complex mediates transcriptional activation and AP-1 inhibition by nuclear receptors. *Cell* 85, 403–414.

Kaufmann, J., and Smale, S. (1994). Direct recognition of initiator elements by a component of the transcription factor IID complex. *Genes Dev.* 8, 821–829.

Kee, B., Arias, J., and Montminy, M. (1996). Adaptor-mediated recruitment of RNA polymerase II to a signal-dependent activator. *J. Biol. Chem.* 271, 2373–2375.

Kelleher, R.J. III, Flanagan, P.M., and Kornberg, R.D. (1990). A novel mediator between activator proteins and the RNA polymerase II transcription apparatus. *Cell* 61, 1209–1216.

Kim, Y.-I., Bjorklund, S., Li, Y., Sayre, M.H., and Kornberg, R.D. (1994). A multiprotein mediator of transcriptional activation and its interaction with the C-terminal repeat domain of RNA polymerase II. *Cell* 77, 599–608.

Kingston, R.E., and Green, M.E. (1994). Modeling eukaryotic transcriptional activation. *Curr. Biol.* 4, 325–332.

Kitajima, S. Chibazakura, T., Yohana, M., and Yasukochi, Y. (1994). Regulation of the human general transcription factor TFIIF by phosphorylation. *J. Biol. Chem.* 269, 29970–29977.

Koh, S.S., Ansari, A.Z., Ptashne, M., and Young, R.A. (1998). An activator target in the RNA polymerase II holoenzyme. *Mol. Cell* 1, 895–904.

Kolb, A., Busby, S., Buc, H., and Adhya, S. (1993). Transcriptional regulation by cAMP and its receptor protein. *Annu. Rev. Biochem.* 62, 749–795.

Koleske, A.J., and Young, R.A. (1994). An RNA polymerase II holoenzyme responsive to activators. *Nature* 368, 466–469.

Kokubo, T., Gong, D.-W., Yamashita, S., Horikoshi, M., Roeder, R.G., and Nakatani, Y. (1993). *Drosophila* 230-kD TFIID subunit, a functional homolog of the human cell cycle gene product, negatively regulates DNA binding of the TATA box-binding subunit of TFIID. *Genes Dev.* 7, 1033–1046.

Kretzschmar, M., Meisterernst, M., Scheidereit, C., Li, G., and Roeder, R.G. (1992). Transcriptional regulation of the HIV-1 promoter by NF-κB *in vitro*. *Genes Dev.* 6, 761–774.

Kretzschmar, M., Meisterernst, M., and Roeder, R.G. (1993). Identification of human DNA topoisomerase I as a cofactor for activator-dependent transcription by RNA polymerase II. *Proc. Natl. Acad. Sci. USA* 90, 11508–11512.

Kretzschmar, M., Stelxer, G., Roeder, R.G., and Meisterernst, M. (1994). RNA polymerase II cofactor PC2 facilitates activation of transcription by GAL4-AH *in vitro*. *Mol. Cell. Biol.* 14, 3927–3937.

Kuchin, S., Yeghiayan, P., and Carlson, M. (1995). Cyclin-dependent protein kinase and cyclin homologs SSN3 and SSN8 contribute to transcriptional control in yeast. *Proc. Natl. Acad. Sci. USA* 92, 4006–4010.

Kuo, M.H., Brownell, J.E., Sobel, R.E., Ranall, T.A., Cook, R.G., Edmondson, D.G., Roth, S.Y., and Allis, C.D. (1996). Transcription-linked acetylation by GCN5p of histone H3 and H4 at specific lysines. *Nature* 383, 269–272.

Kwok, R.P.S., Lundblad, J.R., Chrivia, J.C., Richards, J.P., Bachinger, H.P., Brennan, R.G., Roberts, S.G.E., Green, M.R., and Goodman, R.H. (1994). Nuclear protein CBP is a coactivator for the transcription factor CREB. *Nature* 370, 223–226.

Kwok, R.P.S., Laurance, M.E., Lundblad, J.R., Goldman, P.S., Shih, H., Connor, L.M., Mariott, S.J., and Goodman, R.H. (1996). Control of cAMP-regulated enhancers by the viral transactivator Tax through CREB and the coactivator CBP. *Nature* 380, 642–646.

Li, M. Moyle, H., and Susskind, M.M. (1994). Target of the transcriptional activation function of phage λ cI protein. *Science* 263, 75–77.

Li, Y., Bjorklund, S., Jiang, Y.W., Kim, Y.-J., Lane, W.S., Stillman, D.J., and Kornberg, R.D. (1995). Yeast global transcriptional regulators Sin4 and Rgr1 are components of mediator complex/RNA polymerase II holoenzyme. *Proc. Natl. Acad. Sci. USA* 92, 10864–10868.

Lieberman, P.M., and Berk, A.J. (1994). A mechanism for TAFs in transcriptional activation: activation domain enhancement of TFIID-TFIIA-promoter complex formation. *Genes Dev.* 8, 995–1006.

Lill, N.L., Grossman, S.R., Ginsberg, D., DeCaprio, J., and Livingston, D.M. (1997). Binding and modulation of p53 by p300/CBP coactivators. *Nature* 387, 823–827.

Lin, Y., Nomura, T., Cheong, J., Dorjsuren, D., Iida, K., and Murakami, S. (1997). Hepatitis B virus X protein is a transcriptional modulator that communicates with transcription factor IIB and the RNA polymerase II subunit 5. *J. Biol. Chem.* 272, 7132–7139.

Lundblad, J.R., Kwok, R.P.S., Laurance, M.E., Harter, M.L., and Goodman, R.H. (1995). Adenoviral E1A-associated protein p300 as a functional homologue of the transcriptional co-activator CBP. *Nature* 374, 85–88.

Luo, Y., and Roeder, R.G. (1995). Cloning, functional characterization, and mechanism of action of the B cell-specific transcriptional coactivator OCA-B. *Mol. Cell. Biol.* 15, 4115–4124.

Luo, Y., Fujii, H., Gerster, T. and Roeder, R.G. (1992). A novel B cell-derived coactivator potentiates the activation of immunoglobulin promoters by octamer-binding transcription factors. *Cell* 71, 231–241.

Maldonado, E., Shiekhattar, R., Sheldon, M., Cho, H., Drapkin, R., Rickert, P., Lees, E., Anderson, C.W., Linn, S., and Reinberg, D. (1996). A human RNA polymerase II complex assoicated with SRB and DNA-repair proteins. *Nature* 381, 86–89 (+ erratum: *Nature* 384, 384).

Martinez, E., Chiang, C.-M., Ge, H., and Roeder, R.G. (1994). TATA-binding protein-associated factor(s) in TFIID function through the initiator to direct basal transcription from a TATA-less class II promoter. *EMBO J.* 13, 3115–3126.

Martinez, E., Zhou, Q., L'Etoile, N.D., Oelgeschlager, T., Berk, A.J., and Roeder, R.G. (1995). Core promoter-specific function of a mutant transcription factor TFIID defective in TATA-box binding. *Proc. Natl. Acad. Sci. USA* 92, 11864–11868.

Meisterernst, M., and Roeder, R.G. (1991). Family of proteins that interact with TFIID and regulate promoter activity. *Cell* 67, 557–567.

Meisterernst, M., Roy, A.l., Lieu, H.M., and Roeder, R.G. (1991). Activation of class II gene transcription by regulatory factors is potentiated by a novel activity. *Cell* 66, 981–993.

Mendendez, M., Kolb, A., and Buc, H. (1987). A new target for CRP action at the malT promoter. *EMBO J.* 6, 4227–4234.

Miller, A., Wood, D., Ebright, R.H., Rothman-Denes, L.B. (1997). RNA polymerase β′ subunit: a target of DNA binding-independent activation. *Science* 275, 1655–1657.

Mizushima-Sugano, J., and Roeder, R.G. (1986). Cell type-specific transcription of an immunoglobulin κ ligh chain gene *in vitro*. *Proc. Natl. Acad. Sci. USA* 83, 8511–8515.

Mizzen, C.A., Yang, X.-J., Kokubo, T., Brownell, J.E., Bannister, A.J., Owen-Hughes, T., Workman, J., Wang, L., Berger, S.L., Kouzarides, T., Nakatani, Y., and Allis, C.D. (1996). The $TAF_{II}250$ subunit of TFIID has histone acetyltransferase activity. *Cell* 87, 1261–1270.

Moqtaderi, Z., Yale, J.D., Struhl, K., and Buratowski, S. (1996a). Yeast homologs of higher eukaryotic TFIID subunits. *Proc. Natl. Acad. Sci. USA* 93, 14654–14658.

Moqtaderi, Z., Bai, Y., Poon, D., Weil, P.A., and Struhl, K. (1996b). TBP-associated factors are not generally required for transcriptional activation in yeast. *Nature* 383, 188–191.

Myers, L.C., Gustafsson, C.M., Bushnell, D.A., Lui, M., Erjument-Bromage, H., Tempst, P., and Kornberg, R.D. (1998). The Med proteins of yeast and their function through the RNA polymerase II carboxy-terminal domain. *Genes Dev.* 12, 45–54.

Nakajima, N., Horikoshi, M., and Roeder, R.G. (1988). Factors involved in specific transcription by mammalian RNA polymerase II: purification, genetic specificity, and TATA-box promoter interactions of TFIID. *Mol. Cell. Biol.* 8, 4028–4040.

Nakajima, T., Uchida, C., Anderson, S., Parvin, J., and Montminy, M. (1997a). RNA helicase A mediates association of CBP with RNA polymerase II. *Cell* 90, 1107–1112.

Nakajima, T. Uchida, C., Anderson, S., Parvin, J., and Montminy, M. (1997b). Analysis of a cAMP responsive activator reveals a two-component mechansism for transcriptional induction via signal-dependent factors. *Genes Dev.* 11, 738–747.

Nakatani, Y., Horikoshi, M., Brenner, M., Yamamoto, T., Besnard, F., Roeder, R.G., and Freese, E. (1990). A downstream initiation element required for efficient TATA box binding and i*n vitro* function of TFIID. *Nature* 348, 86–88.

Nielsen, P.J., Georgiev, O., Lorenz, B., and Schaffner, W. (1996). B lymphocytes are impaired in mice lacking the transcriptional coactivator Bob1/OCA-B/OBF1. *Eur. J. Immunol.* 26, 3214–3218.

O'Brien, T., and Tjian, R. (1998). Functional analysis of the human $TAF_{II}250$ N-terminal kinase domain. *Mol. Cell* 1, 905–911.

Oelgeschlager, T., Chiang, C.-H., and Roeder, R.G. (1996). Topology and reorganization of a TFIID-promoter complex. *Nature* 382, 735–738.

Oelgeschlager, T., Tao, Y., Kang, Y.K., and Roeder, R.G. (1998). Transcription activation via enhanced preinitiation complex assembly in a human cell-free system lacking TAF_{II}s. *Mol. Cell* 1, 925–931.

Ogryzko, V.V., Schiltz, R.L., Russanova, V., Howard, B.H., and Nakatani, Y. (1996). The transcriptional coactivators p300 and CBP are histone acetyltransferases. *Cell* 87, 953–959.

Peterson, M.G., Tanese, N., Pugh, B.F., and Tjian, R. (1990). Functional domains and upstream activation properties of cloned human TATA binding protein. *Science* 248, 1625–1630.

Petrij F., Giles R.H., Dauwerse H.G., Saris J.J., Hennekam R.C., Masuno M., Tommerup N., van Ommen G.J., Goodman R.H., Peters D.J. & *et al.* (1995). Rubinstein-Taybi syndrome caused by mutations in the transcriptional co-activator CBP. *Nature* 376, 348–351.

Poon, D., and Weil, P.A. (1993). Immunopurification of yeast TATA-binding protein and associated factors. *J. Biol. Chem.* 268, 15325–15328.

Poon, D., Bai, Y., Cambell, A.M., Bjorklund, S., Kim, Y.J., Zhou, S., Kornberg, R.D., and Weil, P.A. (1995). Identification and characterization of a TFIID-like multiprotein complex from *Saccharomyces cerevisiae*. *Proc. Natl. Acad. Sci. USA* 92, 8224–8228.

Pugh, B.F., and Tjian, R. (1990). Mechanism of transcriptional activation by Sp1: evidence for coactivators. *Cell* 61, 1187–1197.

Puri, P.L., Avantagiatti, M.L., Balsano, C., Sang, N., Graessmann, A., Giordano, A., and Levrero, M. (1997). p300 is specifically required for MyoD-dependent cell cycle arrest and muscle-specific gene transcription. *EMBO J.* 16, 369–383.

Purnell, B.A., Emanuel, P.A., and Gilmour, D.S. (1994). TFIID sequence recognition of the initiator and sequences farther downstream in *Drosophila* class II genes. *Genes Dev.* 8, 830–842.

Qiu, Y., Sharma, A., and Stein, R. (1998). p300 mediates transcriptional stimulation by the basic helix-loop-helix activators of the insulin gene. *Mol. Cell. Biol.* 18, 2957–2964.

Qureshi, S. A., Bell, S. D., and Jackson, S. P. (1997). Factor requirements for transcription in the archaeon *Sulfolobus shibatae*. *EMBO J.* 16, 2927–2936.

Reese, J.C., Apone, L., Walker, S.S., Griffin, L.A., and Green, M. (1994). Yeast TAFIIs in a multisubunit complex required for activated transcription. *Nature* 371, 523–527.

Rhadakrishnan, I., Perez-Alvarado, G.C., Parker, D., Dyson, H.J., Montminy, M.C., and Wright, P.E. (1997). Solution structure of the KIX domain of CBP bound to the transactivation domain of CREB: a model for activator:coactivator interactions. *Cell* 91, 741–752.

Ruppert, S., Wang, E.H., and Tjian, R. (1993). Cloning and expression of human $TAF_{II}250$: a TBP-associated factor implicated in cell-cycle regulation. *Nature* 362, 175–179.

Ruppert, S., and Tjian, R. (1995). $TAF_{II}250$ interacts with RAP74: implications for RNA polymerase II initiation. *Genes Dev.* 9, 2747–2755.

Sauer, F., Hansen, S.K., and Tjian, R. (1995a). Multiple $TAF_{II}s$ directing synergistic activation of transcription. *Science* 270, 1783–1788.

Sauer, F., Hansen, S.K., and Tjian, R. (1995b). DNA template requirement and activator-coactivator requirements for transcriptional synergism by *Drosophila bicoid*. *Science* 270, 1825–1827.

Sawadogo, M., and Roeder, R.G. (1985). Interaction of a gene-specific transcription factor with the adenovirus major late promoter upstream of the TATA box region. *Cell* 43, 165–175.

Schmidt, M.C., Kao, C.C., Pei, R., and Berk, A.J. (1989). Yeast TATA box transcription factor gene. *Proc. Natl. Acad. Sci. USA* 86, 7785–7789.

Schubart, D.B., Rolink, A., Kosco-Vilbois, M.H., Botteri, F., and Matthias, P. (1996). B cell-specific coactivator OBF-1/OCA-B/Bob1 required for immune response and germinal centre formation. *Nature* 383, 538–542.

Segil, N., Guermah, M., Hoffmann, A., Roeder, R.G., and Heitz, N. (1996). Mitotic regulation of TFIID: inhibition of activator-dependent transcription and changes in subcellular localization. *Genes Dev.* 10, 2389–2400.

Shen, W.-S., and Green, M. (1997). Yeast $TAF_{II}145$ functions as a core promoter selectivity factor, not a general coactivator. *Cell* 90, 615–624.

Shi, Y., and Mello, C. (1998). A CBP/p300 homolog specifices multiple differentiation pathways in *Caenorhabditis elegans*. *Genes Dev.* 12, 943–955.

Shikama, N., Lyon, J., and La Thangue, N.B. (1997). The p300/CBP family: integrating signals with transcription factors and chromatin. *Trends Cell Biol.* 7, 230–236.

Simon, M.C., Fisch, T.M., Benecke, B.J., Nevins, J.R., and Heitz, N. (1988). Definition of multiple, functionally distinct TATA elements, one of which is a target in the *hsp70* promoter for E1A regulation. *Cell* 52, 723–729.

Simpson-Brose, M., Treisman, J., and Desplan, C. (1994). Synergy between the Hunchback and Bicoid morphogens is required for anterior patterning in *Drosophila*. *Cell* 78, 855–865.

Smale, S.T. (1997). Transcription initiation from TATA-less promoters within eukaryotic protein-coding genes. *Biochem. Biophys. Acta* 1351, 73–88.

Song, W., Treich, I., Qian, N., Kuchin, S., and Carlson, M. (1996). SSN genes that affect transcriptional repression in *Saccharomyces cerevisiae* encode SIN4, ROX3, and SRB proteins associated with RNA polymerase II. *Mol. Cell. Biol.* 16, 115–120.

Strubin, M., Newell, J.W., and Matthias, P. (1995). OBF-1, a novel B cell-specific coactivator that stimulates immunoglobulin promoter activity through association with octamer-binding proteins. *Cell* 80, 497–506.

Surosky, R.T., Strich, R., and Esposito, R.E. (1994). The yeast UME5 gene regulates the stability of meiotic mRNAs in response to glucose. *Mol. Cell. Biol.* 14, 3446–3458.

Suzuki, Y.Y., Guermah, M., and Roeder, R.G. (1997). The ts13 mutation in the $TAF_{II}250$ subunit (CCG1) of TFIID directly affects transcription of D-type cyclin genes in cells arrested in G1 at the nonpermissive temperature. *Mol. Cell. Biol.* 17, 3284–3294.

Sypes, M.A., and Gilmour, D.S. (1994). Protein/DNA crosslinking of a TFIID complex reveals novel interactions downstream of the transcription start. *Nucl. Acids Res.* 22, 807–814.

Takada, R., Nakatani, Y., Hoffman, A., Kokubo, T., Hasegawa, S., Roeder, R.G., and Horikoshi, M. (1992). Identification of human transcription factor IID components and direct interactions between a 250 kDa polypeptide and the TATA-box binding protein. *Proc. Natl. Acad. Sci. USA* 89, 11809–11813.

Tanese, N., Pugh, B.F., and Tjian, R. (1991). Coactivators for a proline-rich activator purified from the multisubunit human TFIID complex. *Genes Dev.* 5, 2212–2224.

Tang, H.T., Severinov, K., Goldfarb, A., Fenyo, D., Chait, B. and Ebright, R.H. (1994). Location, structure, and function of the target of a transcriptional activator protein. *Genes Dev.* 8, 3058–3067.

Tansey, W.P., and Herr, W. (1997). TAFs: guilt by association? *Cell* 88, 729–732.

Tao, Y., Guermah, M., Martinez, E., Oelgeschlager, T., Hasegawa, S., Takada, R., Yamamoto, T., Horikoshi, M., and Roeder, R.G. (1997). Specific interactions and potential functions of human $TAF_{II}100$. *J. Biol. Chem.* 272, 6714–6721.

Taylor, I.C.A., and Kingston, R.E. (1990). Factor substitution in human HSP70 gene promoter: TATA-dependent and TATA-independent interactions. *Mol. Cell. Biol.* 10, 165–175.

Timmers, H.T.M., and Sharp, P.A. (1991). The mammalian TFIID protein is present in two functional distinct complexes. *Genes Dev.* 5, 1946–1956.

Uesugi, M., Nyanguile, O., Lu, H., Levine, A.J., and Verdine, G.L. (1997). Induced α-helix in the VP16 activation domain upon binding to a human TAF. *Science* 277, 1310–1313.

Verrijzer, P., Yokomori, K., Chen, J.-L., and Tjian, R. (1994). *Drosophila* $TAF_{II}150$: similarity to yeast gene TSM-1 and specific binding to core promoter DNA. *Science* 264, 933–941.

Verrijzer, P., Chen, J.L., Yokomori, K., and Tjian, R. (1995). Binding of TAFs to core elements directs promoter selectivity by RNA polymerase II. *Cell* 81, 1115–1125.

Verrijzer, C.P., and Tjian, R. (1996). TAFs mediate transcriptional activation and promoter selectivity. *Trends Biochem. Sci.* 21, 338–342.

Wahi, M., and Johnson, A.D. (1995). Identification of genes required for $\alpha 2$ repression in *Saccharomyces cerevisiae*. *Genetics* 140, 79–90.

Walker, S.S., Reese, J.C., Apone, L.M., and Green, M.R. (1996). Transcription in cells lacking TAF_{II}s. *Nature* 383, 185–188.

Walker, S.S., Shen, W.-C., Reese, J.C., Apone, L.M., and Green, M.R. (1997). Yeast $TAF_{II}145$ required for transcription of G1/S cyclin genes and regulated by the cellular growth state. *Cell* 90, 607–614.

Wang, E.H., and Tjian, R. (1994). Promoter-selective transcriptional defect in cell cycle mutant ts13 rescued by $hTAF_{II}250$. *Science* 263, 811–814.

Weinzierl, R.O.J., Dynlacht, B.D., and Tjian, R. (1993a). Largest subunit of *Drosophila* transcription factor IID directs assembly of a complex containing TBP and a coactivator. *Nature* 362, 511–517.

Weinzierl, R.O.J., Ruppert, S., Dynlacht, B.D., Tanese, N., and Tjian, R. (1993b). Cloning and expression of Drosophila TAFII60 and human TAFII70 reveal conserved interactions with other subunits of TFIID. *EMBO J.* 12, 5303–5309.

Wirth, T., Staudt, L., and Baltimore, D. (1987). An octamer oligonucleotide upstream of a TATA motif is sufficient for lymphoid-specific promoter activity. *Nature* 329, 174–178.

Wyman, C., Rombel, I., North, A.K., Bustamante, C., and Kustu, S. (1997). Unusual oligomerization required for activity of NtrC, a bacterial enhancer-binding protein. *Science* 275, 1658–1661.

Xie, X., Kokubo, T., Cohen, S.L., Mirza, U.A., Hoffmann, A., Chait, B.T., Roeder, R.G., Nakatani, Y., and Burley, S.K. (1996). Structural similarity between TAFs and the heterotetrameric core of the histone octamer. *Nature* 380, 316–322.

Yang, X.J., Ogryzko, V., Nishikawa, J., Howard, B., and Nakatani, Y. (1996). A p300/CBP-associated factor that competes with the adenoviral ocoprotein E1a. *Nature* 382, 319–324.

You, C.-H., Bolouri, H., and Davidson, E.H. (1998). Genomic *cis*-regulatory logic: experimental and computational analysis of a sea urchin gene. *Science* 279, 1896–1902.

Zhou, Q., Liberman, P.M., Boyer, T.G., and Berk, A.J. (1992). Holo-TFIID supports transcriptional stimulation by diverse activators and from a TATA-less promoter. *Genes Dev.* 6, 1964–1974.

Zou, C., Fujita, N., Igarashi, K., and Ishihama, A. (1992). Mapping the cAMP receptor protein contact site on the alpha subunit of *Escherichia coli* RNA polymerase. *Mol. Microbiol.* 6, 2599–2605.

Chapter 5

Control of RNA Elongation and Termination

Control of transcript initiation is without doubt a major mechanism for the regulation of gene transcription in both prokaryotic and eukaryotic cells. Starting RNA synthesis is, however, merely the first hurdle that needs to be overcome in a multistep process leading up to the production of a mature gene product. Once a transcript has been successfully initiated, RNAPs constantly need to choose from three fundamental options that are available to them at every step: continue with elongation, edit out a misincorporated nucleotide or terminate RNA synthesis altogether. The choice between these alternative pathways (which is irreversible in the case of termination) depends on numerous kinetic factors that are influenced by nucleotide sequences present in the DNA template, availability of nucleoside triphosphates, the structure of the newly synthesized transcript and the presence of various protein factors interacting with RNAPs (reviewed in von Hippel, 1998).

Especially during the last decade the processes influencing transcription elongation rates have been increasingly recognized as significant factors for the transcriptional control of many genes. The logic for controlling the efficiency of transcript elongation is simple: failure to *complete* a transcript is equivalent to failure to *initiate* a transcript and results in both cases in the lack of final gene products (RNA and proteins).

So how can the efficiency of transcript elongation be controlled? During the transcription process there are many opportunities for influencing ongoing

RNA synthesis through mechanisms that range from 'forceful eviction' of elongating RNAPs from the DNA template (e.g. by DNA replication complexes) to very subtle and highly-controlled modifications of RNAPs that alter their enzymatic and kinetic properties. Some of the best examples for the 'forceful eviction' mechanism have come from eukaryotic systems, where there is evidence that gene size is used to control important developmental decisions (O'Farrell, 1992). The *Ultrabithorax* (*Ubx*) and *Antennapedia* (*Antp*) genes are two large transcription units in the fruit fly *Drosophila* (87 and 105 kb, respectively) that give rise, after processing of the primary transcript, to mRNAs of only a few kb in size. These mRNAs encode key transcription factors (homeoproteins) that regulate embryonic development and control regional differentiation events. During the first couple of hours, nuclear divisions in the developing *Drosophila* embryo occur very rapidly (approximately every 10 minutes), and this does not allow enough time for large genes to be completely transcribed, although the *Ubx* and *Antp* promoters are active at that stage and produce short transcripts (e.g. Akam, 1983). The average transcription elongation rate of eukaryotic $RNAP_{II}$s is estimated at around 1.2–1.6 kb/minute (Edwards *et al.*, 1991), which means that full-length transcripts can only emerge from such large genes once the nuclear division time has increased to more than 90 minutes (which happens indeed at the time when complete *Ubx* and *Antp* mRNAs become detectable for the first time).

This mechanism is probably a rare example of a very crude form of controlling gene expression. For many average-sized genes such considerations do not apply and the movement of RNAPs through their coding regions is controlled in more sophisticated ways. In this chapter we will encounter a range of intricate mechanisms that ultimately determine whether RNAPs will be able to transcribe to the end of the coding region of a gene without terminating prematurely. These mechanisms are partially based on the occurance of particular sequence motifs in the transcribed DNA templates, but are even more substantially influenced by the activities of numerous proteins that regulate the processivity and elongation properties of RNAPs in both positive and inhibitory manners.

Ultimately the control of transcript elongation needs to be functionally linked with the various processes leading to stopping transcription altogether. Specific transcription termination events play an important role in regulated gene expression because they allow the functional subdivision of the linear array of genes within the genome into discrete transcription units that can be expressed independently of each other. Just as with transcription elongation, it is easy to understand why specific termination is functionally important. Failure to terminate transcription at specific points downstream of coding regions would allow RNAPs to continue transcription into adjacent genes and lead to the possibly deleterious synthesis of unwanted gene products. Our knowledge of transcript termination mechanisms (and the role of various protein factors influencing them) has reached a fairly advanced stage in bacteria, where there are numerous model systems displaying the various facets of genetic regulation through conditional termination. Less is known about termination in eukaryotic systems, especially by $RNAP_{II}$, but recent breakthroughs have also begun to shed some light on this issue.

5.1. Transcription Elongation: Basic Mechanisms

Much of our current knowledge of elongating RNAPs is based on experimental work carried out on *E. coli* RNAP. The relative ease with which genetic and biochemical experiments can be conducted in bacteria has allowed a detailed dissection of the various structural and kinetic parameters that influence the enzymatic properties of elongating RNAPs. Before we will look at the specialized properties of eukaryotic RNAPs, we will briefly consider some of the most important lessons that have emerged from the studies of bacterial systems. They are of substantial biological interest in their own right and provide a wide range of examples that illustrate the regulatory potential of elongation- and termination control in gene expression programs. Furthermore, there are indications that many of the results obtained from studying *E. coli* RNAP also provide useful guideposts for understanding the elongation properties of the archaeal and eukaryotic RNAPs in more detail.

Models of RNAP Elongation Mechanisms

The simplest model for elongation would be one where a rigid RNAP molecule glides smoothly along the DNA template while incorporating successive rNTPs into the growing RNA chain (e.g. Yager and von Hippel, 1987). Elongating *E. coli* RNAPs have an average footprint of 30 nucleotides (as measured by nuclease protection experiments) and contain a central 18 nucleotide 'transcription bubble,' where the double-stranded DNA template is transiently melted into single-stranded DNA to allow the template-directed formation of the RNA transcript. Specific contacts between the RNAP, the DNA template and the emerging transcript are responsible for maintaining the overall configuration of the ternary transcription complex (Kainz and Roberts, 1992; Uptain *et al.*, 1997). The 3' end of the RNA molecule (i.e. the area where additional nucleotides are added to the chain) is located very close to the leading edge (facing in the transcription direction) of the transcription bubble, suggesting that the protein domains facilitating the unwinding of the DNA template (the 'unwindase' activity) and those responsible for RNA synthesis are located in closely physical proximity to each other (Shi *et al.*, 1988). A major part of the transcription bubble is taken up by newly-synthesized RNA still hybridized to the DNA template strand (approximately 9–12 nucleotides; Sidorenkov *et al.*, 1998). The RNA is actively stripped from the template strand by the 'hybrid separator' activity (Richardson, 1975), which allows the separated DNA strands to join up by base-pairing to reform their original double-helical conformation. The movement of RNAP on DNA resembles that of other cellular motors (such as the components interacting with actin or microtubule filaments), and the force generated is experimentally measureable by sensitive biophysical techniques (Gelles and Landick, 1998).

The relatively static picture painted above, modelling RNAP as a solid workbench on which DNA strands are separated and reclosed while RNA synthesis occurs does not, however, describe adequately all the known facts involved in the RNAP elongation process. The kinetic properties of bacterial RNAPs transcribing different templates vary over many orders of magnitude, suggesting that DNA sequence-specific features have a substantial influence on many aspects of RNAP function (see also further below). Some of these sequence-specific features cause highly localized variations in the RNAP elongation rate. Although constant movement is normally observed over long stretches of sequence, RNAP switches at least occasionally into a discontinuous elongation mode whenever the enzyme encounters particular DNA sequences that slow down the advance of the leading edge of RNAP (Figure 5.1; Nudler *et al.*, 1994). Footprinting studies of artificially-stopped *E. coli* RNAPs display characteristic variations in the footprinting pattern (Figure 5.2), which are also detectable with stopped eukaryotic $RNAP_{II}$s (Rice *et al.*, 1991; Linn and Luse, 1991). These changed footprinting patterns have been rationalized in different ways. The 'inchworm' model postulates that translocation of RNAPs occurs under such circumstances in a stepwise manner with as many as 8 nucleotides becoming incorporated into the elongating RNA chain before the RNAPs become topologically 'strained' and abruptly 'leap' forwards (Krummel and

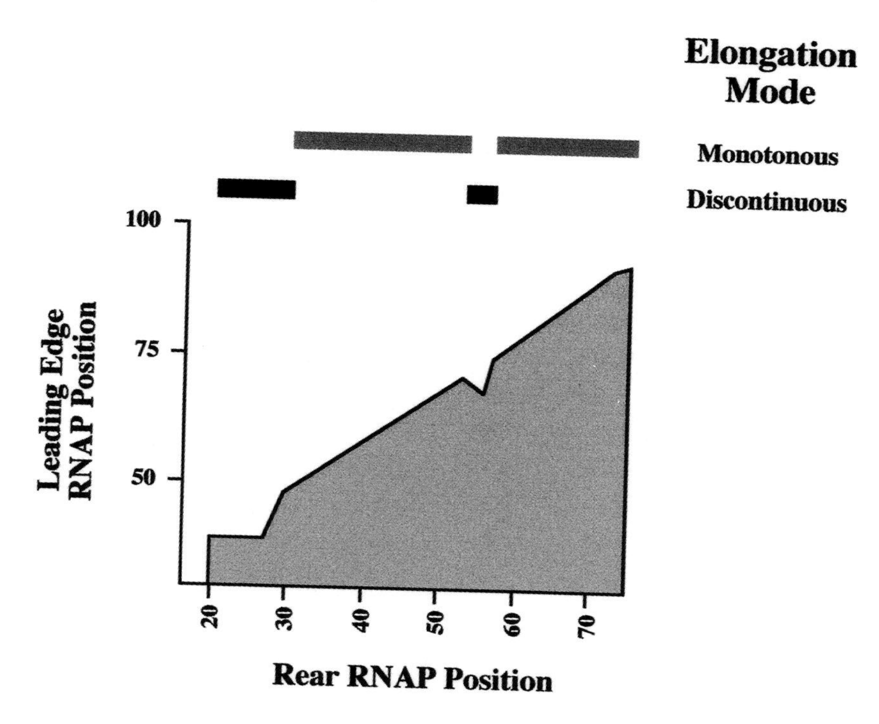

Figure 5.1. Template-Specific Variations in Bacterial RNAP Footprint Size.
The positions of the leading edge and rear end *E. coli* RNAP are plotted against each other as the enzyme moves down a defined DNA template. There are two regions where the RNAP appears to adopt a more compact, contracted state indicating a switch from a continuous into a discontinuous elongation mode. Data plotted from Nudler *et al.* (1994).

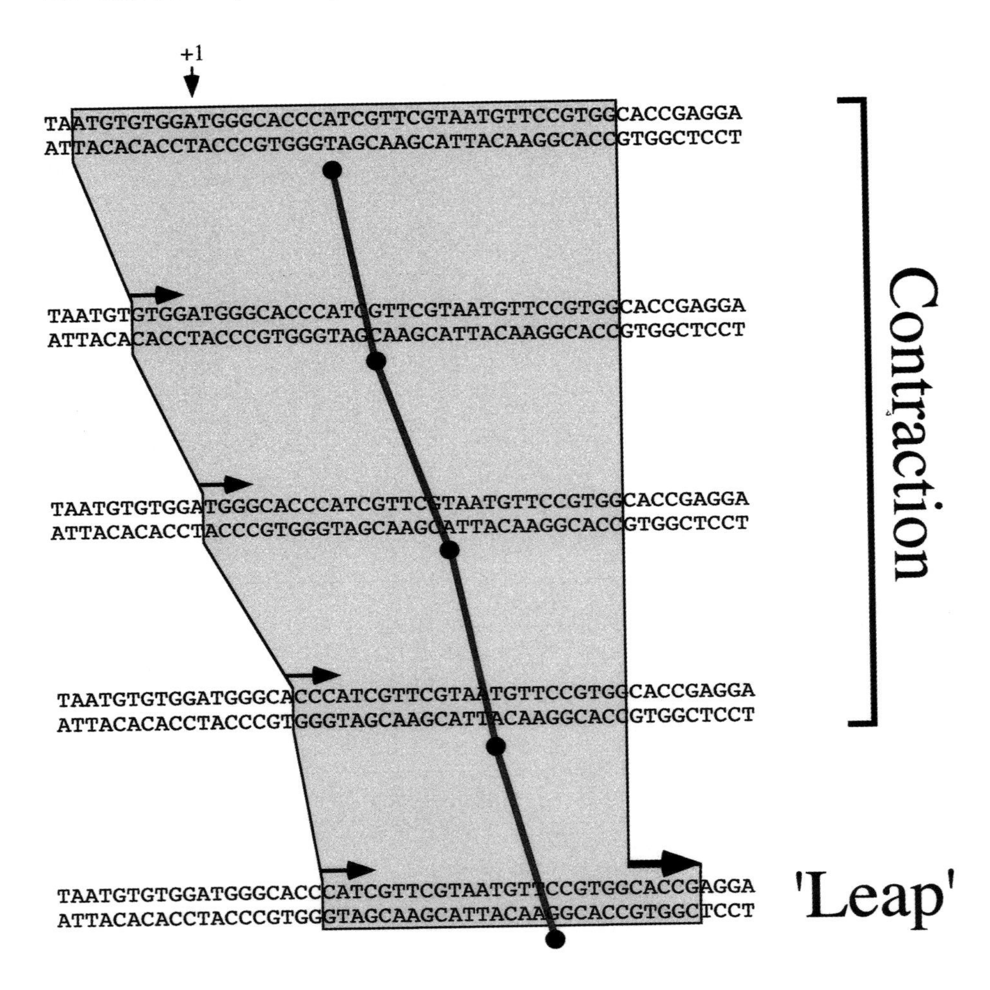

Figure 5.2. DNAase I Footprinting of Paused RNAPs.
A series of *E. coli* RNAP ternary transcription complexes were experimentally stopped at specific positions along identical DNA templates and analyzed by DNAase I footprinting. The extent of the footprints obtained (on the noncoding strand) with the different RNAP/DNA complexes is shown by grey shading. The position of the 3'- nucleotide in the nascent RNA chain is indicated by a black dot and thus defines the approximate position of the catalytic center of RNAP. Note that the 'rear end' and the catalytic center maintain approximately the same distance relative to each other. The 'leading edge' of RNAP remains fixed for a considerable amount of time and then appears to move abruptly forward. Data from Krummel and Chamberlin, 1992.

Chamberlin 1992; Nudler *et al.*, 1994). The discovery of 'loose' and 'tight' RNA binding sites near the active center of RNAP has been used in support of this view. It appears that newly synthesized RNA chains within RNAPs are initially guided to a 'loose' RNA binding site and subsequently threaded through a 'tight' RNA binding site before leaving the enzyme (Nudler *et al.*, 1994). The sequential transfer of RNA between the two sites ('two-stroke' model) could be linked to repetitive conformational changes between contracted and expanded forms of RNAP, similar to the ones seen during inchworming. The two-stroke and inchworming models of bacterial RNAP function thus suggest that these enzymes are highly flexible molecules that undergo substantial topological changes, even under 'normal' elongation conditions, and inchworming could be merely a more extreme version of this routine RNAP working mode.

The inchworming model is not universally accepted and much of the evidence in support of it is compatible with alternative interpretations. One of the most prominent alternative models is derived from a detailed investigation of the structural transitions in bacterial RNAP occurring immediately after the enzyme stops on the DNA template. These studies revealed that in these circumstances the RNAP physically slides backwards along the template strand-RNA hybrid without degrading the transcript (Komissarova and Kashlev, 1997). Such an event removes the 3' end of the RNA (where any new ribonucleotides would be added) from the catalytic site of RNAP, and any further RNA chain elongation becomes temporarily impossible. It has been proposed that on such occasions RNAP 'oscillates' on the template in the immediate vicinity of the stopping point, until the 3' end of the RNA comes back into a suitable position for further elongation. According to this model it would be the oscillation event, rather than a drastic conformational change in RNAP structure, which gives rise to the observed compressions of the footprinting patterns (Figure 5.2). It can be envisaged that DNAase I used in footprinting experiments temporarily gains access to DNA sequences upstream of oscillating RNAPs, resulting in a footprint that reflects an averaged dynamic event rather than static binding. Interpretation of such a dynamic footprint as a static one would then make it appear as if the RNAP was in a topologically contracted state. The currently available data does not allow a clear verification of either the inchworming or the oscillating model. Since they are not mutually exclusive, it is quite possible that eventually elements of both models will contribute to a description of the mechanistic basis of RNAP elongation.

5.2. The Coding Regions of Genes Contain Transcription Pause and Arrest Sites

The synthesis of RNA by RNAPs is a highly processive mechanism. Once the enzyme stops and dissociates from the DNA template, it is not possible for another RNAP to continue extension of the prematurely terminated transcript. For moderately sized genes this high degree of processivity is not usually a problem, because eukaryotic RNAPs can transcribe at a reasonably high rate (1.2–1.6 kb/minute; e.g. Edwards *et al.*, 1991), and thus RNAPs need often only be enzymatically active for several minutes to produce a complete primary transcript. To fully appreciate, however, the scale of the difficulties that RNAPs encounter in certain circumstances, one has to consider that some genes give rise to very large primary transcripts that need to be produced by a *single* $RNAP_{II}$ molecule. The human dystrophin gene is estimated to be more than 2.4 megabases in size and it thus takes up to 40 hours to transcribe a primary transcript, which eventually gives rise to a processed mRNA molecule of 'only' 14 kb in size (Koenig *et al.*, 1987). Some aspects of the remarkable stability of elongating $RNAP_{II}$ molecules can be experimentally demonstrated under *in vitro* conditions because they survive exposure to environments containing high salt- or detergent concentrations without dissociating from their DNA templates. These properties are often used in *in vitro* transcription reactions to abolish transcription re-initiation of new transcripts without disturbing the elongation of already initiated transcription complexes (e.g. in the 'nuclear run-on' assays described further below).

We have already seen in the previous section that there are situations when RNAPs are slowed down (or even get physically stuck), and need to initiate the inchworming or oscillating elongation modes. Such problematic 'blocking' sequences occur in a number of eukaryotic genes in the vicinity of transcription start sites (Figure 5.3; Spencer and Groudine, 1990; Kerppola and Kane, 1991; Christie *et al.*, 1994). We shall see later that these transcriptional 'bottlenecks'

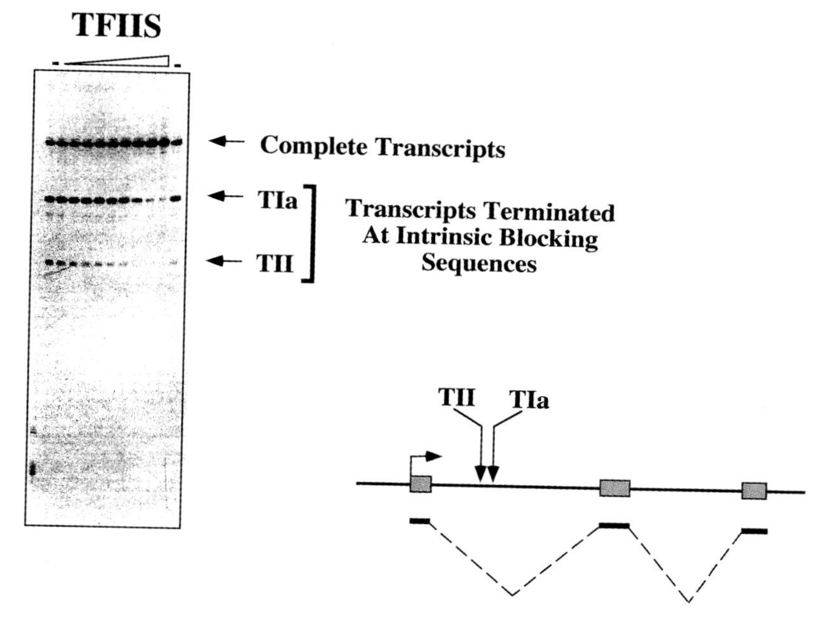

Figure 5.3. The Human Histone H3 Gene Contains Intrinsic Transcription Blocking Sites. The electrophoresis gel on the left hand side shows the result of transcribing the human histone H3 gene with purified yeast $RNAP_{II}$ *in vitro*. Although complete transcripts are transcribed from this gene (near the top of the gel), many of the transcripts terminate at at least two specific sites with different frequencies (50% at 'TIa' and 10% at 'TII') in the absence of the transcription elongation factor TFIIS (left-most and right-most lanes). Incubation of the transcription reactions in the presence of TFIIS for increasing periods (ranging from 5 seconds to 30 minutes; left to right, respectively) results in the release of the majority of the stalled $RNAP_{II}$s from the blocking sites, and increases the quantity of completed transcripts (data from Christie *et al.*, 1994). On the lower right hand side the schematic diagram shows the position of the blocking sites in the first intron relative to the start site. The exons are shown as shaded boxes. After Reines *et al.* (1987).

probably play an important regulatory function by selecting readthrough by specially modified forms of $RNAP_{II}$. Blocking sequences can be functionally divided into 'pause' and 'arrest' sites. As the name implies, transcription pause sites are sequences where RNAPs slow down and sometimes even temporarily cease RNA synthesis altogether. Since RNAPs do not dissociate from the DNA template at pause sites, resumption of transcription occurs spontanously and does not require the assistance of any additional factors. Although the duration of pauses is usually measured in seconds, there are instances where pausing of up to several minutes has been observed (Levin and Chamberlin, 1987; Theissen *et al.*, 1990).

What does a pause site look like on the DNA sequence level? The simple answer is, that we do not know for certain. It is likely that nucleosomes (Izban and Luse, 1991; Protacio and Widom, 1996) have the effect of substantially slowing down the elongation rate of $RNAP_{II}$, and this may account for at least some of the gene-specific pausing events observed *in vivo*. Pausing of transcription elongation can, however, also be observed in biochemically-defined *in vitro* transcription systems using nucleosome-free DNA. It has therefore been suggested that 'unusual' DNA conformations (kinks or 'Z' DNA) may play an important role (Peck and Wang, 1985). This interpretation is undermined by the fact that RNAPs from different organisms transcribing the same DNA template *in vitro* often pause at different sites. Sequences that are troublesome for human $RNAP_{II}$ are often read with ease by *E. coli* RNAP and *vice versa*. It is thus more likely that many pause sites reflect some idiosyncratic differences in RNAPs that have a high degree of species-specificity and may be used for regulatory purposes in evolution by adapting the coding sequences of genes accordingly. The eukaryotic basal transcription factor TFIIF, which is usually found in tight association with $RNAP_{II}$, stimulates the elongation rate by either suppressing $RNAP_{II}$ pausing completely, or by reducing the duration of time spent on individual pause sites (Price *et al.*, 1989; Gu and Reines, 1995).

Elongation Factors Rescue Arrested RNAPs

In addition to temporary pausing, RNAPs also experience more substantial elongation problems that result in an essentially irreversible 'arrested' situation. Such emergencies require the assistance of additional transcription factors to release the arrested RNAPs. In general, the presence of arrested RNAPs within the coding region of a gene poses a serious problem for the transcriptional machinery. Any arrested RNAP molecule is a physical obstacle that prevents other enzymes from completing their transcription cycle, and causes a pile-up of elongating RNAPs behind the arrested molecule. What can be done to transform an arrested RNAP back into an actively elongating one?

Apart from the catalytic center required for RNA synthesis, RNAPs also contain an exonuclease activity that allow them to chew off the 3' -most nucleotides of the currently synthesizing transcript. Although the physiological role of this activity is not yet entirely understood, it is likely to play a key role in rescuing arrested RNAPs. Activation of the exonuclease occurs spontanously (Surratt *et al.*, 1991; Orlova *et al.*, 1995), but is greatly stimulated by the temporary recruitment of a family of transcription elongation factors. In *E. coli*, the two proteins GreA and GreB induce cleavage of the nascent transcript in arrested RNAP complexes, and thus antagonize the action of arrest sites (Borukhov *et al.*, 1993). The crystal structure of GreA reveals a distinct α-helical extension (Figure 5.4), and the positively-charged tip of this structure can be crosslinked to the nascent RNA molecule (Stebbins *et al.*, 1995). It is therefore likely that GreA stimulates the transcript cleavage reaction by 'poking' its finger-like extension into the RNAP catalytic site. GreB is similar in primary sequence organization to GreA, but molecular modelling studies suggest that the positively-charged tip of GreB is more extensive in size (Koulich *et al.*, 1997). This difference correlates well with the documented functional effects of GreA and GreB on the extent of the RNA cleavage reaction (GreA: 2–3 nucleotides; GreB up to 9 nucleotides; Borukhov *et al.*, 1993). By enhancing the cleavage rate of the nascent transcript, elongation factors provide RNAPs with another chance to read across the arrest site which will in most cases eventually allow a successful elongation event.

Eukaryotic cells contain a diverse range of elongation factors that may carry out comparable functions (see Table 5.1; e.g. Kassavetis and Chamberlin, 1981; Sluder *et al.*, 1989; Arndt and Chamberlin, 1990). One of the best-studied elongation factor, TFIIS, rescues arrested $RNAP_{II}$ by stimulating the RNA cleavage reaction, before any readthrough across the arrest site occurs (Reines, 1992; Reines *et al.*, 1992; Izban and Luse, 1992; Wang and Hawley, 1993; Rudd *et al.*, 1994). TFIIS homologs have also recently been found in archaeal

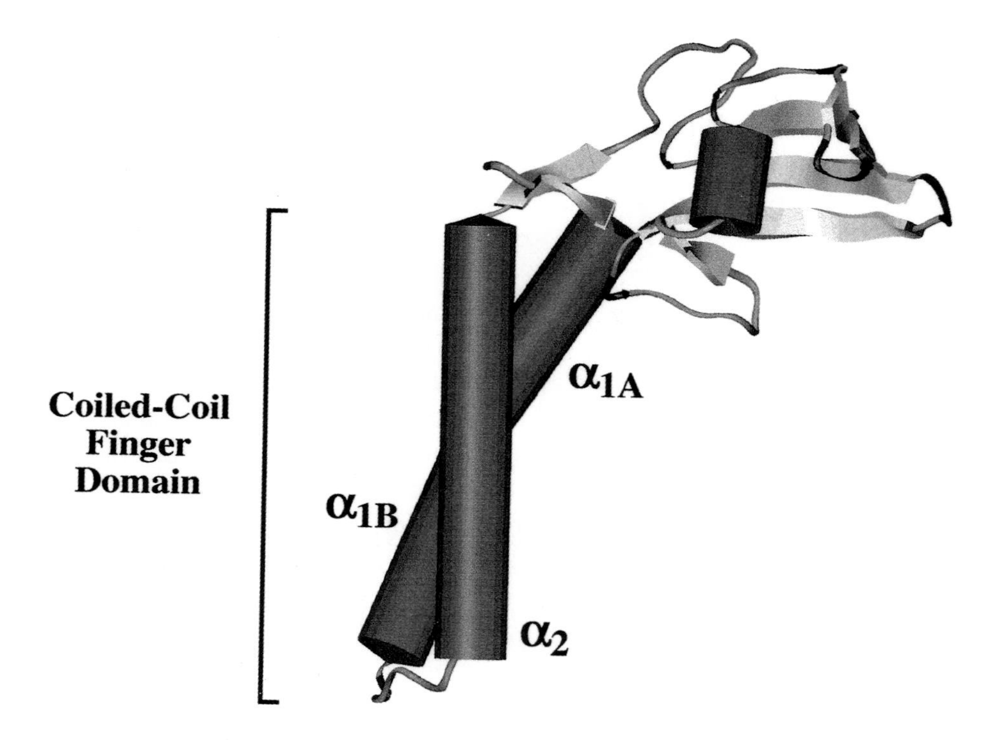

Figure 5.4. Structure of GreA Elongation Factor from *E. coli*.
The N-terminal domain consists mostly of an α-helical coiled-coil domain that comes in close contact with the catalytic center of RNAP to stimulate exonuclease-mediated cleavage of the nascent transcript in arrested RNAPs. The C-terminal domain is made up mostly of β-sheets and is implicated in the recruitment of GreA to RNAP. (PDB Access No. 1GRJ)

Table 5.1. **A Selection of Proteins Associated with $RNAP_{II}$ during the Elongation Phase.**

Protein	Proposed Function	Occurance
TFIIF	Suppresses $RNAP_{II}$ Pausing	All Organisms
TFIIS (SII)	Prevents $RNAP_{II}$ Arresting	All Organisms
Elongin	Suppresses $RNAP_{II}$ Pausing	All Organisms
P-TEFb	Prevents $RNAP_{II}$ Arresting	Mammals/Yeast?
ELL	Suppresses $RNAP_{II}$ Pausing	Mammals

genomes, hinting at the ancient evolutionary origin of this protein family (Langer and Zillig, 1993; Kaine *et al.*, 1994; Bult *et al.*, 1996). All TFIIS homologs contain a highly-conserved zinc-binding motif ('zinc ribbon') with four appropriately-spaced cysteine residues (Figure 5.5; Qian *et al.*, 1993). The results of mutagenesis studies suggest that this zinc ribbon is a cryptic nucleic acid-binding motif, which may bind in the vicinity of the DNA-RNA hybrid and could be activated through allosteric changes in protein structure (Aggarwal *et al.*, 1991). It remains unclear at this stage how such a mechanism could lead to a stimulation of the $RNAP_{II}$ cleavage reaction. Yeast cells lacking TFIIS are viable, but display an increased sensitivity to 6-azauracil (6-azauracil lowers the intracellular amounts of UTP and GTP, and thus causes problems for transcription elongation because elongating RNAPs have to struggle to recruit the NTP precursors into the catalytic core). Conversely, increasing the amount of TFIIS by genetic means overcomes certain mutations in the largest $RNAP_{II}$ subunit, RPB1, that cause sensitivity to 6-azauracil, implying that TFIIS interacts directly with RPB1 (Archambault et *al.*, 1992).

5.3. Control of Elongation 'Competence' of $RNAP_{II}$

Many Eukaryotic Promoters have 'Stalled' $RNAP_{II}$ at the Transcription Start Site

The examples of the pause and arrest sites discussed above show that transcriptional elongation through the coding region of a gene is not necessarily

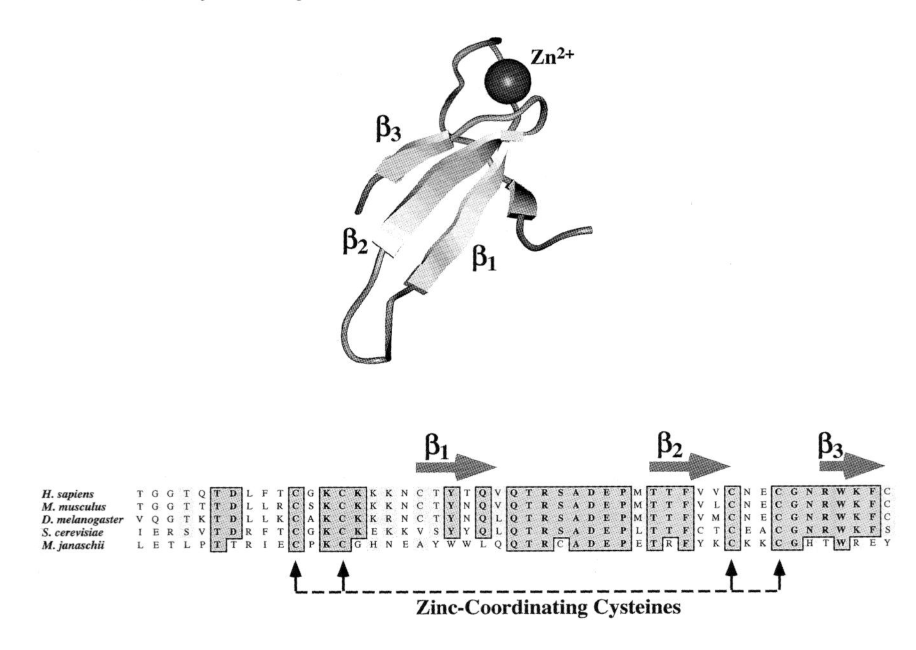

Figure 5.5. Structure of the Highly Conserved TFIIS Zinc Ribbon.
Top: Solution structure of the zinc-ribbon of human TFIIS (PDB Access No. 1TFI). *Bottom*: Sequence alignment of one archaeal (*M. janaschii*) and several eukaryotic TFIIS zinc ribbons. Note the absolute conservation of all four zinc-binding cysteine residues and the high degree of sequence similarity in other parts of the motif.

a smooth and continuous process. In this section we are going to investigate in more detail how such sites can be used, in conjunction with post-translationally modified $RNAP_{II}$s, to regulate eukaryotic gene expression. The control of transcript elongation efficiency provides another chance of controlling transcription after $RNAP_{II}$ has left the initiation complex (see Chapter 2 for more details on the mechanism allowing $RNAP_{II}$ to go into elongation mode, and the recycling of basal factors).

It has been known for a considerable time that many eukaryotic genes (including various *Drosophila* heatshock genes, the human proto-oncogene *c-myc*, the α- and β-tubulin genes etc.) contain 'stalled' $RNAP_{II}$ molecules near the transcription start site (Bentley and Groudine, 1986; Rougvie and Lis,

et al., 1986; Cadena *et al.*, 1987). Although it was initially not clear whether the 'poised' $RNAP_{II}$s found on the heat-shock promoters were representative of the general situation on other cellular promoters, numerous studies on a diverse range of model systems have since confirmed a strict correlation between the phosphorylation events of the $RNAP_{II}$-CTD and its transcriptional elongation competence. Some of the best examples include gene-specific activation events mediated by the viral transactivators HIV-Tat, E1A, E2F and VP16 and cellular transcription factors such as p53 (Yankulov *et al.*, 1994; Blau *et al.*, 1996; Parada and Roeder, 1996). The general importance of the phosphorylation of the largest $RNAP_{II}$ subunit during transcript elongation is underlined by the finding that a similar phosphorylation even occurs during elongation in evolutionarily highly diverged eukaryotic RNA polymerases that completely lack a recognizable CTD motif (such as the one found in trypanosomes; Chapman and Agabian, 1994). This observation suggests that phosphorylation is a universal mechanisms used in all eukaryotic species to form elongation competent $RNAP_{II}$, which evolved at an early stage during the divergence of the eukaryotic domain.

TFIIH and SRBs are Implicated in CTD-Phosphorylation

As we have seen above, the phosphorylation of the $RNAP_{II}$ CTD is directly responsible for generating an elongation-competent form of polymerase in response to activation by (at least some) gene-specific transcription factors. This observation raises the question which kinase activity might be responsible. Casting back our minds back to Chapter 2, we see that we have already come across two possible candidates. The basal factor TFIIH contains at least one kinase activity and mounting evidence suggests that this activity is mainly responsible for carrying out the CTD-phosphorylation (Gileadi *et al.*, 1992; Lu *et al.*, 1992; Serizawa *et al.*, 1992). Injection of antibodies directed against the kinase subunit of TFIIH specifically inhibits activator-dependent elongation in *Xenopus* oocytes (Yankulov *et al.*, 1996), and inhibitors of TFIIH-kinase affect elongation efficiency under activated (but not basal) transcription conditions (Serizawa *et al.*, 1993; Yankulov *et al.*, 1995). This direct biochemical evidence is strongly supported by genetic data obtained from studying transcription elongation in yeast cells that are mutant for a variety of genes (Akhtar *et al.*, 1996). A temperature-sensitive mutation in the KIN28 gene, encoding the yeast TFIIH kinase subunit, prevents efficient GAL4-stimulated elongation at non-permissive temperatures (Figure 5.9; Valay *et al.*, 1995; Akhtar *et al.*, 1996). Even more dramatic results can be obtained in cells carrying deletions of the SRB2 or SRB10 genes that encode components of the mediator complex within the yeast holoenzyme. Nuclear run-on assays suggest a massive failure of GAL4-dependent $RNAP_{II}$ elongation in ΔSRB2 (Figure 5.5) and ΔSRB10 mutant cells, and thus strongly implicate the SRBs as key components of the pathway controlling the formation of elongation-

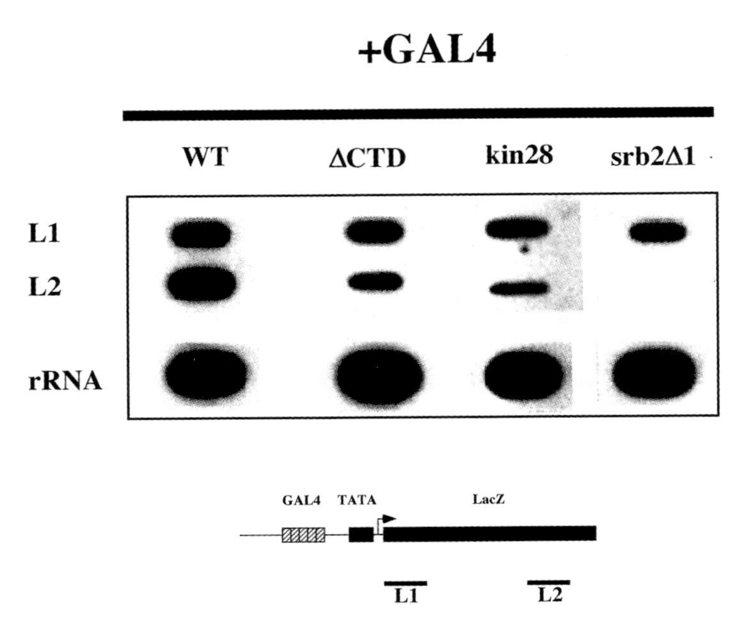

Figure 5.9. $RNAP_{II}$ CTD, TFIIH Kinase and SRB2 are Crucially Involved in Determining $RNAP_{II}$ Processivity.
These nuclear run-on assays demonstrate that in yeast cells containing an $RNAP_{II}$ with a truncated CTD only 19% of $RNAP_{II}$ molecules (relative to the wildtype) reach the 3' end of the reporter gene in presence of GAL4 activator (as monitored by the L1/L2 ratio in run-on assay). Mutations in the kinase unit of TFIIH (*kin28ts* at nonpermissive temperature), or the SRB2 gene (a component of the holoenzyme mediator complex), similarly reduce transcript elongation at least 10-fold. Figure compiled from data shown in Akthar *et al.* (1996).

competent $RNAP_{II}$. These results fit previous data suggesting that the SRB-mediator complex has a substantial structural and functional influence on the operation of the $RNAP_{II}$-CTD (e.g. Kim *et al.*, 1994), and that the SRB-containing mediator complex stimulates the kinase activity of TFIIH (Chapter 2; Myers *et al.*, 1998). The kinase activity present in SRB10 may also be directly involved in CTD phosphorylation (Liao *et al.*, 1995). While the TFIIH kinase is mainly reponsible for stimulating transcription by transforming $RNAP_{II}$ into an elongation-competent version, genetic evidence suggests that SRB10 acts as a negative regulator (Suroski *et al.*, 1994; Kuchin *et al.*, 1995; Wahi and Johnson, 1995). This observation raises the intriguing question, how CTD phosphorylation by the two different kinases can have such an opposite effect on transcriptional efficiency. The answer lies in the observation that, although the activities of both SRB10 and TFIIH kinases result in a biochemically indistinguishable phosphorylated CTD (Figure 5.10), the timing of this event is very different: SRB10-mediated phosphorylation occurs predominantly before the formation of the preinitiation complex, which has a repressive effect on the efficiency of transcript initiation. On the other hand, the stimulating CTD phosphorylation effect of TFIIH occurs at the transition from the transcript initiation to elongation phase (Hengartner *et al.*, 1998). At the moment the factors regulating the kinase activity of SRB10 are unknown, but a simple model based on competition between SRB10 and TFIIH kinases for the productive or non-productive phosphorylation of the $RNAP_{II}$-CTD could explain the previously observed effects of SRB10 on gene expression.

The phosphorylation status of the CTD is not entirely under the exclusive control of kinases. Serine and threonine residues present in the CTD are often highly glycosylated with O-linked N-acetylglucosamine (O-GlcNAc) moieties, which have to be specifically removed before phosphorylation of the same residues can take place. This suggests a reciprocal relationship between regulatory glycosylation and phosphorylation events (Hart, 1997). Also, once phosphorylated, the CTD-domain needs to be extensively dephosphorylated before entering new transcription initiation complexes. A CTD-specific phosphatase, whose activity is stimulated by TFIIF, may be responsible for the 'recycling' of elongation-competent into initiation-competent $RNAP_{II}$ after termination of the transcript (Chambers and Kane, 1996; Archambault *et al.*,

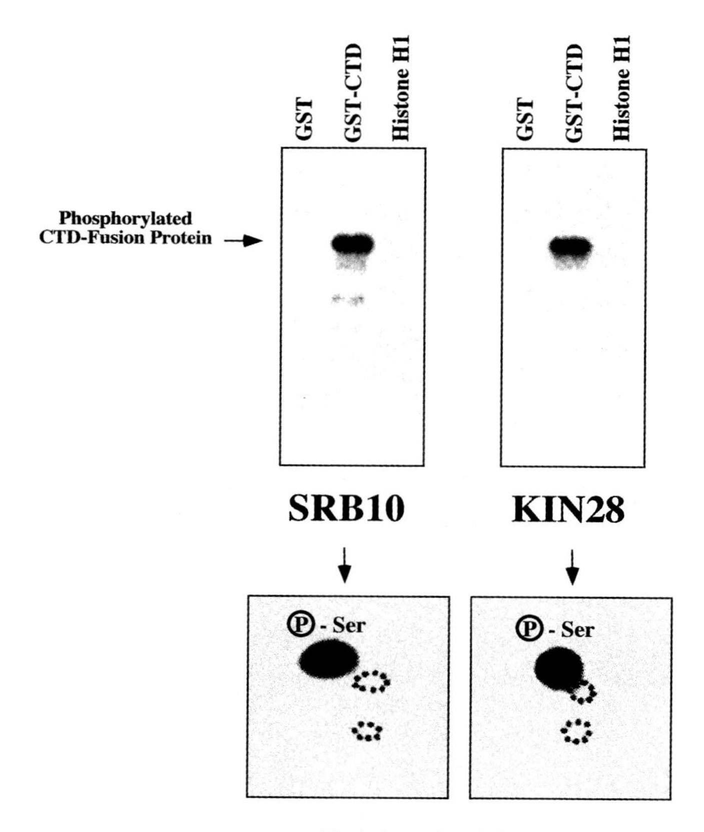

Figure 5.10. SRB10- and KIN28 (TFIIH)-Mediated Phosphorylation of RNAPII-CTD is Biochemically Indistinguishable.

Top: Three different proteins were incubated with recombinant SRB10/SRB11 and KIN28/CCL1 kinases in the presence of γ^{32}P-ATP *in vitro.* A glutathione-S-transferase (GST) fusion protein containing the $RNAP_{II}$-CTD domain is radioactively labelled to a similar extent by both kinases. GST alone and histone H1 are not labelled under these conditions, indicating that the two kinases display a high degree of substrate specificity. *Bottom*: Analysis and identification of the amino acid residues of the CTD domain that underwent phosphorylation by either SRB10 or KIN28 kinase by two dimensional thin-layer electrophoresis. The specifically-phosphorylated amino acids were subsequently visualized by autoradiography. Both kinases phosphorylate exclusively serine residues, whereas threonine and tyrosine (positions marked by dotted circles) remain unmodified. Data from Hengartner e*t al.*, 1998.

1997). Although the regulatory significance of the glycosylation/deglycosylation enzymes and the CTD-phosphatase could be of equal importance to the CTD-kinases for obtaining a better understanding of the $RNAP_{II}$ initiation/elongation/termination cycle (summarized in Figure 5.11), very little is currently known about these activities and the role they play in regulating various aspects of gene-specific transcription.

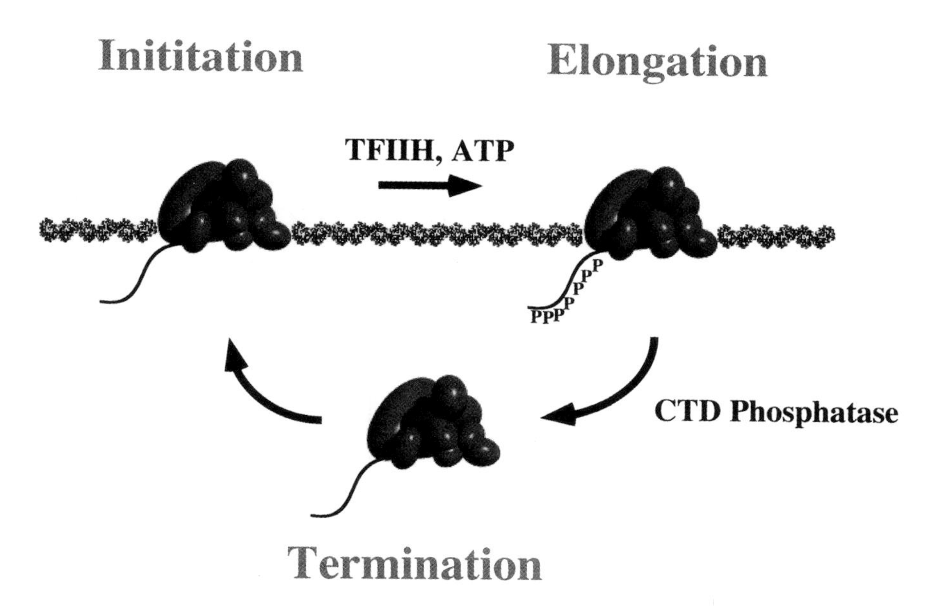

Figure 5.11. Overview of the Transcription Cycle of $RNAP_{II}$.
See main text for more information.

How widespread is the use of elongation-competent $RNAP_{II}$s to control the expression of different types of genes? The data obtained by Akhtar *et al.* (1996), based on several yeast genes, clearly demonstrate the effectiveness of elongation control as a mechanism for controlling gene expression. Nevertheless, the fact that cells carrying deletions of the SRB2 or SRB10 genes, which crucially affect the elongation competence of $RNAP_{II}$, are viable does suggest that many other important 'house-keeping' genes do not depend on elongation control (or bypass the missing polypeptides because of functional redundancy within the holoenzyme). It is likely that many genes that encode common cell components are probably transcribed at a fairly constant rate without any elaborate elongation control mechanisms. The task ahead is to classify the elongation properties of $RNAP_{II}$ on a larger sample of eukaryotic genes in order to obtain a clearer picture about this fascinating aspect of eukaryotic gene expression.

5.4. Transcription Termination in Bacteria

The control of transcription termination plays an important regulatory role in bacteria. Many well-known genetic phenomena affecting the expression of individual genes, such as polarity, retroregulation, or attenuation are a direct consequence of various proteins and RNAs working together in a concerted manner to influence the processivity of RNAP. In bacteriophages, conditional termination signals and the carefully-timed expression of 'antiterminators' provide successive checkpoints for the controlled immediate-early, delayed early and late gene expression programs.

Unconditional Transcriptional Termination Mechanisms

Approximately 50% of all bacterial genes contain DNA sequence motifs at the 3' end that terminate transcription without the need of specific protein termination factors. RNAP stops elongation whenever it encounters such sequences, releases the complete transcript from the ternary transcription complex, and dissociates from the DNA template. Strong, unconditional terminator sequences usually include a palindromic G/C-rich region, followed by a series of T residues (see e.g. Carafa *et al.*, 1991). The RNA transcribed from the G/C-rich region is thought to take up a specific hairpin formation, which causes (through an unknown mechanism) a substantial slow-down of the RNAP elongation rate and, eventually, a massive destabilization of the RNAP/DNA/RNA transcription complex (Figure 5.12; Wilson and von Hippel, 1994 and 1995). One of the components of the destabilization process could be the exceptionally unstable interactions between the T-residues present in

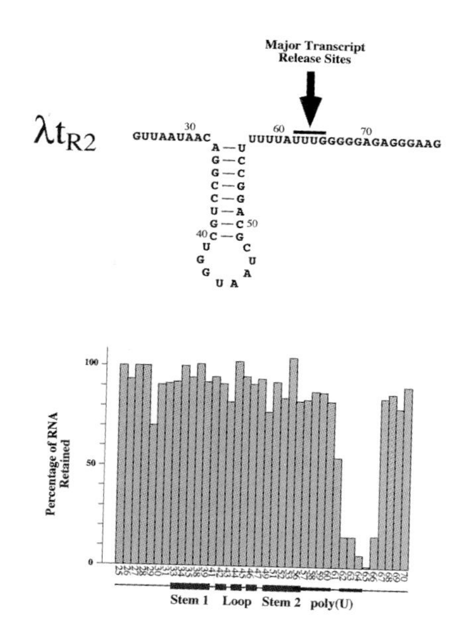

Figure 5.12. Sequence-Dependent Transcript Release from an ρ-Independent Terminator. *Top:* Schematic illustration of the structural elements present in a transcript derived from the λt_{R2} ρ-independent terminator. Note the presence of an inverted repeat structure (capable of forming a hairpin structure through basepairing between complementary nucleotides from positions 33 to 56), and an U-rich region surrounding position 60. Under normal circumstances the majority of transcripts would specifically terminate and be released from the transcription complex in the region indicated by the arrow. *Bottom:* The association of the RNA transcript with the transcription complex is substantially lowered within a termination zone encompassing six nucleotides (positions 61–66). A series of transcription complexes were stalled at various positions along the template DNA (as indicated by number along the horizontal axis) and left for 10 minutes. The percentage of the transcripts retained in transcription complexes after this incubation period was measured and is shown in the graph in a position-dependent manner. Note that the RNAs present in the majority of transcription complexes are specifically retained. The RNA molecules associated with the transcription complexes located over nucleotides 61–66 are, however, efficiently released under these circumstances. This result demonstrates that this region of λt_{R2} has a substantial destabilization effect on the ternary transcription complex, which could be due to the decreased base-pairing stability of the U-rich transcript with the template DNA strand. The position of the hairpin stem, the loop elements and the poly(U) stretch is schematically drawn underneath the nucleotide positions, which correspond to the ones used in the top half of the figure. Data from Wilson and von Hippel, 1994.

the DNA template strand and the newly-synthesized complementary RNA stretch, which may faciliate the spontanous dissociation of the transcript from the template DNA (Farnham and Platt, 1980; Martin and Tinoco, 1980). It has also been suggested that the RNA hairpin structure may interact directly with RNAP and induce a conformational change which would cause an active disruption of the transcription complex (Cheng *et al.*, 1991). Sequences located downstream of the terminator influence the efficiency of termination, indicating that contacts between the DNA template and the leading edge of RNA polymerase may also play an important role in determining overall terminator efficiency (Telesnitsky and Chamberlin, 1989; Reynolds and Chamberlin, 1992).

The asymmetric structure of unconditional terminators suggests that they function, just like promoters, in a directional manner. Only RNAPs encountering the G/C-rich region first and the sequence of T residues afterwards will terminate transcription. Interestingly however, there are several known examples of terminators that have a more symmetric structure with a run of A-residues preceeding the G/C-rich region and are thus able to function in a bidirectional manner (e.g. Pinkham and Platt, 1983). Such sites could provide a space-saving transcriptional barrier between adjacent oppositely-transcribed transcription units and prevent the formation of inhibitory antisense transcripts.

Regulation of Gene Expression Through Attenuation

The structure and functional elements of unconditional bacterial terminators are fixed through the genomic DNA sequence, and it therefore seems as if there were no apportunities to use them to control gene expression in any subtle control mechanisms. Two particular features of terminating sequences, i.e. the importance of the formation of an RNA-hairpin structure and the fact that transcription and translation are closely coupled in bacterial cells, allow nevertheless a rather unusual way of transcriptional control generally referred to as 'attenuation' (Platt, 1981; Yanofski, 1981). Several different operons that encode the enzymes for the biosynthesis of various amino acids (including leucine, isoleucine, histidine, threonine, tryptophan and phenylalanine) contain a 'leader' region within the first 100–200 nucleotides of the transcribed mRNA. This leader region contains a short open reading frame that allows the cell to sample the abundance of key amino acids. In one of the best-understood examples, the leader region of the tryptophan operon (trp) contains two adjacent tryptophan codons which requires high levels of the relatively rare amino acid tryptophan to be present in order to be efficiently translated. If tryptophan levels are low in the cell, translation would either stop or become substantially delayed at this point. So how can the position of such potential translation barriers influence transcription? The key for understanding attenuation mechanisms is based on the fact that the leader RNA can take up several alternative hairpin formations governed by the position of the translating ribosomes (Figure 5.13). Efficient readthrough of the translational machinery across the region containing the adjacent tryptophan codons allows the formation of an RNA hairpin structure next to a series of T residues in the DNA template, causing the formation of a *bona fide* terminator structure. If, on the other hand, the ribosomes stall over the tryptophane codons due to a shortage of tryptophan in the cell, an alternative RNA hairpin is formed that is wrongly positioned relative to the T residue series and fails to terminate transcription. Under these conditions, RNAP continues transcription into the *trp* operon coding regions, which leads to the production of the enzymes required for tryptophane biosynthesis. The dependance of attenuation mechanisms on the close coupling between transcription and translation precludes similar mechanisms from operating in eukaryotic cells, where the two processes are strictly separated by physical barriers (nuclear membrane) and the requirement for sophisticated post-transcriptional RNA processing and transport mechanisms (see Chapter 8).

Factor-Dependent Transcriptional Termination Mechanisms

While unconditional terminators contain all the necessary information to stop transcription, a variety of other bacterial terminator sequences require the assistance of additional protein termination factors to bring RNA synthesis to a halt. The bacterial transcription termination factor ρ ('rho') plays a prominent role in terminating transcription of many genes at ρ-dependent terminator sites. ρ attaches itself specifically to the newly synthesized RNA molecule emerging from elongating bacterial RNAP via a specific sequence motif present on these transcripts, the 'ρ-loading site.' Comparison of various ρ-loading sites reveals

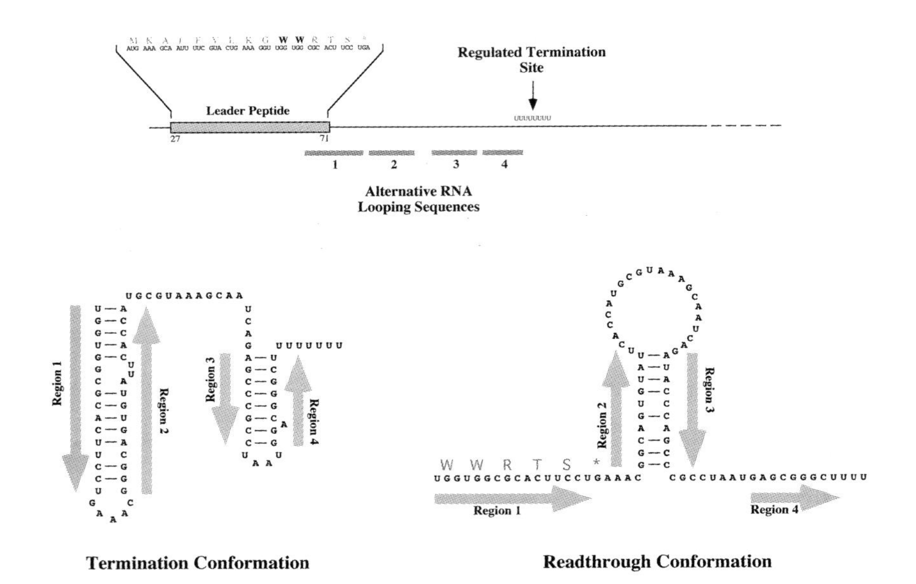

Figure 5.13. The Tryptophane Operon Attenuator Model.
Top: The 5'-most portion of the transcript from the *trp* operon is shown schematically. Between nucleotides 27 and 71 a short open reading frame contains two adjacent codons encoding tryptophan (W; shown in bold). Immediately downstream of the leader peptide coding sequence there are four RNA sequences capable of participating in alternative hairpin structures. *Bottom*: Detailed depiction of two alternative hairpin structures formed by differential basepairing. *Left*: Hybridization of regions 1/2 and 3/4 generate an RNA hairpin characteristic of a ρ-independent terminator and leads to termination of further transcription within the U-rich sequence. *Right*: A delay in the translation of the leader peptide because of tryptophane shortage arrests the ribosome over the trp-codons and therefore prevents region 1 from basepairing with region 2 (note the spatial relationship between the position of the two trp-codons and region 1!). Instead, regions 2 and 3 hybridize to form a hairpin structure. This hairpin is not close enough to the U-rich region to form an efficient termination structure and allows transcription to continue.

that they are not necessarily defined by a specific nucleotide primary sequence, but by sharing the common features of being rich in cytosine residues, varying in length between 60–100 nucleotides and lacking extensive secondary structure (Morgan *et al.*, 1985; Chen and Richardson, 1987; Alifano *et al.*, 1991; Walstrom *et al.*, 1998). The binding of ρ to single-stranded RNA molecules stimulates an endogenous ATP hydrolase activity within ρ (Lowery-Goldhammer and Richardson, 1974; Richardson, 1982), which allows ρ to translocate processively (in a 5' to 3' direction) along the transcript towards RNAP. The 'kinetic coupling' model postulates that the translocation of ρ towards RNAP and the elongation rate of RNAP creates a 'race' situation between RNAP and ρ to catch up with each other (Jin *et al.*, 1992). At the normal RNAP elongation rate, ρ is too slow to make physical contact with the polymerase. If, however, RNAP slows down due to the presence of pause sites or ρ-dependent terminator sequence motifs, then ρ will be able to catch up with RNAP and induce a specific transcript termination event within RNAP. It is likely that the release of the completed transcript from the DNA template involves the DNA/RNA helicase activity known to be present in ρ (Brennan *et al.*, 1987). The kinetic coupling model has been used to explain the so-called 'polarity' mutants observed in certain operons (Newton *et al.*, 1965). Briefly, the introduction of a premature translation stop codon in one of the upstream open reading frames of a polycistronic mRNA often also diminishes the level of expression of the open reading frames located further downstream, which should not be affected in their expression levels. It is likely, however, that the presence of a premature stop codon in an upstream open reading frame causes a large stretch of the mRNA to be free of ribosomes, leading to the exposure of cryptic ρ- binding sites that would normally be hidden by translating ribosomes. Such a situation would allow ρ to terminate transcription of the operon at internal RNAP pause site, and thus reduce the amount of full-length mRNA produced. This model is supported by observations showing that polarity effects can be suppressed by mutations in either ρ itself, or by the presence of the anti-termination factors, such as N and Q encoded by bacteriophage λ (see below; Greenblatt, 1981; Forbes and Herskowitz, 1982).

What is the biochemical basis of ρ function? Electron microscopical studies have revealed that ρ is a ring-shaped hexameric complex assembled from six identical subunits with a central opening measuring approximately 25Å in diameter (Figure 5.14; Gogol *et al.*, 1991). Each of the 47 kDa subunits consists of an RNA-binding domain, located near the N-terminus, and a C-terminal domain that displays a significant degree of sequence homology to the mitochondrial F_1-ATPase (reviewed in Richardson, 1996). Recent structural investigations of the ρ RNA-binding motif revealed a conserved RNA recognition motif in the form of an 'OB' domain (Allison *et al.*, 1998; Briercheck *et al.*, 1998). Some of the most crucial amino acid residues required

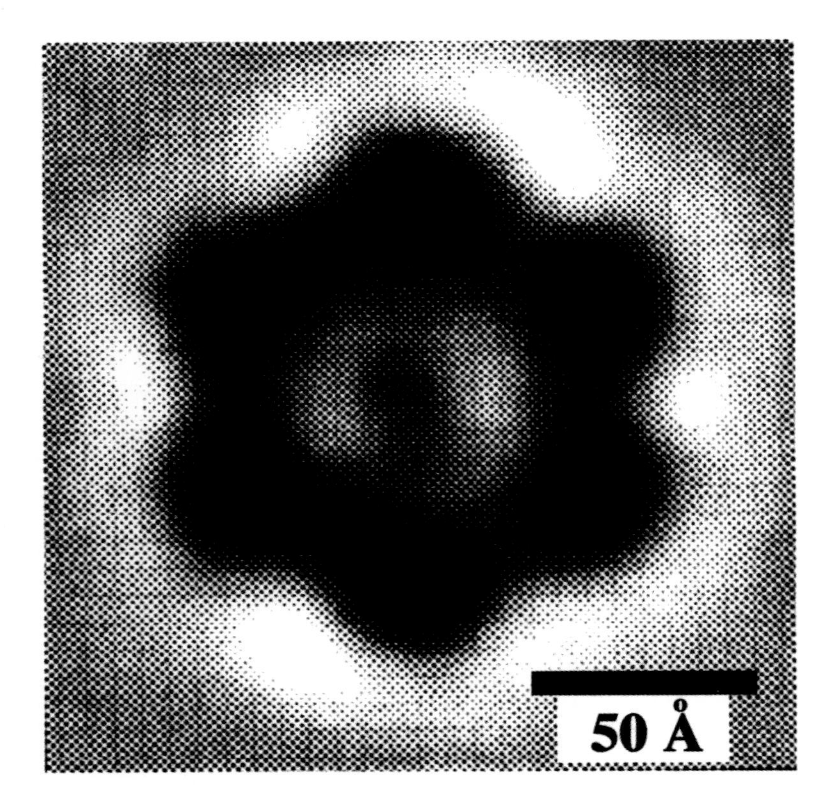

Figure 5.14. Electron Micrograph of the ρ Bacterial Termination Factor.
Six copies of ρ form a hexameric complex which slides along the transcript powered by continuous ATP hydrolysis. From Gogol *et al.* (1991).

for the specific RNA binding to the OB motif appear to be two highly conserved phenylalanine residues that are thought to bind to nucleotides via hydrophobic stacking forces (Brennan and Platt, 1991; see also Oubridge *et al.*, 1994). This type of RNA-protein interaction probably explains the preference of ρ for binding to single-stranded RNA, because such contacts can not be made if the nucleotides are in a base-paired conformation.

Although ρ plays a central role in the bacterial transcription termination process, there are several other factors that assist it in this task and/or regulate its function during antitermination. One of these factors, NusG, is an abundant 21 kDa protein that stimulates ρ-dependent termination greatly, especially under conditions that are suboptimal for ρ function (Sullivan and Gottesman, 1992; Li *et al.*, 1993). While NusG has no direct influence on termination by itself, various *in vitro* studies indicate that NusG facilitates ρ-dependent termination if a terminator sequence is located very close to a promoter, or for termination of transcripts synthesized by RNAP at high elongation rates (Burns and Richardson, 1995).

Regulation of Gene Expression Through Specific Antitermination

The temporally-precise expression of various genes is a particularly important feature during bacteriophage infection. In bacteriophage λ, many of the 'early' gene products encode regulatory transcription factors (such as the *cI* and *cro* proteins), that play an important role in controlling the lysis/lysogeny decision pathway. The presence of conditional terminators, and the controlled synthesis of the the bacteriophage-encoded antitermination proteins N and Q, play a major role in the temporal unfolding of the gene expression program from the bacteriophage genome and represents one of the most thoroughly-studied model systems for understanding antitermination control mechanisms. Shortly after infection, transcription from two early λ promoters gives rise to two distinct mRNA molecules that terminate specifically at the terminator sequences, t_{L1} and t_{R1} (Figure 5.15). One of the protein products encoded by these immediate early mRNAs is the N protein (Roberts 1969; Luzzati, 1970). N binds sequence-specifically to the RNA containing the 'N utilization site' (*nut* sites; see Figure 5.15; Horwitz *et al.*, 1987; Nodwell and Greenblatt, 1991). A relatively short portion (22 amino acids) of the N-terminal domain of protein

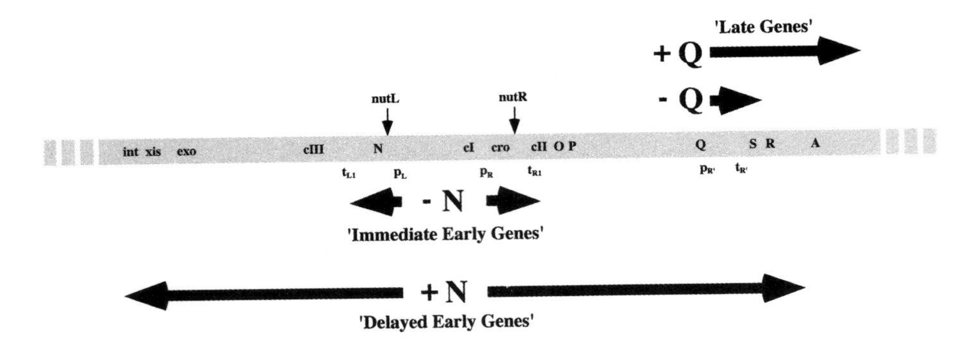

Figure 5.15. Regulation of Gene Expression by Antitermination in Bacteriophage λ. Transcription of the 'immediate early' genes stops at terminators t_{L1} and t_{R1}. One of these mRNAs encodes protein N, which interacts with the nutL and nutR sites to overide the terminators and thus allows the expression of the 'delayed early' genes. Another antiterminator protein, Q, regulates the expression of late genes in a similar manner.

N is sufficient for RNA-binding via the major groove formed by the transcribed region of the *nut* sites (Figure 5.16; Cai *et al.*, 1998; Legault *et al.*, 1998; Mogridge *et al.*, 1998). This N-RNA$_{nut}$ complex associates with a variety of other bacterial proteins, such as NusA, NusB (Das and Wolska, 1984), NusG (Li *et al.*, 1992; Sullivan *et al.*, 1992), the ribosomal (!) protein S10 (Friedman *et al.*, 1981), and RNAP (Goda and Greenblatt, 1985), and allows the RNAP complex to transcribe through the terminators t_{L1} and t_{R1} without prematurely terminating transcription. NusA, S10 and NusG are known to bind directly to RNAP (Mason and Greenblatt, 1991a), and many of the factors also interact with each other to form a stable 'N-modified' RNAP elongation complex (Mason and Greenblatt, 1991b; Mogridge *et al.*, 1998).

The antitermination control strategy of bacteriophage λ may appear to be a highly specialized and unusual way of regulating gene expression. Interestingly though, the human HIV virus uses a very similar stratgey. The majority of the early transcripts, that are specifically initiated from a promoter sequence located within the 5' long-terminal repeat (5' LTR), terminate shortly after initiation, and only a small proportion of transcripts reads through the protein-coding portions of the viral genome (Figure 5.17). The virus-encoded factor HIV-1

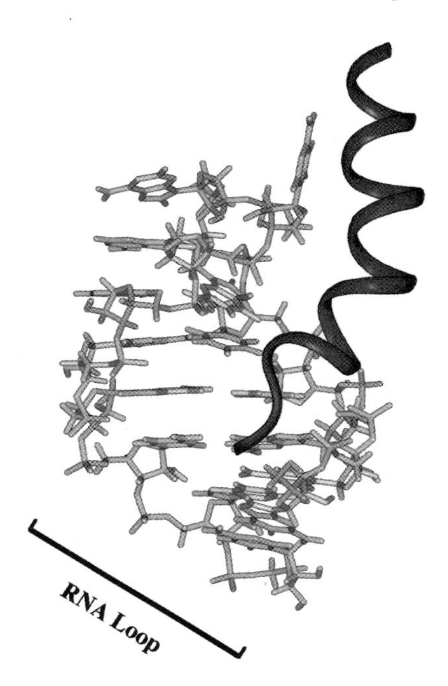

Figure 5.16. Solution Structure of the Aminoterminal Domain of Bacteriophage P22 Protein 'N' Bound to 'N' RNA.
The α-helical aminoterminal domain of protein N binds to the major groove of a double-stranded RNA loop formed by the 'BoxB' sequence element. After Cai *et al.* (1998).

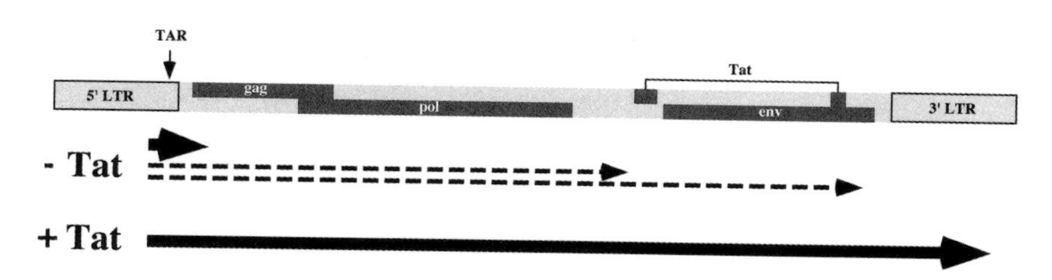

Figure 5.17. Antitermination Strategy in Human Immunodeficiency Virus 1 (HIV-1).
Most of the transcripts promoted by the 5' LTR terminate shortly after initiation, but a few continue elongation through the viral genome and result in the production of the Tat antitermination protein. Tat protein interacts via the TAR site to produce high levels of full-length transcripts.

Tat produced from some of these full-length transcripts stimulates the elongation properties of $RNAP_{II}$ (by enhancing the phosphorylation of the $RNAP_{II}$-CTD domain via TFIIH; see above for details; Parada and Roeder, 1996; Garcia-Martinez *et al.*, 1997), and thus leads to an increased level of the various proteins required to complete the infectious cycle.

5.5. Specific Termination of Transcription in Eukaryotes

Termination of $RNAP_{II}$ Transcripts is Generally Coupled with mRNA 3' End Formation

We have seen at the beginning of this chapter that some intrinsic DNA sequences slow down transcription by $RNAP_{II}$ ('pause sites'), or even make it grind to an irreversible halt ('arrest sites'). From these observation one might conclude that most eukaryotic genes would have such sequence motifs, maybe in an even more potent form, downstream of the coding region to act as a transcriptional terminator motifs similar to the ρ-independent sequences found in bacteria. Some $RNAP_{II}$-transcribed genes, like the ones encoding human histone H3, have indeed sequences of 5–8 consecutive thymine residues at the 3' end that are thought to be involved in transcription pausing, and may cause DNA to bend (Figure 5.18; Dedrick *et al.*, 1987; Reines *et al.*, 1988). Similar clusters play an important role in transcription termination by $RNAP_{III}$, and the rather simple sequence motif $^{5'}$GCAAAAGC$^{3'}$ is indeed sufficient to terminate 5S rRNA gene transcription in *Xenopus borealis* (Bogenhagen and Brown, 1981; Cozarelli *et al.*, 1983). On the other hand, the sequences from the mouse β-globin gene, that have been shown to be implicated in specifying transcriptional termination *in vivo* fail to impede $RNAP_{II}$ transcription under *in vitro* conditions (Reines *et al.*, 1987). It is thus highly likely that most genes depend on specific protein termination factors, rather than DNA sequences, to initiate the termination event of the transcription cycle (see also Heidmann *et al.*, 1992). This view recently received experimental support from Xie and Price (1996), who purified a specific ATP-dependent transcript release factor that is probably crucially involved in the termination process. One of the problems associated with studying termination under defined *in vitro* conditions is, however, due to the fact that most primary transcripts extend beyond the 3'

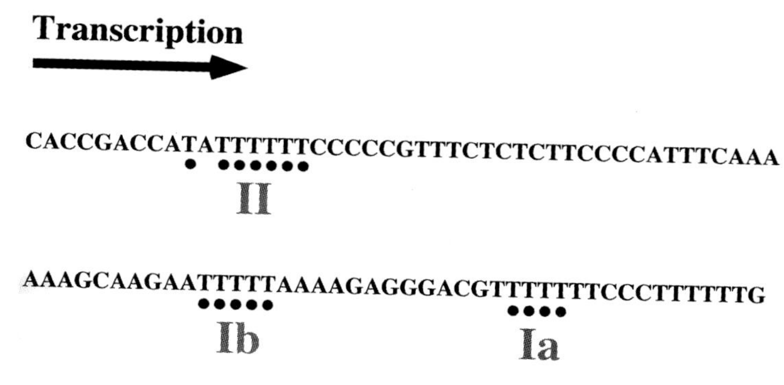

Figure 5.18. Intrinsic Transcription Pause Sequences Contain Clusters of Consecutive Thymidine Residues.
The positions of the three pause sites (Ia, Ib, and II) are indicated. Clusters of T residues coinciding with them are marked with black dots. After Reines *et al.* (1987).

end of the mature mRNA (Darnell, 1982) and termination seems to be functionally tightly linked with 3' end processing (Manley, 1983; Edwalds-Gilbert *et al.*, 1993). It is probably for this reason that the termination signals for the $RNAP_I$ and $RNAP_{III}$-transcribed genes are relatively straightforward, because they do not undergo such an event (although the resulting RNAs are usually subject to post-transcriptional cleavage and modification processes).

The model postulating tight coupling between transcription termination and post-transcriptional processing in $RNAP_{II}$-transcribed genes has recently received strong experimental support from genetic studies in yeast. Birse *et al.* (1998) studied the effect of several temperature-sensitive mutants in various components involved in the 3' end formation of mRNA on the termination efficiency of the *CYC1* gene under permissive and non-permissive conditions. Using a transcription run-on assay, similar to the one described earlier, it was possible to demonstrate normal termination of the mRNA transcripts in wildtype cells at a point closely downstream of the polyadenylation signal present at the 3' end of *CYC1* (Figure 5.19). Mutations in the genes encoding factors involved in various aspects of polyadenylation (e.g. poly A-polymerase [*pap1*] and various factors interacting with it, such as *fip1* and *yth1*) exert little or no discernible effects on the efficiency of transcript termination. In contrast,

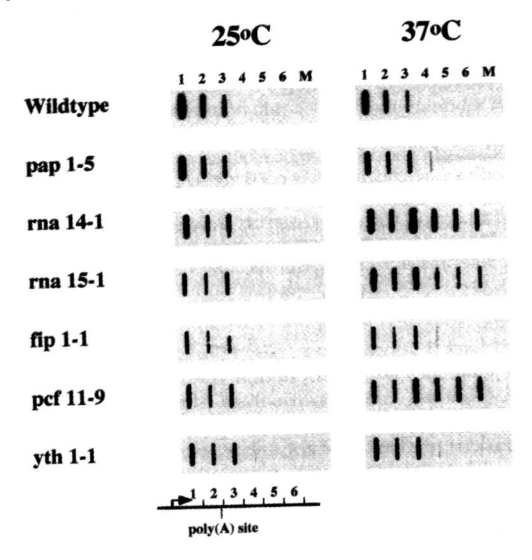

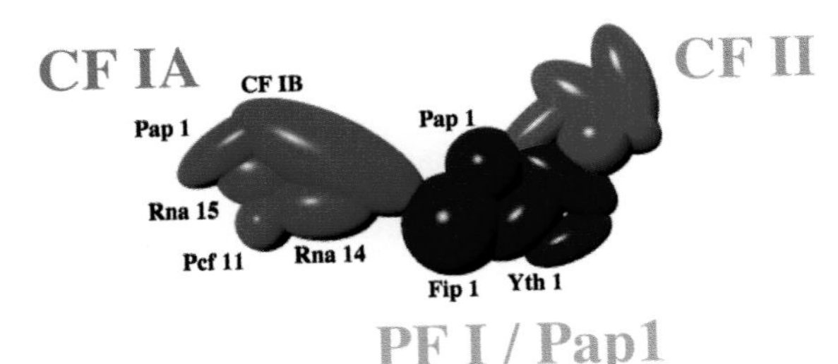

Figure 5.19. The 3' Cleavage Factor CF 1A Determines Transcription Termination by RNAP$_{II}$.

Top: Results of nuclear run-on studies near the polyadenylation and transcript termination sites of the CYC1 gene. Each membrane contains six distinct probes (+ a negative control 'M') and the polyadenylation site is present in probe 2. In wildtype cells the majority of transcripts terminate specifically immediately downstream of the polyadenylation site within probe 3. Transcription termination also occurs at the same position in cells harbouring temperature-sensitive mutants in various components of the 3' end cleavage factor complex CF 1A (Rna14, Rna15 and Pcf11) or the polyadenylation machinery (Pap1, Fip1 and Yth1) at permissive temperature (25°C). Raising the temperature to the non-permissive level (37°C) results in lack of specific termination in the cells containing CF 1A mutants (note the presence of hybridization signals in the positions containing probes 4, 5 and 6), but mutants in the polyadenylation machinery terminate (almost) normally. This proves that cleavage factor CF 1A, and not the polyadenylation machinery, is required for normal transcript termination by RNAP$_{II}$. Data from Birse *et al.* (1998). *Bottom*: Schematic diagram of the various subunits present in CF 1A and the polyadenylation machinery.

mutations in the subunits of CF 1A, the cleavage factor catalyzing the first step in mRNA 3' end processing, prevent efficient termination, and extensive transcription beyond the normal termination site can be detected in such cells (Figure 5.8). It is therefore likely that the sequence-specific recognition of the polyadenylation signal, followed by an endonucleolytic cleavage of the nascent mRNA by CF 1A, constitutes an important primary signal for $RNAP_{II}$ to terminate transcription. In mammalian systems the cleavage factor is found in association with the CTD-domain of the largest $RNAP_{II}$ subunit, suggesting that it associates with elongating $RNAP_{II}$ molecules and is therefore in a position to recognize the polyadenylation signal as soon as it emerges from the polymerase (McCracken *et al.*, 1997; Dantonel *et al.*, 1997). The specific recognition event between CF1A and the polyadenylation signal may then trigger allosteric changes which are transmitted to $RNAP_{II}$ and causes it to specifically terminate transcription.

5.6. Conclusions

For many genes the rate of transcript elongation constitutes an important control point for determining the rate of production of full-length transcripts by RNAPs. Experimental evidence suggests that transcript elongation rates vary in a template-dependent manner which cause RNAPs to switch between continuous and discontinuous translocation modes. On elongation arrest sites, RNAPs need the assistance of elongation factors that stimulate an endogenous nuclease activity to overcome such transcriptional barriers. In eukaryotic cells, the presence of sequence-intrinsic pause sequences in the coding regions of many genes impede the progress of unmodified $RNAP_{II}$ molecules and prevent such genes from being actively transcribed in the absence of suitable gene-specific activation. Particular types of activation domains present in gene-specific transcriptional activators are especially potent in stimulating the formation of elongation-competent $RNAP_{II}$. The C-terminal domain of the largest subunit of elongation-competent $RNAP_{II}$ is highly-phosphorylated, and numerous genetic and biochemical studies suggest that the kinase subunit present in the basal transcription factor TFIIH, in conjunction with components of the mediator complex, is responsible for this modification. Transcriptional termination in bacteria can be carried out by RNAP alone on ρ-independent terminators, or with the assistance of ρ and other termination factors on ρ-dependent terminators. Several strategies for the selective expression of particular genes are based on conditional termination. Relatively little is known about transcript termination by eukaryotic RNAPs, but many lines of evidence indicate that $RNAP_{II}$ termination is tightly linked to the post-transcriptional mRNA processing events (3' end formation).

References

Agarwal, K., Baek, K.H., Jeon, C.J., Miyamoto K., Ueno, A. & Yoon, H.S. (1991). Stimulation of transcript elongation requires both the zinc finger and RNA polymerase II binding domains of human TFIIS. *Biochemistry* 30, 7842–7851.

Akam, M.E. (1983). The location of *Ultrabithorax* transcripts in *Drosophila* tissue sections. *EMBO J.* 2, 2075–2084.

Akhtar, A., Faye, G., and Bentley, D.L. (1996). Distinct activated and non-activated RNA polymerase II complexes in yeast. *EMBO J.* 15, 4654–4664.

Alifano, P., Rivellini, F., Limauro, D., Bruni, C.B., and Carlomagno, S.M. (1991). A consensus motif common to all rho-dependent prokaryotic transcription terminators. *Cell* 64, 553–563.

Allison, T.J., Wood, T.C., Briercheck, D.M., Rastinejad, F., Richardson, J.P., and Rule, G.S. (1998). Crystal structure of the RNA-binding domain from transcription termination factor rho. *Nature Struct. Biol.* 5, 352–356.

Archambault, J., Lacroute, F., Ruet, A., and Friesen, J.D. (1992). Genetic interaction between transcription elongation factor TFIIS and RNA polymerase II. *Mol. Cell. Biol.* 12, 4142–4152.

Archambault, J., Chambers, R.S., Kobor, M.S., Ho, Y., Cartier, M., Bolotin, D., Andrews, B., Kane, C.M. and Greenblatt, J. (1997). An essential component of a C-terminal domain phosphatase that interacts with transcription factor IIF in *Saccharomyces cerevisiae*. *Proc. Natl. Acad. Sci. USA* 94, 14300–14305.

Arndt, K.M., and Chamberlin, M.J. (1990). RNA chain elongation by *Escherichia coli* RNA polymerase. Factors affecting the stability of elongating ternary complexes. *J. Mol. Biol.* 213, 79–108.

Bartholomew, B., Dahmus, M.E., and Meares, C.F. (1986). RNA contacts subunits IIo and IIc in HeLa RNA polymerase II transcription complexes. *J. Biol. Chem.* 261, 14226–14231.

Bentley, D.L., and Groudine, M. (1986). A block to elongation is largely reponsible for decreased transcription of *c-myc* in differentiated HL60 cells. *Nature* 321, 702–706.

Birse, C.E., Minvielle-Sebastia, L., Lee, B.A., Keller, W., and Proudfoot, N.J. (1998). Coupling termination of transcription to messenger RNA maturation in yeast. *Science* 280, 298–301.

Blau, J., Xiao, H., McCracken, S., O'Hare, P., Greenblatt, J., and Bentley, D. (1996). Three functional classes of transcriptional activation domains. *Mol. Cell. Biol.* 16, 2044–2055.

Bogenhagen, D.F., and Brown, D.D. (1981). Nucleotide sequences in *Xenopus* 5S DNA required for transcription termination. *Cell* 24, 261–270.

Borukhov, S., Sagitov, V., and Goldfarb, A. (1993). Transcript cleavage factors from *E. coli*. *Cell* 459–466.

Brennan, C.A., and Platt, T. (1991). Mutations in an RNP1 consensus sequence of rho protein reduce RNA binding affinity but facilitate helicase turnover. *J. Biol. Chem.* 266, 17296–17305.

Brennan, C.A., Dombroski, A.J., and Platt, T. (1987). Transcription termination factor ρ is an RNA-DNA helicase. *Cell* 48, 945–952.

Briercheck, D.M., Wood, T.C., Allison, T.J., Richardson, J.P., and Rule, G.S. (1998). The NMR structure of the RNA binding domain of *E. coli* rho factor suggests possible RNA-protein interactions. *Nature Struct. Biol.* 5, 393–399.

Brown, S.A., Weirich, C.S., Newton, E.M., and Kingston, R.E. (1998). Transcriptional activation domains stimulate initiation and elongation at different times and via different residues. *EMBO J.* 17, 3146–3154.

Bult, C. J., *et al.*, (1996). Complete genome sequence of the methanogenic archaeon, *Methanococcus jannaschii*. *Science* 273, 1058–1073.

Cadena, D.L., and Dahnums, M.E. (1987). Messenger RNA synthesis in mammalian cells is catalyzed by the phosphorylated form of RNA polymerase II. *J. Biol. Chem.* 262, 12468–12474.

Cai, H., and Luse, D.S. (1987). Transcription initiation by RNA polymerase II *in vitro:* properties of preinitiation, initiation and elongation complexes. *J. Biol. Chem.* 262, 298–304.

Cai, Z., Gorin, A., Frederick, R., Ye, X., Hu, W., Majumdar, A., Kettani, A., and Patel, D.J. (1998). Solution structure of P22 transcriptional antitermination N peptide boxB RNA complex. *Nature Struct. Biol.* 5, 203–212.

Carafa, Y.D., Brody, E., and Thermes, C. (1991). Prediction of rho-independent *Escherichia coli* transcription terminators. A statistical analysis of their RNA stem-loop structures. *J. Mol. Biol.* 216, 835–858.

Chambers, R.S., and Kane, C.M. (1996). Purification and characterization of an RNA polymerase II phosphatase from yeast. *J. Biol. Chem.* 271, 24498–24504.

Chapman, A.B., and Agabian, N. (1994). *Trypanosoma brucei* RNA polymerase II is phosphorylated in the absence of carboxy-terminal domain heptapeptide repeats. *J. Biol. Chem.* 269, 4754–4760.

Chen, C.-Y. A., and Richardson, J.P. (1987). Sequence elements essential for rho-dependent transcription termination at lambda t_{R1}. *J. Biol. Chem.* 262, 11292–11299.

Cheng, S., Lynch, E.C., Leason, K.R., Court, D.L., Shapiro, B.A., and Friedman, D.I. (1991). Functional importance of sequence in the stem-loop of a transcription terminator. *Science* 254, 1205–1297.

Christie, K.R., Awrey, D.E., Edwards, A.M., and Kane, C.M. (1994). Purified yeast RNA polymerase II reads through intrinsic blocks to elongation in response to the yeast TFIIS analogue, P37. *J. Biol. Chem.* 269, 936–943.

Cozarelli, N.R., Gerrard, S.P., Schlissl, M., Brown, D.D., and Bogenhagen, D.F. (1983). Purified RNA polymerase III accurately and efficiently terminates transcription of 5S RNA genes. *Cell* 34, 829–835.

Dantonel, J.-C., Murthy, K.G.K., Manley, J.L., Tora, L. (1997). Transcription factor TFIID recruits factor CPSF for formation of 3' end of mRNA. *Nature* 389, 399–402.

Darnell, J.E. (1982). Variety in the level of gene control in eukaryotic cells. *Nature* 297, 365–373.

Das, A., and Wolska, K. (1984). Transcription antitermination *in vitro* by lambda *N* gene product: requirement for a phage *nut* site and the products of host *nusA*, *nusB*, and *nusE* genes. *Cell* 38, 165–173.

Dedrick, R.L., Kane, C.M., and Chamberlin, M.J. (1987). Purified RNA polymerase II recognizes specific termination sites during transcription *in vitro*. *J. Biol. Chem.* 262, 9098–9108.

Edwalds-Gilbert, G., Prescott, J., and Falck-Pedersen, E. (1993). 3' RNA processing efficiency plays a primary role in generating termination-competent RNA polymerase II elongation complexes. *Mol. Cell. Biol.* 13, 3472–3480.

Edwards, A.M., Kane, C.M., Young, R.A., and Kornberg, R.D. (1991). Two dissociable subunits of yeast RNA polymerase II stimulate the initiation of transcription at a promoter *in vitro*. *J. Biol. Chem.* 266, 71–75.

Elion, E.A., and Warner, J.R. (1986). An RNA polymerase I enhancer in *Saccharomces cerevisiae*. *Mol. Cell. Biol.* 6, 2089–2097.

Farnham, P.J., and Platt, T. (1980). A model for transcription termination suggested by studies of the *trp* attenuator *in vitro* using base analogs. *Cell* 20, 739–748.

Forbes, D., and Herskowitz, I. (1982). Polarity suppression by the Q gene product of bacteriophage λ. *J. Mol. Biol.* 160, 549–569.

Friedman, D.I., Schauer, A.T., Baumann, M.R., Baron, L.S. and Adhya, S.L. (1981). Evidence that ribosomal protein S10 participates in the control of transcription termination. *Proc. Natl. Acad. Sci. USA* 78, 1115–1118.

Garcia-Martinez, L.F. Mavankal, G., Neveu, J.M., Lane, W.S., Ivanov, D., and Gaynor, R.B. (1997). Purification of Tat-associated kinase reveals a TFIIH complex that modulates HIV-1 transcription. *EMBO J.* 16, 2836–2850.

Gelles J., and Landick, R. (1998). RNA polymerase as a molecular motor. *Cell* 93, 13–16.

Gileadi, O., Feaver, W.J., and Kornberg, R.D. (1992). Cloning of a subunit of RNA polymerase II transcription factor β and CTD kinase. *Science* 257, 1389–1392.

Giniger, E., and Ptashne, M. (1987). Transcription in yeast activated by a putative amphipathic α helix linked to a DNA-binding unit. *Nature* 330, 670–672

Goda, Y., and Greenblatt, J. (1985). Efficient modification of *E. coli* RNA polymerase *in vitro* by the *N* gene transcription antitermination protein of bacteriophage lambda. *Nucl. Acids Res.* 13, 2569–2582.

Gogol, E.P., Seifried, S.E., and von Hippel, P. (1991). Structure and assembly of the *Escherichia coli* transcription termination factor rho and its interaction with RNA. I. Cryoelectron microscopic studies. *J. Mol. Biol.* 221, 1127–1138.

Greenblatt, J. (1981). Regulation of transcription termination by the *N* gene protein of bacteriphage λ. *Cell* 24, 8–9.

Gu, W., and Reines, D. (1995). Identification of a decay in transcription potential that results in elongation factor dependence of RNA polymerase II. *J. Biol. Chem.* 270, 11238–11244.

Hart, G.W. (1997). Dynamic O-linked glycosylation of nuclear and cytoskeletal proteins. *Annu. Rev. Biochem.* 66, 315–335.

Heidmann, S., Obermaier, B., Vogel, K., and Domdey, H. (1992). Identification of pre-mRNA polyadenylation sites in *Saccharomyces cerevisiae*. *Mol. Cell. Biol.* 12, 4215–4229.

Hengartner, C.J., Myer, V.E., Liao, S.-M., Wilson, C.J., Koh, S.S., and Young, R.A. (1998). Temporal regulation of RNA polymerase II by SRB10 and Kin28 cyclin-dependent kinases. *Mol. Cell* 2, 43–53.

Horwitz, R.J., Li, J., and Greenblatt, J. (1987). An elongation control particle containing the *N* gene transcription antitermination protein of bacteriophage lambda. *Cell* 51, 631–641.

Izban, M.G., and Luse, D.S. (1991). Transcription on nucleosome templates by RNA polymerase II *in vitro*: inhibition of elongation with enhancement of sequence-specific pausing. *Genes Dev.* 5, 683–696.

Izban, M.G., and Luse, D.S. (1992). The RNA polymerase II ternary complex cleaves the nascent transcript in a 3'–5' direction in the presence of elongation factor SII. *Genes Dev.* 6, 1342–1356.

Jin, D.J., Burgess, R.R., Richardson, J.P., and Gross, C.A. (1992). Termination and efficiency of rho-dependent terminators depends on kinetic coupling between RNA polymerase and rho. *Proc. Natl. Acad. Sci. USA* 89, 1453–1457.

Kaine, B.P., Mehr, I.J., and Woese, C.R. (1994). The sequence, and it evolutionary implications, of a *Thermococcus celer* protein associated with transcription. *Proc. Natl. Acad. Sci. USA* 91, 3854–3856.

Kainz, M., and Roberts, J. (1992). Structure of transcription elongation complexes *in vivo*. *Science* 255, 838–841.

Kassavetis, G.A., and Chamberlin, M.J. (1981). Pausing and termination of transcription within the early region of bacteriophage T7 DNA *in vitro*. *J. Biol. Chem.* 256, 2777–2786.

Kerppola, T.K., and Kane, C.M. (1991). RNA polymerase: regulation of transcript elongation and termination. *FASEB J.* 5, 2833–2841.

Kim, Y.-J., Bjorklund, S., Li, Y., Sayre, M., and Kornberg, R. D. (1994). A multiprotein mediator of transcriptional activation and its interaction with the C-terminal repeat domain of RNA polymerase II. *Cell* 77, 599–608.

Koenig, M., Hoffmann, E.P., Bertelson, C.J., Monaco, A.P., Feener, C., and Kunkel, L.M. (1987). Complete cloning of the Duchenne Muscular Dystrophy (DMD) cDNA and preliminary genomic organization of the DMD gene in normal and affected individuals. *Cell* 50, 509–517.

Komissarova, N., and Kashlev, M. (1997). RNA polymerase switches between inactivated and activated states by translocating back and forth along the DNA and the RNA. *J. Biol. Chem.* 272, 15329–15338.

Koulich, D., Orlova, M., Malhotra, A., Sali, A., Darst, S.A., and Borukhov, S. (1997). Domain organziation of *Escherichia coli* transcript cleavage factors GreA and GreB. *J. Biol. Chem.* 272, 7201–7210.

Krumm, A., Meulia, T., Brunvand, M., and Groudine, M. (1992). A block to transcriptional elongation within the human *c-myc* gene is determined in the promoter-proximal region. *Genes Dev.* 6, 2201–2213.

Krumm, A., Hickey, L., and M., G. (1995). Promoter-proximal pausing of RNA polymerase II defines a general rate-limitingstep after transcription initiation. *Genes Dev.* 9, 559–572.

Krummel. B., and Chamberlin, M.J. (1992). Structural analysis of ternary complexes of *Escherichia coli* RNA polymerase. Deoxyribonuclease I footprinting of defined complexes. *J. Mol. Biol.* 225, 239–250.

Kuchin, S., Yeghiayan, P., and Carlson, M. (1995). Cyclin-dependent protein kinase and cyclin homologs SSN3 and SSN8 contribute to transcriptional control in yeast. *Proc. Natl. Acad. Sci. USA* 92, 4006–4010.

Langer, D., and Zillig, W. (1993). Putative tfIIs gene of *Sulfolobus acidocaldarius* encoding an archaeal transcription elongation factor is situated directly downstream of the gene for a small subunit of DNA-dependent RNA polymerase. *Nucl. Acids Res.* 21, 2251.

Levin, J.R., and Chamberlin, M.J. (1987). Mapping and characterization of transcriptional pause sites in the early genetic region of bacteriophage T7. *J. Mol. Biol.* 196, 61–84.

Liao, S.-M., Zhang, J., Jeffery, D.A., Koleske, A.J., Thompson, C.M., Chao, D.M., Viljoen, M., van Vuuren, H.J.J., and Young, R.A. (1995). A kinase-cyclin pair in the RNA polymerase holoenzyme. *Nature* 374, 193–196.

Li, J., Mason, S.W., and Greenblatt, J. (1993). Elongation factor NusG interacts with termination factor ρ to regulate termination and antitermination of transcription. *Genes Dev.* 7, 161–172.

Linn, S.C., and Luse, D.S. (1991). RNA polymerase II elongation complexes paused after the synthesis of 15- or 35-base transcripts have different structures. *Mol. Cell. Biol.* 11, 1508–1522.

Lis, J., and Wu, C. (1993). Protein traffick on the heat shock promoter: parking, stalling and trucking along. *Cell* 74, 1–4.

Lowery-Goldhammer, C. and Richardson, J.P. (1974). An RNA-dependent nucleoside triphosphate phosphohydrolase (ATPase) associated with rho termination factor. *Proc. Natl. Acad. Sci. USA* 71, 2003–2007.

Lu, H., Zawel, L., Fisher, L., Egly, J.M., and Reinberg, D. (1992). Human general transcription factor IIH phosphorylates the C-terminal domain of RNA polymerase II. *Nature* 358, 641–645.

Luzzati, D. (1970). Regulation of λ exonuclease synthesis: role of the *N* gene product and λ repressor. *J. Mol. Biol.* 49, 525–539.

Manley, J.L. (1983). Analysis of the expression of genes encoding animal mRNA *in vitro* techniques. *Progr. Nucl. Acids Res.* 30, 195–244.

Martin, F.H., and Tinoco, I. (1980). DNA-RNA hybrid duplexes containing oligo (dA:rU) sequences are exceptionally unstable and may facilitate transcriptional termination. *Nucl. Acids Res.* 8, 2295–2299.

Mason, S.W., and Greenblatt, J. (1991a). Assembly of transcription elongation complexes containing N protein of bacteriophage λ and the *Escherichia coli* elongation factors NusA, NusB, NusG and S10. *Genes Dev.* 5, 1504–1512.

Mason, S.W., and Greenblatt, J. (1991b). A direct interaction between two *Escherichia coli* transcription antitermination factors, NusB and ribosomal protein S10. *J. Mol. Biol.* 223, 55–66.

McCracken, S., Fong, N., Yankulov, K., Ballantyne, S., Pan, G., Greenblatt, J., Patterson, S.D., Wickens, M., and Bentley, D.L. (1997). The C-terminal domain of RNA polymerase II couples mRNA processing to transcription. *Nature* 385, 357–361.

Mogridge, J., Legault, P., Li, J., Van Oene, M.D., Kay, L.E., and Greenblatt, J. (1998). Independent ligand-induced folding of the RNA-binding domain and two functionally distinct antitermination regions in the pahge lambda N protein. *Mol. Cell* 1, 265–275.

Morgan, W.D., Bear, D.G., Litchman, B.L., and von Hippel, P.H. (1985). RNA sequence and secondary structure requirements for rho-dependent transcription termination. *Nucl. Acids Res.* 13, 3739–3754.

Myers, L.C., Gustafsson, C.M., Bushnell, D.A., Lui, M., Erjument-Bromage, H., Tempst, P., and Kornberg, R.D. (1998). The Med proteins of yeast and their function through the RNA polymerase II carboxy-terminal domain. *Genes Dev.* 12, 45–54.

Newton, W.A., Beckwith, J.R., Zipser, D., and Brenner, S. (1965). Nonsense mutants and polarity in the *lac* operon of *Escherichia coli. J. Mol. Biol.* 14, 290–296.

Nodwell, J.R., and Greenblatt, J. (1991). The *nut* site of bacteriophage λ is made of RNA and is bound by transcription antitermination factors on the surface of RNA polymerase. *Genes Dev.* 5, 2141–2151.

Nudler, E., Goldfarb, A., and Kashlev, M. (1994). Discontinuous mechanism of transcription elongation. *Science* 265, 793–796.

O'Brien, T., Hardin, S., Greenleaf, A., and Lis, J.T. (1994). Phosphorylation of RNA polymerase II C-terminal domain and transcriptional elongation. *Nature* 370, 75–77.

O'Farrell, P.H. (1992). Developmental biology. Big genes and little genes and deadlines for transcription. *Nature* 359, 366–367.

Orlova, M., Newlands, J., Das, A., Goldfarb, A., and Borukhov, S. (1995). Intrinsic transcript cleavage activity by RNA polymerase. *Proc. Natl. Acad. Sci. USA* 92, 4596–4600.

Oubridge, C., Ito, N., Evans, P.R., Teo, C.-H., and Nagai, K. (1994). Crystal structure at 1.92Å resolution of the RNA-binding domain of the U1A spliceosomal protein complexed with an RNA hairpin. *Nature* 372, 432–438.

Parada, C.A., and Roeder, R.G. (1996). Enhanced processivity of RNA polymerase II triggered by Tat-induced phosphorylation of its carboxy-terminal end. *Nature* 384, 375–378.

Peck, L.J., and Wang, J.C. (1985). Transcriptional block caused by a negative supercoiling induced structural change in an alternating CG sequence. *Cell* 40, 129–137.

Pinkham, J.L., and Platt, T. (1983). The nucleotide sequence of the rho gene of *E. coli* K-12. *Nucl. Acids Res.* 11, 3531–3545.

Platt, T. (1981). Termination of transcription and its regulation in the tryptophan operon of *E. coli*. *Cell* 24, 10–23.

Price, D.H., Sluder, A.E., and Greenleaf, A.L. (1989). Dynamic interaction between a *Drosophila* transcription factor and RNA polymerase II. *Mol. Cell. Biol.* 9, 1465–1475.

Protacio, R.U., and Widom, J. (1996). Nucleosome transcription studied in real-time synchronous system: test of the lexosome model and direct measurement of effects due to histone octamer. *J. Mol. Biol.* 256, 458–472.

Qian, X., Gozani, S.N., Yoon, H., Jeon, C., Agarwal, K., and Weiss, M.A. (1993). Novel zinc finger motif in the basal transcriptional machinery: three-dimensional NMR studies of the nucleic acid binding domain of transcriptional elongation factor TFIIS. *Biochemistry* 32, 9944–9959.

Reines, D., Wells, D., Chamberlin, M.J., and Kane, C.M. (1987). Identification of intrinsic termination sites *in vitro* for RNA polymerase II within eukaryotic gene sequences. *J. Mol. Biol.* 196, 299–312.

Reines, D., Ghanouni, P., Li, Q., and Mote, J. (1992). The RNA polymerase II elongation complex. Factor-dependent transcription elongation involves nascent RNA cleavage. *J. Biol. Chem.* 267, 15516–15522.

Reines, D. (1992). Elongation factor-dependent transcript shortening by template-engaged RNA polymerase II. *J. Biol. Chem.* 267, 3795–3800.

Reynolds, R., and Chamberlin, M.J. (1992). Parameters affecting transcription termination by *Escherichia coli* RNA polymerase. II. Construction and analysis of hybrid terminators. *J. Mol. Biol.* 224, 53–63.

Rice, G.A., Kane, C.M., and Chamberlin. M.J. (1991). Footprinting analysis of mammalian RNA polymerase II along its transcript: an alternative view of transcription elongation. *Proc. Natl. Acad. Sci. USA* 88, 4245–4249.

Richardson, J.P. (1975). Attachment of nascent RNA molecules to superhelical DNA. *J. Mol. Biol.* 98, 565–579.

Richardson, J.P. (1982). Activation of rho protein ATPase requires simultanous interaction at two kinds of nucleic acid-binding sites. *J. Biol. Chem.* 257, 5760–5766.

Richardson, J.P. (1996). Structural organization of transcription termination factor rho. *J. Biol. Chem.* 271, 1251–1254.

Roberts, J.W. (1969). Termination factor for RNA synthesis. *Nature* 224, 1168–1174.

Rougvie, A.E., and Lis, J.T. (1988). The RNA polymerase II molecule at the 5' end of the uninduced hsp70 gene of *D. melanogaster* is transcriptionally engaged. *Cell* 54, 795–804.

Rudd, M.D., Izban, M., G., and Luse, D.S. (1994). The active site of RNA polymerase II participates in transcript cleavage within arrested ternary complexes. *Proc. Natl. Acad. Sci. USA* 91, 8057–8061.

Serizawa, H., Conaway, R.C., and Conaway, J.W. (1992). A carboxy-terminal domain kinase associated with RNA polymerase II transcription factor delta from rat liver. *Proc. Natl. Acad. Sci. USA* 89, 7476–7480.

Serizawa, H., Conaway, J.W., and Conaway, R.C. (1993). Phosphorylation of C-terminal domain of RNA polymerase II is not required in basal transcription. *Nature* 363, 371–374.

Shi, Y.B., Gamper H., Van Houten B., and Hearst J.E. (1988). Interaction of *Escherichia coli* RNA polymerase with DNA in an elongation complex arrested at a specific psoralen crosslink site. *J. Mol. Biol.* 199, 277–293.

Sidorenkov, I., Komissarova, N., and Kashlev, M. (1998). Crucial role of the RNA:DNA hybrid in the processivity of transcription. *Mol. Cell* 2, 55–64.

Sluder, A.E., Greenleaf, A.L., and Price, D.H. (1989). Properties of a *Drosophila* RNA polymerase II elongation factor. *J. Biol. Chem.* 264, 8963–8969.

Spencer, C.A., and Groudine, M. (1990). Transcription elongation and eukaryotic gene regulation. *Oncogene* 5, 777–785.

Stebbins, C.E., Norukhov, S., Orlova, M., Polyakov, A., Goldfarb, A., and Darst, S.A. (1995). Crystal structure of the GreA transcript cleavage factor from *Escherichia coli*. *Nature* 373, 636–640.

Sullivan, S.L., and Gottesman, M.E. (1992). Requirement for *E. coli* NusG protein in factor-dependent transcription termination. *Cell* 68, 989–994.

Surosky, R.T., Strich, R., and Esposito, R.E. (1994). The yeast UME5 gene regulates the stability of meiotic mRNAs in response to glucose. *Mol. Cell. Biol.* 14, 3446–3458.

Surratt, C.K., Milan, S.C., and Chamberlin, M.J. (1991). Spontaneous cleavage of RNA in ternary complexes of *Escherichia coli* RNA polymerase and its significance for the mechanism of transcription. *Proc. Natl. Acad. Sci. U.S.A.* 88, 7983–7987.

Telesnitsky, A.P.W., and Chamberlin, M.J. (1989). Terminator-distal sequences determine the *in vitro* efficiency of the early terminators of bacteriophages T3 and T7. *Biochemistry* 28, 5210–5218.

Theissen, G., Pardon, B., and Wagner, R. (1990). A quantitative assessment for transcriptional pausing of DNA-dependent RNA polymerase *in vitro*. *Anal. Biochem.* 189, 254–261.

Uptain, S.M., Kane, C.M., and Chamberlin, M.J. (1997). Basic mechanisms of transcription elongation and its regulation. *Annu. Rev. Biochem.* 66, 117–172.

Valay, J.-G., Simon, M., Dubois, M.-F., Bensaude, O., Facca, C., and Faya, G. (1995). The KIN28 gene is required for RNA polymerase II-mediated transcription and phosphorylation of the RPB1 CTD. *J. Mol. Biol.* 249, 535–544.

von Hippel, P.H. (1998). An integrated model of the transcription complex in elongation, termination and editing. *Science* 281, 660–665.

Wahi, M., and Johnson, A.D. (1995). Identification of genes required for alpha 2 repression in *Saccharomyces cerevisiae*. *Genetics* 140, 79–90.

Walstrom, K.M., Dozono, J.M., and von Hippel, P.H. (1998). Effects of reaction conditions on RNA secondary structure and on the helicase activity of *Escherichia coli* transcription termination factor rho. *J. Mol. Biol.* 279, 713–726.

Wang, D.G., and Hawley, D.K. (1993). Identification of a 3'–5' exonuclease activity associated with human RNA polymerase II. *Proc. Natl. Acad. Sci. USA* 90, 843–847.

Weeks, J.R., Hardin, S.E., Shen, J.J., Lee, J.M., and Greenleaf, A.L. (1994). Locus-specific variation in phosphorylation state of RNA polymerase II *in vivo* — correlations with gene activity and transcript processing. *Genes Dev.* 7, 2329–2344.

Wilson, K.S., and von Hippel, P.H. (1995). Stability of *Escherichia coli* transcription complexes near an intrinsic terminator. *J. Mol. Biol.* 244, 36–51.

Wilson, K.S., and von Hippel, P.H. (1995). Transcription termination at intrinsic terminators: the role of the RNA hairpin. *Proc. Natl. Acad. Sci. USA* 92, 8793–8797.

Xie, Z., and Price, D.H. (1996). Purification of an RNA polymerase II transcript release factor from *Drosophila*. *J. Biol. Chem.* 271, 11043–11046.

Yager, T.D., and von Hippel, P. (1987). Transcript elongation and termination in *Escherichia coli*. In '*Escherichia coli* and *Salmonella typhimurium*: Cellular and Molecular Biology (Ed.: Neidhardt, F.C.), American Society of Microbiology, Washington DC.

Yankulov, K., Blau, J., Purton, T., Roberts, S., and Bentley, D.L. (1994). Transcriptional elongation by RNA polymerase II is stimulated by transactivators. *Cell* 77, 749–759.

Yankulov, K., Yamashita, K., Roy, R., Egly, J.-M., and Bentley, D. (1995). The transcriptional elongation inhibitor 5,6-dichlor-1-β-D-ribofuranosylbenzimidazole inhibits transcription factor IIH-associated protein kinase. *J. Biol. Chem.* 270, 23922–23925.

Yankulov, K.Y/. Pandes, M., McCracken, S., Bouchard, D., and Bentley, D.L. (1996). TFIIH functions in regulating transcriptional elongation by RNA polymerase II in *Xenopus* oocyctes. *Mol. Cell. Biol.* 16, 3291–3299.

Yanofski, C. (1981). Attenuation in the control of expression of bacterial operons. *Nature* 289, 751–758.

Chapter 6

$RNAP_I$ and $RNAP_{III}$ Transcriptional Machineries

Much of this book has up to now focused on the control mechanisms employed by the $RNAP_{II}$-transcriptional machinery. Many additional interesting features about eukaryotic transcription mechanisms have emerged from similar studies carried out in the $RNAP_I$- and $RNAP_{III}$ transcription systems. This knowledge complements the insights obtained in the $RNAP_{II}$ system, and extends our understanding of eukaryotic gene expression in a much more comprehensive manner. The comparison of the regulatory features in all three transcription systems has allowed common, and often surprising, themes to emerge.

6.1. Evolutionary History of the Compartmentalization of Eukaryotic Transcriptional Machinery

In two of the three evolutionary domains, Bacteria and Archaea, a single type of RNA polymerase transcribes all the genes that are present in the genome. In eukaryotes three distinct nuclear RNA polymerases (plus a group of distinct basal factors for each of them) share the task of transcribing various types of genes from the genome in an essentially non-overlapping manner. The $RNAP_I$ transcription system is responsible for the exclusive transcription of the 'large' (28S, 18S and 5.7S) rRNA genes, and the $RNAP_{III}$ system for the transcription of a number of small RNAs involved in various aspects of post-transcriptional RNA processing and protein translation.

At present we do not understand, even in the barest outline, the evolutionary pressures that lead to the splitting the eukarytic transcriptional machinery into the three distinct RNAP systems during the evolution of eukaryotic cells. These systems are relatively independent of each other, and most of the transcription factors they employ are type-specific and used only in conjunction with one of the three different RNAPs. Based on our current understanding it looks as if the transcriptional compartmentalization is essentially based on a 'division of control' concept. According to this line of thought, the split is a consequence of an event occuring at an early stage of eukaryotic evolution that resulted in separating the capacity for controlling the expression of protein-encoding genes away from the systems that are involved in the production of RNAs with mainly structural and catalytic functions. The $RNAP_{I}$- and $RNAP_{III}$ systems involve a relatively small number of basal/gene-specific transcription factors and appear somewhat primitive in their narrowly-specialized functions. The $RNAP_{II}$ system, with its vast array of gene-specific and basal transcription factors, elongation control and chromatin remodelling mechanisms, may have been set free by the compartmentalization process to evolve into the complex transcriptional apparatus capable of controlling the sophisticated transcriptional programs necessary for the development of multicellular organisms. The task of the $RNAP_{I}$ system, on the other hand, is to transcribe a single gene type at high levels. Although rRNA-encoding genes generally account for only 1% or so of a genome, they are responsible for up to 40% of total RNA biosynthesis that occurs in actively growing cells. It is therefore evident that the evolution of a sophisticated control system for regulating $RNAP_{II}$ could have depended at some stage of pre-eukaryotic evolution on releasing the system from the 'heavy manual' work involved in rRNA and tRNA biosynthesis.

As we will see below, there are still many traces of a common evolutionary ancestry that are easily detectable in the three eukaryotic nuclear transcription systems. The TATA-binding protein (TBP) participates in promoter-selective basal factors in all three systems, and five different RNAP subunits occur in all three polymerases and presumably participate in analogous functions. We will also look at the transcription of the snRNA genes, which is an area of significant overlap between the $RNAP_{II}/RNAP_{III}$ systems and which are therefore of great interest for understanding the evolutionary history of eukaryotic transcription systems. The molecular characterization of the transcriptional machineries of evolutionary-diverged eukaryotic lineages, such as ciliates and trypanosomes, may provide a better experimental basis for speculations about this topic in the future.

6.2. Structure and Function of the $RNAP_I$ Transcriptional Machinery

Genomic and Nuclear Arrangement of rRNA Genes

The genes encoding the three 'large' ribosomal RNAs (28S, 18S, and 5.8S) are organized as single transcription units that are tandemly repeated and arranged as large clusters. In eukaryotes each of these these clusters usually contains between 100 and 1,000 individual repeat units (Figure 6.1; reviewed in Long and Dawid, 1980; Paule, 1994; Moss and Stefanovsky, 1995). Every repeat unit produces a complete primary transcript that is neither capped nor polyadenylated, but is specifically methylated at approximately 100 ribose moieties through the action of snRNA-containing ribonucleoproteins (reviewed in Maden and Hughes, 1997), before it is post-transcriptionally processed into the 18S, 5.8S and 28S mature rRNAs found in ribosomes. In the human genome there are more than 200 rRNA gene copies that are are dispersed into several distinct clusters ('nucleolar organizers') located on several chromosomes.

The rate of transcription of rRNA from the tandem repeats is so high that a specialized organelle, the nucleolus, forms around them to create a spatially distinct region with the nucleus. Even single rRNA genes have an intrinsic ability to initiate the formation of a specific nucleolar structure (Karpen *et al.*, 1988). $RNAP_I$ transcription in intact cells can be visualized in permeabilized nuclei by incubating them with labelled nucleotide precursors in presence of α-amanitin (in contrast to $RNAP_{II}$ and $RNAP_{III}$, $RNAP_I$ is essentially resistant to this drug). This reveals approximately 25 distinct nucleolar foci that are still detectable, even if the surrounding chromatin is removed, suggesting that they are structurally independent nuclear components (Jackson *et al.*, 1993; Hozak

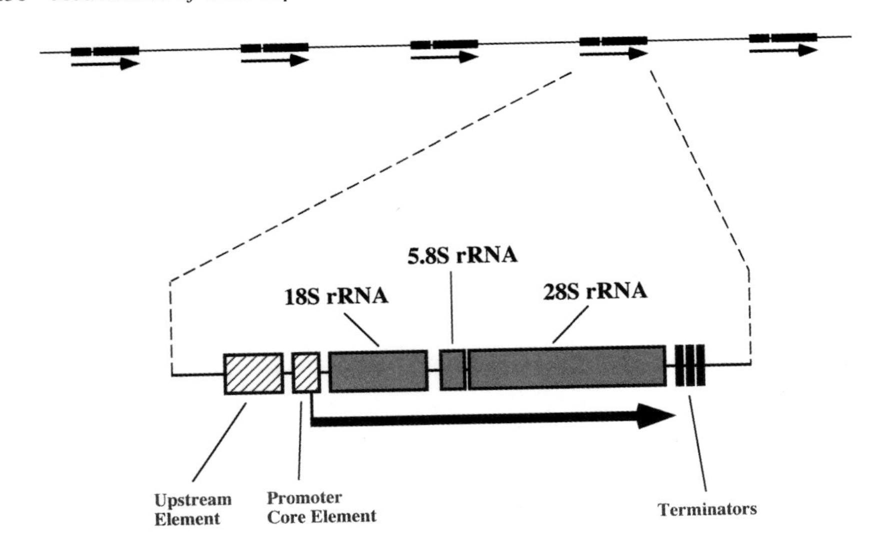

Figure 6.1. Molecular Organization of Genes Encoding the 18S-5.8S-28S rRNAs. Three different types of rRNAs are transcribed from single, tandemly-repeated transcription units. Expression of each individual transcription unit is controlled from intergenic spacer elements and individual promoter elements that are recognized by the basal factors of the $RNAP_I$ transcriptional machinery. The rRNA primary transcripts are specifically terminated by a row of terminator motifs.

et al., 1994; reviewed in Scheer and Weisenberger, 1994). The nucleolus breaks down and reforms in most cells during each mitotic cycle and is thus a highly dynamic structure. Interestingly, nucleoli have also been implicated in the storage and transport of specific mRNA molecules, such as the transcripts from the *c-myc*, *N-myc* and *myoD* genes (Bond and Wold, 1992), indicating that these structures are not solely involved in $RNAP_I$-mediated transcription and rRNA processing events, but may also play a (yet not understood) role in the control of mRNA export from the nucleus.

Ribosomal Gene Promoters

The lack of obvious sequence similarity between rRNA gene promoters from different species, even those that are closely related, has for a long time frustrated efforts to define any functionally relevant sequence elements (see e.g. Moss and Stefanovsky, 1995). Most point mutations have relatively minor effects on the efficiency of rRNA gene promoters, and generally only the absolute identity of two (!) nucleotides seems to be important (Read *et al.*, 1992). The application of promoter deletion- and 'linker scanning' experiments (see Chapter 3) eventually led to the identification of two functionally important motifs in the human rRNA gene promoter, namely the 'core' and the 'upstream control' elements (UCE; Figure 6.2; Haltiner *et al.*, 1986). The spacing between the core element and UCE is critical, because deletions/insertions that change the distance between them negatively affect the overall activity of the rRNA promoter. The two elements are present in a number of rRNA gene promoters from higher eukaryotes (Miller *et al.*, 1995b; Windle and Sollner-Webb, 1986), but lower eukaryotes seem to have only a single regulatory element flanking the position where the transcript initiates (Kwonin *et al.*, 1985; Tyler *et al.*, 1985).

'Upstream Binding Factor' Guides the Assembly of $RNAP_I$ Promoter Complexes

The $RNAP_I$ transcriptional machinery has been most extensively characterized in vertebrate systems, where it has been shown to consist of at least three major biochemically-distinct transcription factors. These include 'Upstream Binding Factor' (UBF), SL1 and $RNAP_I$, all of which are necessary to reconstitute specific $RNAP_I$ transcription from rRNA gene promoters under *in vitro* conditions (Figure 6.3; Bell *et al.*, 1988). UBF and SL1 are the basal factors that recognize the specific promoter motifs and facilitate the recruitment and positioning of $RNAP_I$ over the transcript initiation site.

Human UBF contains six distinct DNA-binding domains revealing a high degree of sequence homology to motifs that were previously characterized in the chromatin-associated proteins HMG1 and HMG2 (Figure 6.4; Jantzen *et al.*, 1990). Each of these 'HMG' boxes binds up to 20 nucleotides and introduces a kink of approximately 130° (Read *et al.*, 1993; Weir *et al.*, 1993). Sequence comparisons between a number of vertebrate UBF homologs have shown that each of the HMG boxes is more closely related to the box in the equivalent position of UBF from another species, rather than to any adjacent HMG boxes within the same UBF molecule. A particular HMG box cannot be

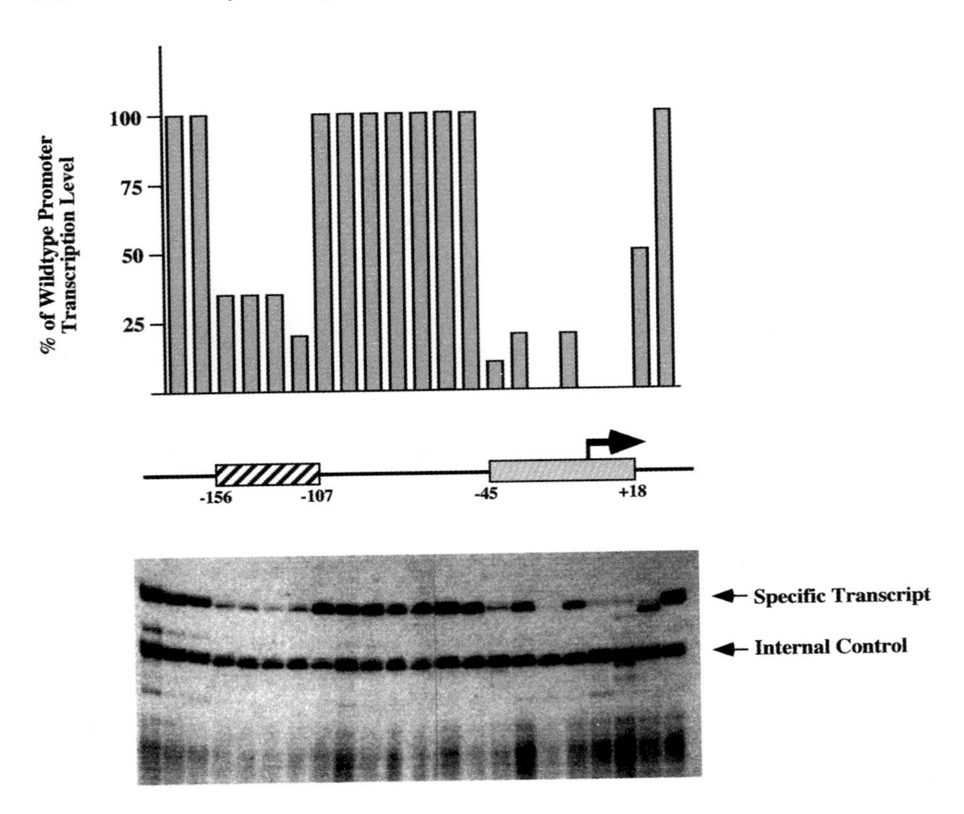

Figure 6.2. Linker Scanning Mutagenesis of the Human rRNA Gene Promoter.
A series of evenly-spaced 'linker-scanning' mutants spanning 200 nucleotides upstream of the transcript initiation site of the human rRNA gene promoter was analyzed in an *in vitro* transcription system. *Top*: In comparison to the unmodified ('wildtype') promoter, insertions of small linker sequences in most parts of the promoter have no detectable effects on the amount of specifically-initiated transcripts produced. Mutagenesis of a region between −156/−107 ('Upstream Control Element', 'UCE') and −45/+18 ('Core') leads to a substantial loss of promoter activity, indicating the presence of control sequences that are required for high-level transcription. *Bottom*: *In vitro* transcription assay result of the various linker-scanning promoter mutants. Each experiment contained an internal positive control template giving rise to a transcript of a different size from the ones initiated on the linker-scan mutant promoters. The internal control transcript remains largely invariant, whereas the transcripts from the mutant promoters reflect the presence/absence of the crucial control elements. Data from Haltiner *et al.* (1986).

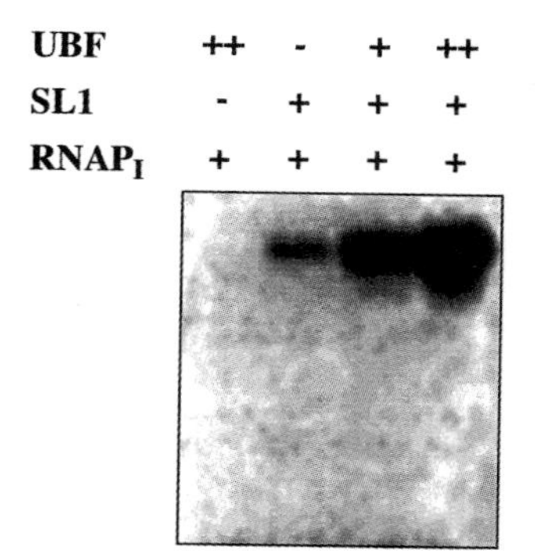

Figure 6.3. Human UBF Stimulates Transcription of the RNAP$_I$/SL-1 Initiation Complex *In Vitro*. In the human system SL1 is essential and increasing amounts of UBF substantially stimulate the rate of transcription of an RNAP$_I$ test promoter. Data from Bell *et al.* (1988).

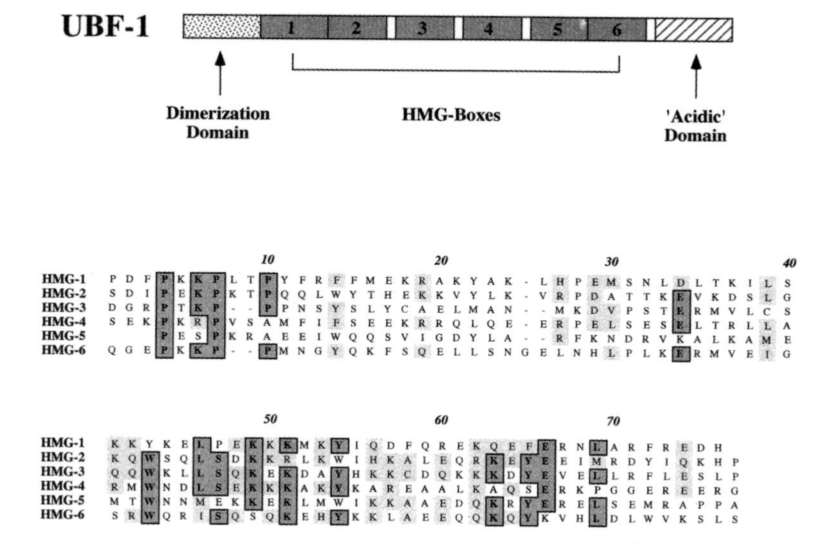

Figure 6.4. Structural Organization of Human UBF.
Apart from unique N- and C-terminal domains, the major part of the protein is made up of the repetitive, DNA-binding HMG-domains. The sequence of the six HMG domains are aligned in the lower half of the figure to reveal the location of several highly conserved amino acids.

replaced with another one derived from another position in UBF, but can be functionally replaced by an equivalent one from another species (Cairns and McStay, 1995). This observation suggests that each HMG box has a rather precise structural role to fulfill, which may at least partially explain the degree of species-specificity that prevents UBFs from distantly related vertebrates to substitute functionally for each other (e.g. Jantzen *et al.*, 1992; Cairns and McStay, 1995).

Since the binding of UBF is probably the first step towards the assembly of stable $RNAP_I$ basal transcription complexes (especially during mitosis; Figure 6.5), there is a need for UBF to specifically recognize rRNA promoters. Interestingly, point mutations and deletions in the promoter regions containing UBF recognition sites have little overall effect on UBF binding (Read *et al.*, 1992; LeBlanc *et al.*, 1993). These results suggests that the interaction is not particularly sequence-specific, but may depend on structural features of DNA, such as DNA-curving. Recent structural studies of UBF-DNA complexes have shed more light on this mystery: UBF usually occurs as a homodimer capable of binding up to 200 nucleotides of DNA. Images obtained by electron spectroscopic imaging (ESI) techniques suggest that the DNA is wound round the outside of the UBF dimer in a single complete turn, which packs 180 nucleotides into an 'enhancesome' (Figure 6.6; Bazett-Jones *et al.*, 1994). The architectural role of UBF in forming the promoter loop structure explains the previously observed importance of the spacing between the promoter core and upstream elements, since alterations in the length of the DNA separating these elements would disrupt their precise spatial arrangement relative to each other when looped around the UBF homodimer (Le Blanc *et al.*, 1993; Hu *et al.*, 1994).

SL1

UBF is the apparently the major $RNAP_I$-basal factor that displays sequence-specific binding to DNA and therefore plays a major role in promoter recognition. The other basal factor, SL1, has little or no sequence-specific affinity for ribosomal promoters. Once UBF is stably bound, however, SL1 binds with greatly enhanced cooperativity (Clos *et al.*, 1986; Bell *et al.*, 1988; Schnapp and Grummt, 1991).

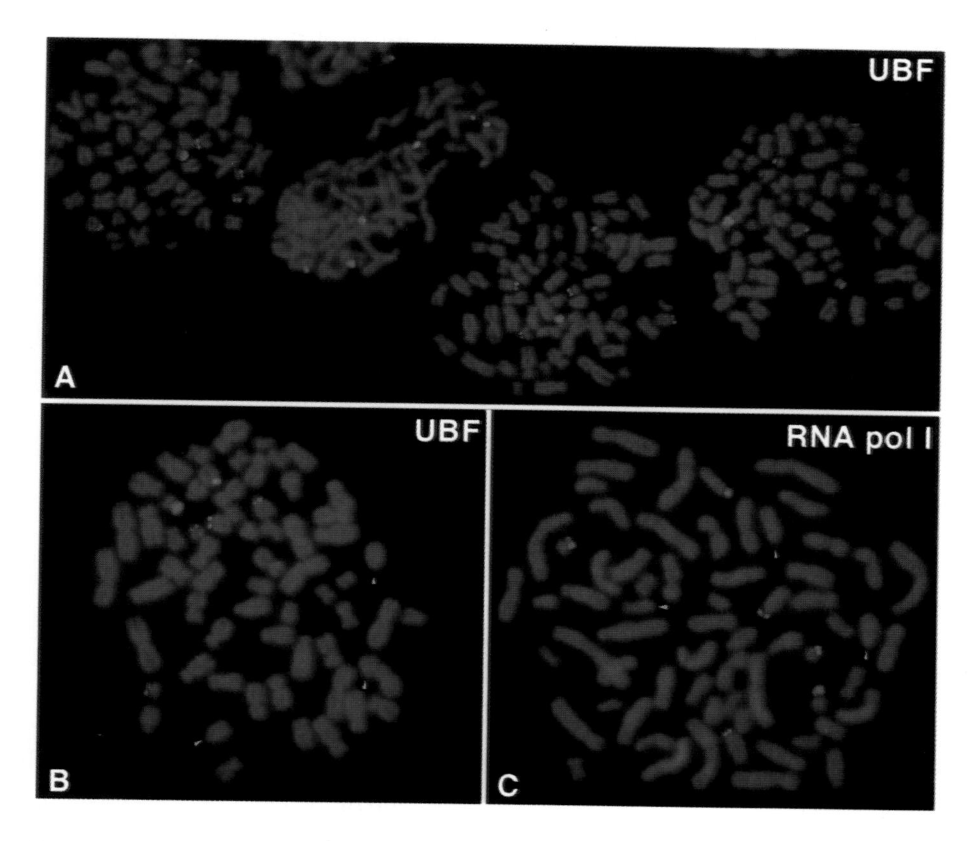

Figure 6.5. Immunolocalization of UBF and $RNAP_I$ on Mitotic Human Chromosomes. The $RNAP_I$ transcriptional machinery remains associated with specific nucleolar organizer regions (NORs) during all mitotic phases. Chromosomes are stained with DAPI (blue), and either UBF or $RNAP_I$ are labelled with fluorescent antibodies (red). Six acrocentric chromosomes that are known to bear NORs are labelled to a similar extent with antibodies directed against both $RNAP_I$ and UBF, suggesting that they contain fully assembled $RNAP_I$ transcription inititation complexes. The assembly of such transcription complexes preceeding mitosis may determine the activity of NORs at the beginning of the next cell cycle. From Roussel *et al.* (1996).

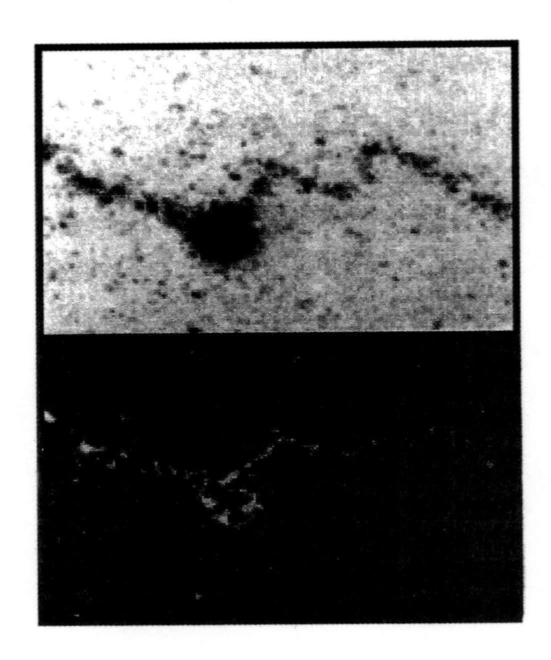

DNA + protein

Phosphorus

Figure 6.6. The RNAP$_I$ 'Enhancesome'.
Top: Transmission electron micrograph of a UBF-promoter complex. *Bottom*: Selective visualization of phosphorus within the UBF-promoter complex supports the notion that UBF-homodimers wind DNA around them in a 360° loop. From Bazett-Jones *et al.* (1994).

SL1 is a TFIID-like transcription factor consisting of TATA-binding protein (TBP) and three distinct TBP-associated factors (TAF$_I$s; Comai *et al.*, 1992; Eberhard *et al.*, 1993; Rudloff *et al.*, 1994). The TAF$_I$s are completely distinct from the TAF$_{II}$s found in the TFIID complex, and transcriptionally active SL1 can be reconstituted from recombinant proteins (Comai *et al.*, 1994; Zomerdjik *et al.*, 1994; reviewed in Zomerdijk and Tjian, 1998). Interestingly, the TBP present in SL1 does not seem to be directly involved in sequence-specific recognition of promoter elements, which is reminscent of the role of TBP on TATA-less promoters (Chapter 2). The genes encoding rRNA do not have a TATA-like sequence near the transcription initiation site (with the exception of some plant rRNA genes!), and *in vitro* transcription efficiency is not impaired by the presence of competitor TATA-containing oligonucleotides or TBP-specific inhibitor proteins (Radebaugh *et al.*, 1994; White *et al.*, 1994). On all promoters transcribed by $RNAP_I$ (and, as we will see later, most promoters transcribed by $RNAP_{III}$), the TBP-containing transcription complex seems to become recruited into the preinitiation complex predominantly via protein-protein interactions. TBP is also utilized in different ways in the various transcription systems because specific TBP mutants have been characterized that affect transcription by the $RNAP_I$-, $RNAP_{II}$-, and $RNAP_{III}$-specific transcription systems to different extents (Cormack *et al.*, 1992), and distinct areas of TBP are essential for transcription by $RNAP_{III}$ (Cormack *et al.*, 1993). TBP acquires probably most of its specific properties trough association with specific sets of TBP-associated factors (TAFs) which are specific for the $RNAP_I$-, $RNAP_{II}$-, and $RNAP_{III}$-systems. We have already seen in Chapter 4 how the TAF_{II}s that are part of the TFIID complex play a major role in conveying transcriptional regulatory signals from promoter-specific transcription factors to other basal transcription factors and $RNAP_{II}$. Less is known about the functional role of TAF_Is and TAF_{III}s, but it seems very likely that they also participate crucially in template commitment and signal processing in these systems. TAF_Is and TAF_{II}s bind to TBP in a mutually exclusive type-specific manner, thus avoiding the formation of hybrid (and possibly nonfunctional) complexes (Comai *et al.*, 1994).

$RNAP_I$ Recruitment Step

The assembled UBF/SL1 complex recruits $RNAP_I$ onto the rRNA gene promoters through protein-protein interactions and supports multiple rounds of transcription (Schnapp *et al.*, 1991; Comai *et al.*, 1992). This step is probably directly comparable to the recruitment of $RNAP_{II}$ to the TFIID/TFIIB complex bound to the TATA box in the $RNAP_{II}$ transcription system (Chapter 2). Observations from several laboratories suggest that purified $RNAP_I$ is present in two different versions, one that can initiate transcription accurately, and another one which transcribes only non-specifically. Although it has been suggested that differential modification might be the underlying cause for the existence of the two forms of polymerase (Bateman and Paule, 1986; Tower and Sollner-Webb, 1987), another likely explanation is the existence of another transcription factor that assists $RNAP_I$ in specifically recognizing the preassembled UBF/SL1 complex. Due to the inability to reconstitute active eukaryotic RNAPs from recombinant subunits, all currently available transcription systems utilize RNAPs purified from various endogenous sources. Schnapp and Grummt (1991) describe the presence of two additional general $RNAP_I$ transcription factors that are necessary for the specific *in vitro* transcription from mouse rRNA promoters. It is possible that these factors also exist in the human system, but are more tightly associated with $RNAP_I$ and therefore not easily separated from the enzyme during the purification procedure.

Genetic screens in yeast have provided strong evidence fot the presence of an additional transcription factor, RRN3, which appears to be an entity distinct from the UBF and SL-1 homologs in yeast. RRN3 stimulates the recruitment of $RNAP_I$ to the promoter by direct association with the polymerase (Yamamoto *et al.,* 1996). This interaction is superficially reminiscent of the interaction between TFIIF and $RNAP_{II}$ (Chapter 2), but lack of sequence homology to TFIIF and differences in functional properties makes it likely that RRN3 plays a very different role from TFIIF. At this stage it still remains to be seen whether a similar factor is also present in higher eukaryotic organisms.

Role of the Intergenic 'Spacer' $RNAP_I$-Promoters

In all eukaryotes the tandemly-arranged rRNA transcription units are separated by 'intergenic spacer' ('IGS') elements. While the size and sequences of the rRNA coding regions are largely invariant across the whole eukaryotic range (which turns them into valuable tools for measuring the evolutionary relationships between different species), the length and sequences of the IGSs are often highly variable from species to species. The human ribosomal rRNA gene repeat unit, with 14 kb one of the longest ones known, is twice the size of the one found in yeast, and this difference is almost entirely due to variation in the IGS length. The IGSs separating the pre-rRNA transcription units contain one or several 'spacer promoters' that direct transcription in the same direction as the direction of the rRNA precursors. The function of these intergenic transcripts is currently still somewhat unclear, but some observations suggests that they may play an important role in the regulation of transcription of rRNA genes. A terminator sequence located approximately 150–200 nucleotides upstream of the rRNA promoter (T_0; Figure 6.1) has been shown to have a stimulatory effect on rRNA transcription. This terminator is the target site for a nucleolar transcription termination factor, TTF1 (Grummt, 1986). Recent studies demonstrated that, apart from possibly shielding the rRNA promoter from transcripts initiated in the intergenic region and acting as a replication barrier (Gerber *et al.*, 1997), TTF1 is a key component involved in 'opening up' chromatin domains and making them available to the $RNAP_I$ transcriptional machinery (Langst *et al.*, 1997). TTF1 binds to transcriptionally inactive, nucleosome-packed rRNA genes, destabilizes nucleosome structures and creates a defined promoter architecture containing a precisely-positioned nucleosome (Langst *et al.*, 1998; see also Chapter 7). Actively transcribed rRNA genes in yeast are nucleosome-free (Dammann *et al.*, 1993).

6.3. Structure and Function of the $RNAP_{III}$ Transcriptional Machinery

The $RNAP_{III}$ Transcriptional Machinery Transcribes a Diverse Set of Small RNA-Encoding Genes

In the $RNAP_I$ system there is a high degree of homogenity in the promoter structure of all genes encoding the large rRNAs. In contrast, the genes transcribed by the $RNAP_{III}$ systems (Figure 6.7) are rather diverse in their structure, function and promoter organization. This diversity is reflected in a variety of basal factors required at different $RNAP_{III}$ promoters (reviewed in Willis, 1993). The original characterization of a specific $RNAP_{III}$ gene type, utilizing a template containing the 5S rRNA-promoter, led to the preliminary definition of three distinct, biochemically-defined fractions, TFIIIA, TFIIIB and TFIIIC (Segall *et al.*, 1980). All these factors are required, in addition to $RNAP_{III}$, for accurate transcription of the 5S promoter under *in vitro* conditions. We will see, however, that most other genes transcribed by $RNAP_{III}$ differ in many important details from this rather specialized model system and these discoveries have led to a renewed appreciation of the regulatory diversity of other $RNAP_{III}$ transcribed genes. The organization of regulatory elements of several $RNAP_{III}$ promoters is schematically illustrated in Figure 6.8. Unusually, many of these promoters contain sequence motifs that are located within the

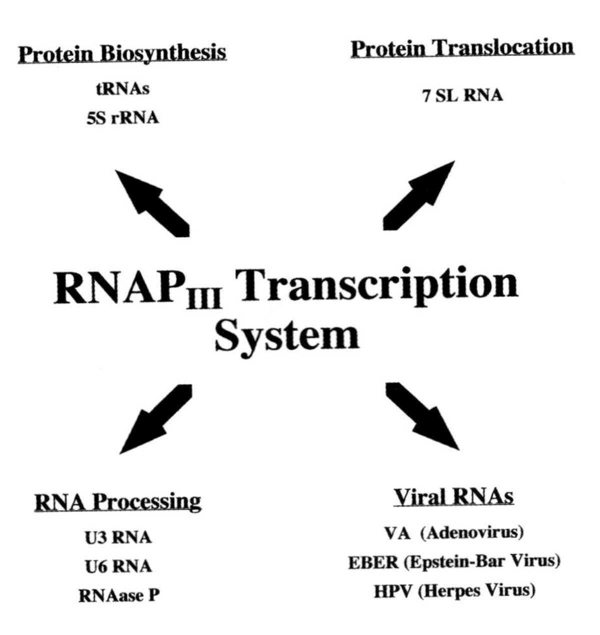

Figure 6.7. Overview of Genes Transcribed by the RNAP$_{III}$ System.
RNAP$_{III}$ is responsible for the production of a diverse set of transcripts that are essential components of various cellular macromolecular processing machineries and small RNAs required for viral infection.

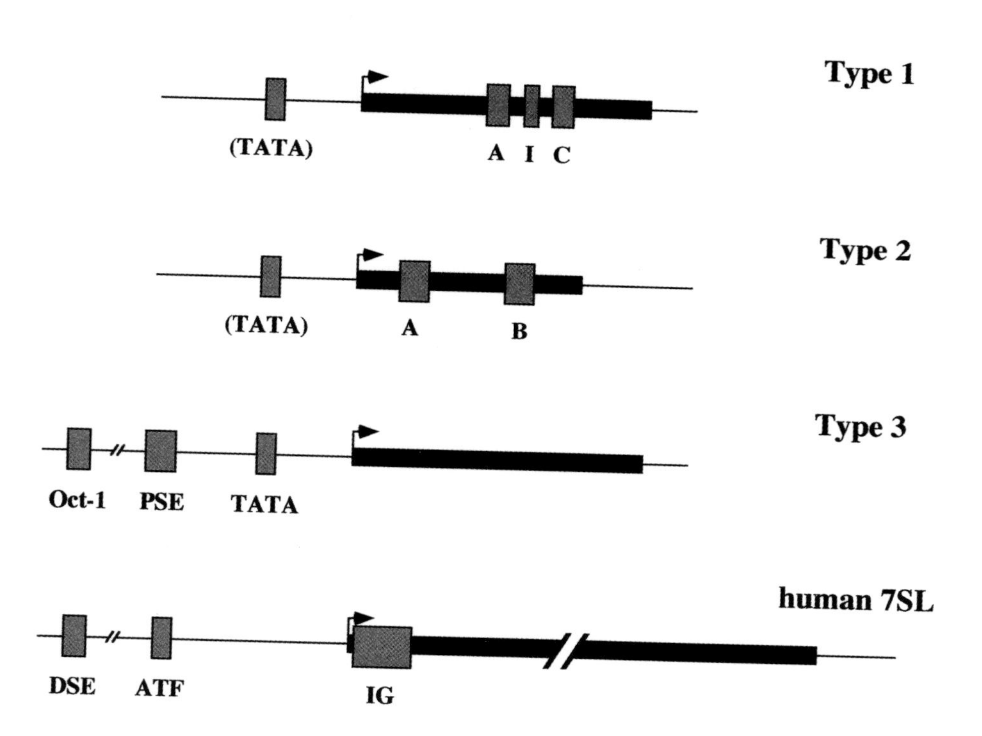

Figure 6.8. Schematic Diagram Showing the Arrangement of Transcriptional Control Elements in Various $RNAP_{III}$-Transcribed Genes.
Genes transcribed by the $RNAP_{III}$ system differ substantially in the type and arrangement of individual promoter elements. The transcribed region of each gene is shown in black, and the inititation site/transcript orientation is indicated by an arrow. Individual regulatory elements are shown as grey boxes. Note the presence of many of these sites within the transcribed region of various genes ('intragenic' elements'). DSE = distal sequence element; PSE = proximal sequence element; TATA = TATA-box; 'A', 'B', 'C' and 'I' = various sequence motifs recognized by the TFIIIC and TFIIIA basal factors. Modified from Willis (1993).

transcribed region and genuine TATA-boxes. Below we will investigate in more detail the functional significance of these promoter elements and how they are recognized by various gene-specific and basal transcription factors.

TBP-containing Multiprotein Complex TFIIIB is the Key $RNAP_{III}$ Basal Transcription Factor

Although the early experimental work on the 5S rRNA promoter suggested the presence of three essential basal factors assisting $RNAP_{III}$ recrutiment, it is now clear that this is not the generic minimal configuration of proteins required for all $RNAP_{III}$-transcribed genes. In fact, TFIIIB is the only basal factor that is universally required to transcribe all types of $RNAP_{III}$ genes (Kassavetis *et al.*, 1990), and TFIIIB is therefore a major target for cellular processes that influence the rate of $RNAP_{III}$ transcription (see below). In some cases, such as the yeast U6 snRNA gene (encoding an RNA required for pre-mRNA splicing), TFIIIB is actually the only basal factor required to correctly guide $RNAP_{III}$ to its initiation site under certain artificial *in vitro* conditions (Margottin *et al.*, 1991). TFIIIB binds sequence-specifically to sequences located upstream of the start site and its footprint covers approximately 50 nucleotides of genes encoding tRNAs and 5S rRNA (Kassavetis *et al.*, 1989). The binding of TFIIIB to its target site is kinetically rate-limiting, depends strongly on the presence of a TATA motif (Margottin *et al.*, 1991), and results in DNA-bending in a manner reminiscent of the binding of TBP to TATA-boxes in the $RNAP_{II}$ system (Leveillard *et al.*, 1991).

The biochemical characterization of TFIIIB revealed a multiprotein complex containing TATA-binding protein (TBP) and several TAF_{III}s, one of which has substantial homology to the $RNAP_{II}$ basal transcription factor TFIIB ('TFIIB-related factor' or 'BRF'; Buratowski and Zhou, 1992; Colbert and Hahn, 1992; Lopez-de-Leon *et al.,* 1992; Khoo *et al.*, 1994; Wang and Roeder, 1995). Although this homology initially suggested that BRF might interact with TBP using a mechanism comparable to that used in the TBP/TFIIB complex in the $RNAP_{II}$ system, subsequent mutagenesis studies have clearly shown that the interaction between TBP and BRF has very different structural requirements. BRF deletion variants lacking the region homologous to TFIIB retain their ability to interact with DNA-bound TBP (Kassavetis *et al.*, 1997), and point mutations in TBP abolishing recognition by BRF map to the top surface of the TBP saddle structure (Figure 6.9; Shen *et al.*, 1998). This work suggests that BRF interacts with TBP like a TAF, rather than as a protein stabilizing the TBP-induced kink as described for the $RNAP_{II}$-specific TBP/TFIIB/DNA complex (Chapter 2; Nikolov *et al.*, 1995). High resolution hydroxyl radical footprinting and competition experiments involving BRF and TFIIB support the idea that BRF binds to the top surface of TBP, but also suggest that BRF

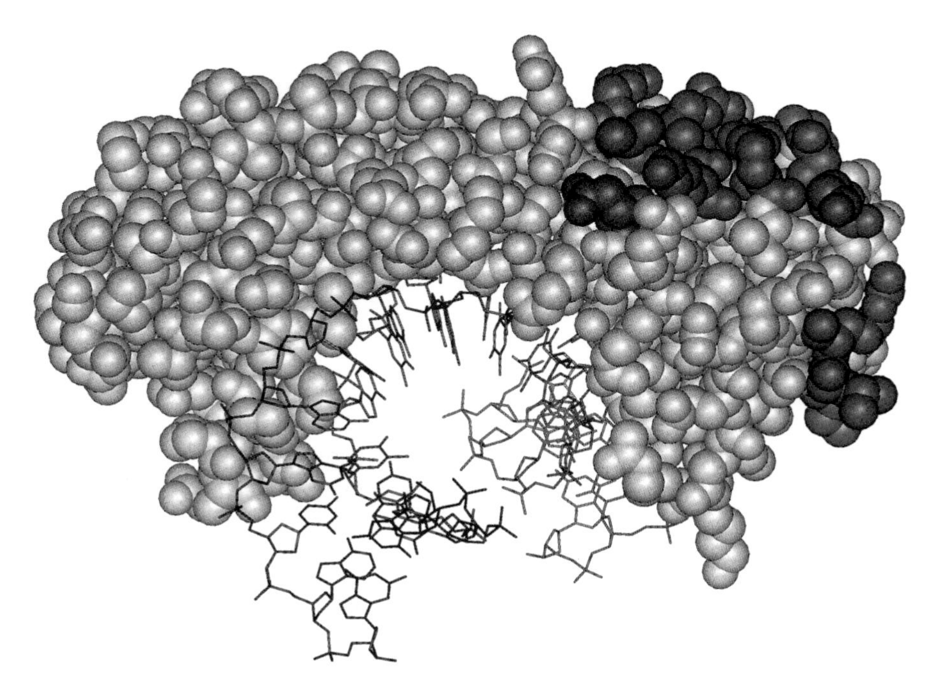

Figure 6.9. Location of Specific Amino Acid Residues in TPB Involved in the Interaction with BRF.

Spacefilling sideview of TBP with the amino acid residues identified as functionally active in the interaction with BRF shown in black. Contrary to previous expectations, BRF interacts with TBP via a tightly-clustered positively-charged 'patch' on the top of the TBP saddle structure. This mode of interaction is very different from the structural requirements for the formation of stable complexes of TFIIB and TBP on the TATA-box in $RNAP_{II}$ promoters. Based on data from Shen *et al.* (1998).

and another TFIIIB subunit, B", stabilize TBP binding by surrounding both faces of the TBP-DNA complex (Colbert *et al.*, 1998).

The subunit composition of mammalian TFIIIB has not been yet completely elucidated, but is probably very similarly organized and structurally related to the yeast factor (Wang and Roeder, 1996; Teichmann *et al.*, 1997). Similarly to TFIID and SL1, it has been possible to successfully reconstitute TFIIIB from recombinant subunits in yeast (e.g. Kassavetis *et al.*, 1995). TFIIIB is thought to be the only basal transcription factor that is capable of recruting $RNAP_{III}$ through direct protein:protein contacts with certain subunits of the $RNAP_{III}$-specific cluster, including RPC34, RPC31 and RPC82 (Figure 6.10; Bartholomew *et al.*, 1993; Werner *et al.*, 1993). The mapping of the molecular interactions between TFIIIB and these distinct $RNAP_{III}$ subunit represents a unique insight into the protein:protein contacts between eukaryotic basal factors and RNAPs because we still do not know at this stage in comparable detail how $RNAP_I$ and $RNAP_{II}$ interact with their basal factors.

TFIIIC Recruits TFIIIB to Promoters of $RNAP_{III}$-Transcribed Genes

The tRNA encoding genes require TFIIIC in addition to TFIIIB and $RNAP_{III}$ for accurate transcription *in vitro*. TFIIIC participates in the sequence-specific recognition of two distinct promoter elements that, similarly to the 5S rRNA genes, are located within the coding region and correspond to the DNA sequences encoding the highly conserved D and TψC-loop motifs present in all tRNAs ('Box A' and 'Box B'; Figure 6.8; Geiduschek and Tocchini-Valentini, 1988). When TFIIIC is stably bound to the intragenic tRNA gene promoter elements, it is then able to recruit TFIIIB via direct protein-protein interactions to create a platform suitable for the recruitment of $RNAP_{III}$ (Figure 6.10). Once TFIIIB is specifically bound to its target sites on DNA, it becomes remarkably stable and resistant to exposure to high salt and negatively charged polyanions. Under such extreme *in vitro* conditions TFIIIC, which was originally responsible for facilitating TFIIIB recruitment into the initiation complex, dissociates and leaves TFIIIB behind. TFIIIC is therefore often referrend to as an 'assembly factor' (e.g. Geiduschek and Kassavetis, 1995), and its main function is probably only the correct positioning of TFIIIB on the promoter. Genomic footprinting studies on yeast tRNA genes show little

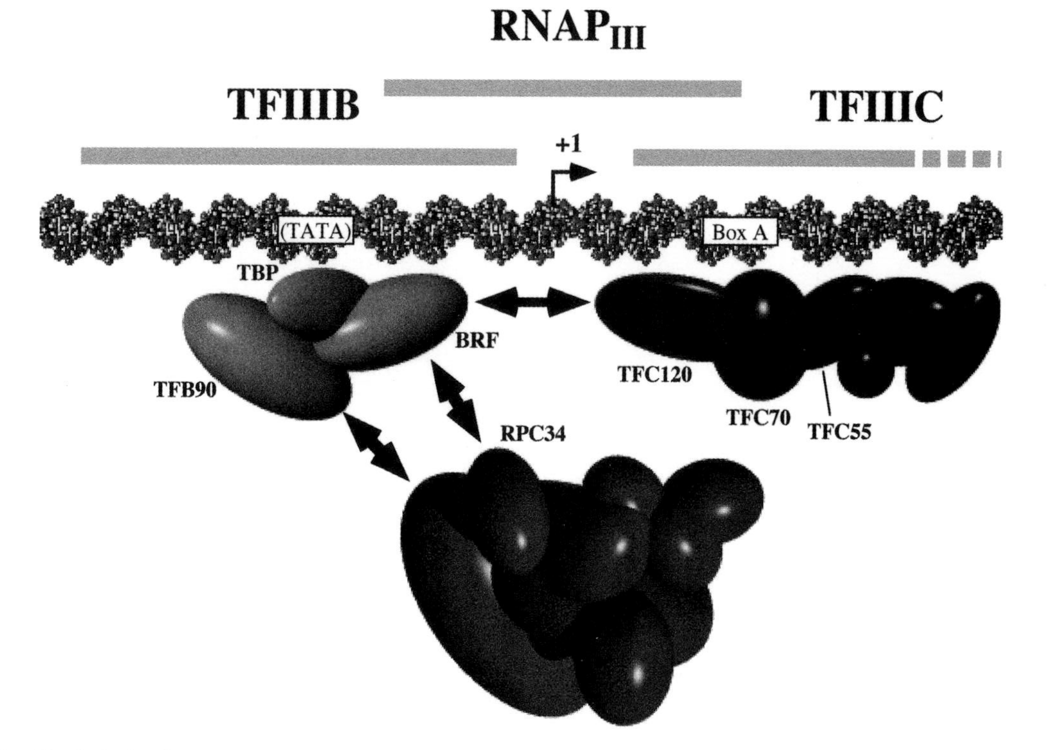

Figure 6.10. Typical Configuration of $RNAP_{III}$ Transcription Complexes on a Class II tRNA Gene.
TFIIIC binds to the intragenic promoter elements Box A and Box B (not illustrated) and recruits TFIIIB through specific contacts involving the BRF subunit of TFIIIB and the TFC120 subunit of TFIIIC (protein interactions shown by black double-headed arrows). $RNAP_{III}$ makes several specific protein contacts with at least two of the TFIIIB subunits, including BRF and TFB90, and is thus precisely positioned over the start site ('+1'). After Geiduschek and Kassavetis (1995).

evidence for the existence of stable protein complexes over the internal A- and B-boxes and therefore support the view that TFIIIC may also become displaced from genes *in vivo* once it has fulfilled its primary assembly role (Huibregtse and Engelke, 1989).

On the biochemical level, TFIIIC has been defined as a large multisubunit complex consisting of six distinct subunits forming a large DNA-protein complex that covers almost the entire coding region of the tRNA genes (Figure 6.11; Camier *et al.*, 1985; Baker *et al.*, 1987; Bartholomew *et al.*, 1991).

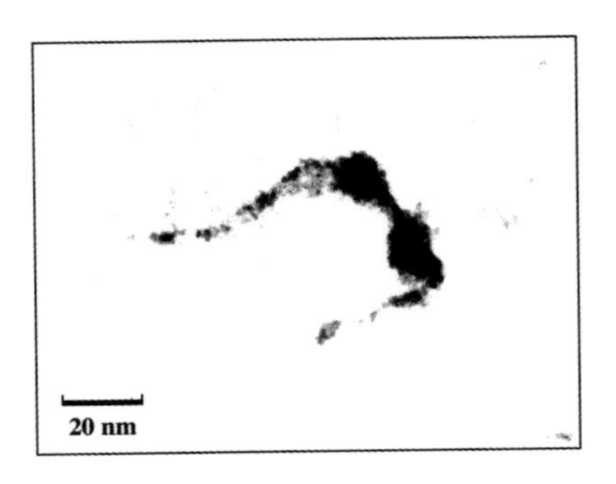

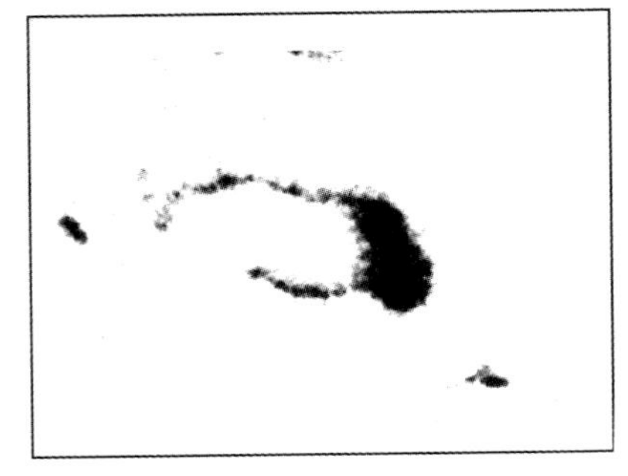

Figure 6.11. Scanning Transmission Electron Micrographs of TFIIIC-DNA Complexes. Purified TFIIIC was incubated with the $tRNA_3^{Leu}$ gene to form specific *in vitro* complexes. Most of the complexes detected (66%) were similar to the dumb-bell-shaped complexes illustrated above. Comparison of the shapes of TFIIIC complexes formed on tRNA genes with variable distances between the A and B boxes showed that the two halves of the dumb-bell complexes represent separate globular domains of TFIIIC specifically bound to the A and B boxes. From Schultz *et al.* (1989).

Comparison of cloned TFIIIC subunits from several eukaryotic species revealed no substantial sequence homology between yeast and human proteins, suggesting that the $RNAP_{III}$ system evolved very quickly into a system displaying a high degree of species-specificity (L'Etoile *et al.*, 1994; Lagna *et al.*, 1994; Sinn *et al.*, 1995). Moreover, even some of the TFIIIB factors that do display detectable homology are only 30% identical between yeast and humans (Wang and Roeder, 1995). This observation is reminiscent of the low degree of sequence conservation of the $RNAP_I$-basal factors and supports the general interpretation that many of the components of the RNAP$_{I/III}$ systems are freely coevolving at a much higher rate than their RNAP$_{II}$ counterparts. Due to their much more highly specialized functions, most structural changes in RNAP$_I$/RNAP$_{III}$ basal factors could be compensated for by mutations in their molecular interaction partners without loss of function. In contrast, the high degree of conservation of the majority of RNAP$_{II}$-basal factors across the entire eukaryotic range shows clearly that many of these proteins are structurally constrained, presumably because they need to function in a number of standardized configurations on a large variety of differently organized RNAP$_{II}$-promoters.

TFIIIA is a 5S rRNA Gene-Specific Transcription Factor

One of the most important basal factors involved in the synthesis and subsequent processing of 5S rRNA is TFIIIA. TFIII-A is exclusively required for transcription of the intragenically-located 5S rRNA promoters, where it is involved in the specific recognition of sequence elements spread over 45 nucleotides within the coding region of the 5S rRNA gene (elements 'A', 'I', and 'C'; Figure 6.8). Strictly speaking, TFIIIA should therefore be classified as a gene-specific transcription factor, but for historic reasons TFIIIA remains currently categorized as an RNAP$_{III}$ basal factor in the research literature and textbooks.

The molecular characterization of the TFIIIA protein revealed a repetitive arrangement of nine individual 'zinc-fingers', each 27 amino acid residues in length, that are involved in DNA-binding (Figure 6.12; Miller *et al.*, 1985). The subsequent high-resolution structural analysis of individual zinc-finger motifs has led to a detailed understanding of the functional contribution of individual finger elements to the recognition of specific nucleotide sequences under different conditions during the many roles that TFIIIA plays (see Chapter 3; Lee *et al.,* 1989; Pavletich and Pabo, 1991; Clemens *et al.*, 1992; Clemens *et al.*, 1993; reviewed in Pieler and Theunissen, 1993). The initial binding of TFIIIA to the 5S rRNA promoter initiates a sequentially-ordered recruitment of TFIIIC, TFIIIB and RNAP$_{III}$ in a manner strongly reminiscent of the sequential assembly process seen with RNAP$_{II}$ basal factors under *in vitro* conditions (Chapter 2).

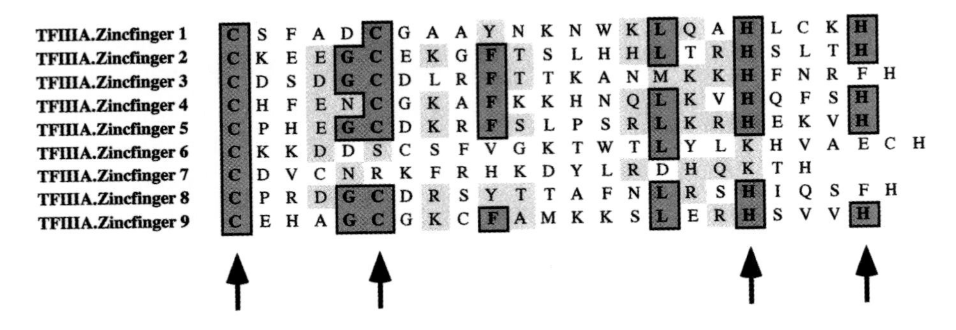

Figure 6.12. Zinc Finger Motifs Found in *Xenopus* TFIIIA.
TFIIIA contains zinc fingers of the characteristic 'C_2H_2' structure (note the high degree of conservations and arrangement of the cysteine ['C'] and histidine ['H'] residues indicated with arrows).

6.4. The 'Grey Area' Between the $RNAP_{II}$ and $RNAP_{III}$ Transcription Systems: snRNA Genes in Higher Eukaryotes

snRNA Genes Use Extragenic Promoters

Conventional thinking has always put very clearly defined boundaries between the classes of genes transcribed by the three different eukaryotic RNAP transcription systems. The discovery of common RNAP subunits shared between $RNAP_I$, $RNAP_{II}$ and $RNAP_{III}$ (Chapter 1), and the presence of TBP in basal factors for all three systems (in SL1, TFIID and TFIIIB) provided strong molecular evidence for the common evolutionary origin of the three nuclear RNAP transcription systems. The only clear example where the class-specific compartmentalization is somewhat blurred is in the case of small nuclear RNA ('snRNA')-encoding genes. Some of them (e.g. snRNA U1 and U2) are transcribed by the $RNAP_{II}$ system, whereas others (e.g. snRNA U6) are transcribed by the $RNAP_{III}$ machinery.

A comparative analysis of the promoter structures of these two types of genes has revealed surprisingly few differences (Figure 6.13). Both the $RNAP_{II}$-transcribed U2 and the $RNAP_{III}$-transcribed U6 genes contain the fairly general gene-specific transcriptional activator, Oct-1, bound to the 'distal sequence element' ('DSE') located around 200 basepairs upstream of the transcription

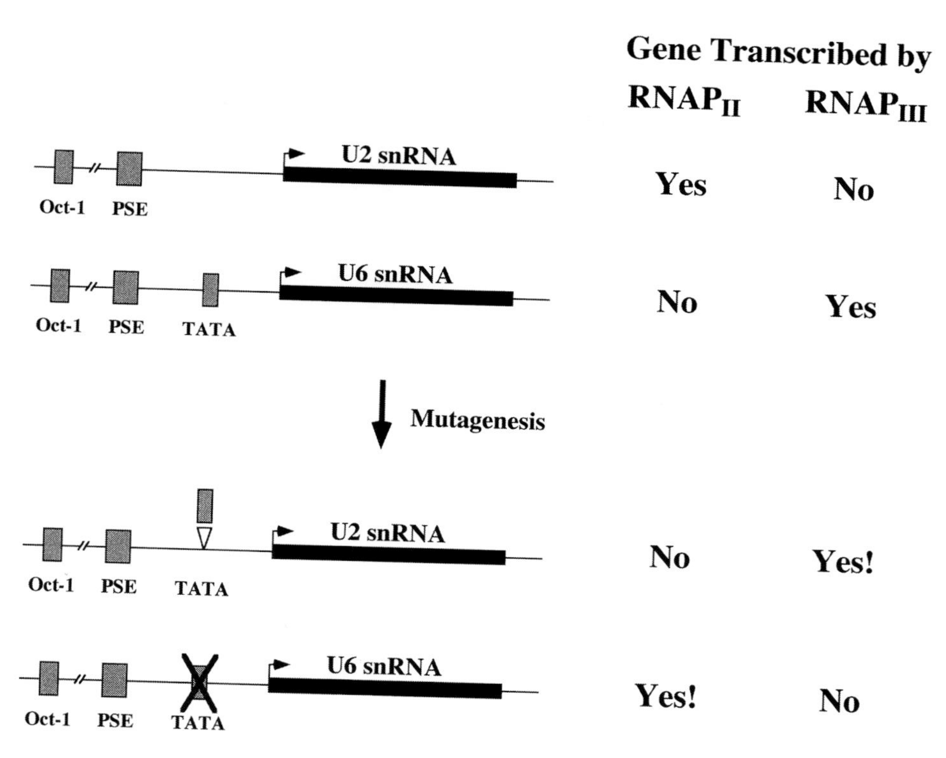

Figure 6.13. Promoter Organization of the U2 and U6 snRNA Genes.
The promoter of the U2 snRNA gene is very similar to the promoter of the gene encoding U6 snRNA although they are transcribed by very different transcriptional machineries. A small promoter-proximal region is crucially involved in selecting either the $RNAP_{II}$ or $RNAP_{III}$ system.

initiation site (some DSEs also contain Sp1 binding sites; Hernandez, 1992; Das *et al.*, 1995). In addition, the U2 and U6 gene also share a very similar 'proximal sequence element' ('PSE'), which serves as the target sequence for another TBP-containing basal transcription factor, $SNAP_c$. $SNAP_c$ consists of TBP and at least four additional protein subunits, some of which (SNAP50 and SNAP190) are mainly responsible for the sequence-specific recognition of the PSE consensus sequence (Figures 6.14 and 6.15; Sadowski *et al.*, 1993; Henry *et al.*, 1995; Yoon *et al.*, 1995; Wong *et al.*, 1998). $SNAP_c$ interacts

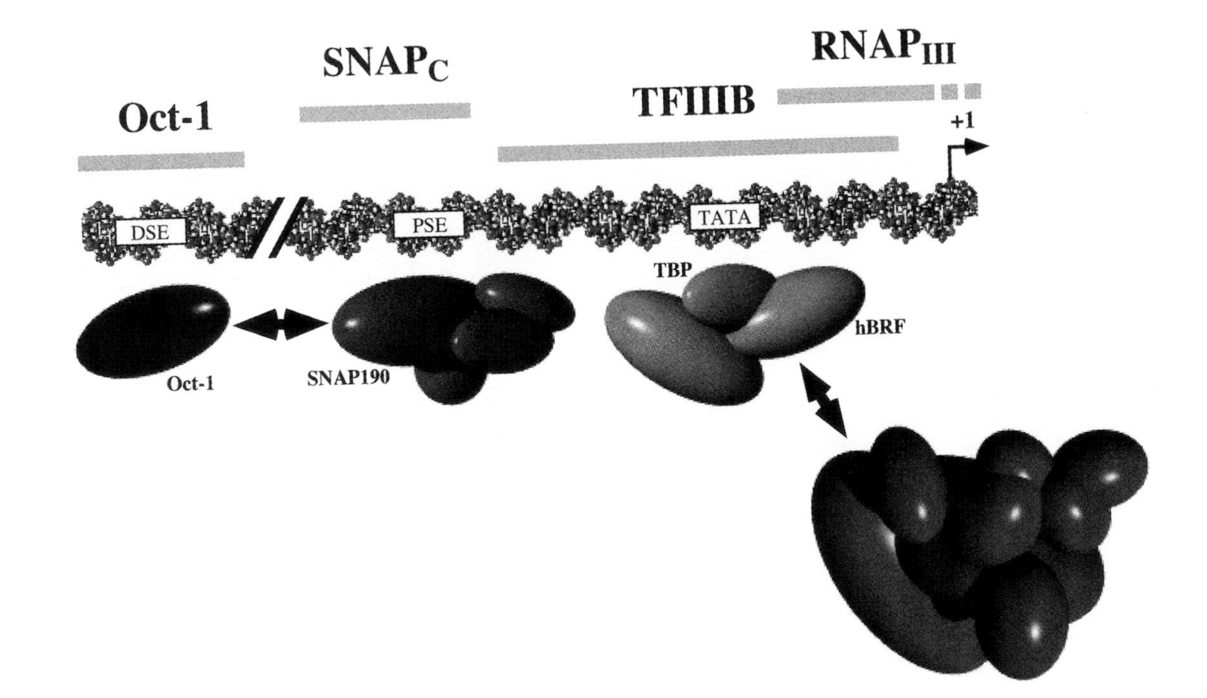

Figure 6.14. Configuration of $RNAP_{III}$ Transcription Complexes on a Vertebrate Class III U6 snRNA Gene.
In contrast to $RNAP_{III}$-class I and $RNAP_{III}$-class II promoters, the $RNAP_{III}$-class III promoter region illustrated here consists entirely of sequence elements located upstream of the transcription initiation site ('+1'). The $SNAP_c$ complex is recruited to the proximal sequence element ('PSE'; located around position –50) through specific protein:protein contacts with the DNA-binding domain of the transcriptional activator Oct-1 bound to the distal sequence element ('DSE'; located approximately 200 nucleotides upstream of the transcription initiation site). UV-crosslinking studies implicate SNAP190 and SNAP50 in the sequence-specific recognition of the proximal sequence element ('PSE'). $RNAP_{III}$ is probably specifically recruited to the U6 snRNA promoter via TFIIIB(?) bound to a TATA-element.

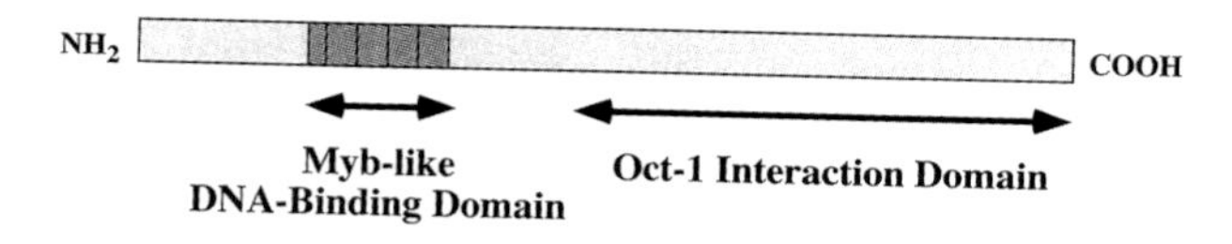

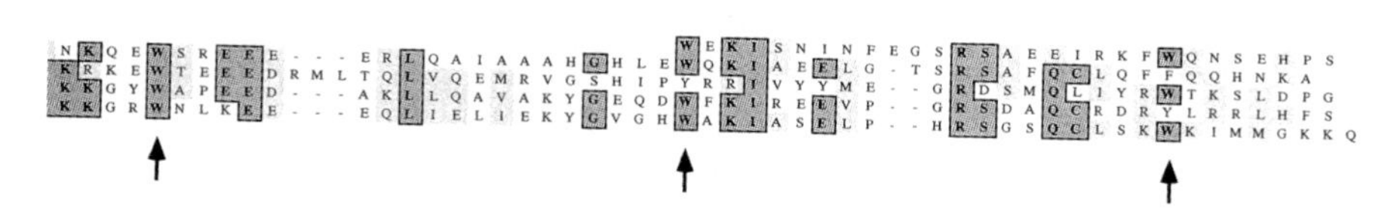

Figure 6.15. Domain Organization of the Largest Human $SNAP_c$ Subunit, SNAP190. The N-terminal domain contains four complete and one incomplete DNA-binding Myb-domains that are involved in the sequence recognition of the proximal sequence element ('PSE'). In the lower half of the figure the aligned sequences of the individual Myb domains are shown. Three highly conserved hydrophobic residues (usually W, i.e. tryptophane), that are hallmarks of Myb domains, are highlighted with arrows. Deletion studies of SNAP190 have revealed that the C-terminal half of the protein is sufficient for specific interactions with the Oct-1 activator protein bound to the distal sequence element. After Wong *et al.*, 1998.

cooperatively with the DSE-bound Oct-1 transcription factor through protein-protein contacts that mainly involve the 'POU' DNA-binding domain of Oct-1 (Figure 6.16; Murphy *et al.*, 1992; Mittal *et al.*, 1996; Ford and Hernandez, 1997; Wong *et al.*, 1998). This Oct-1/$SNAP_c$ complex is thought to be capable of specific interactions with both $RNAP_{II}$ *and* $RNAP_{III}$ basal factors. Although it is possible that different types of $SNAP_c$ complexes with varying subunit compositions will be discovered in the future, it seems likely that the true selection between the $RNAP_{II}$/$RNAP_{III}$ systems actually involves a small region located approximately 30 nucleotides upstream of the snRNA transcription start site. The only functional sequence element that distinguishes the U2 and U6 promoters is a TATA-element that is absolutely required for the $RNAP_{III}$-mediated transcription. Deletion of the TATA-element from the U6 promoter changes the transcription specificity from the $RNAP_{III}$- to the $RNAP_{II}$-system. Similarly, introduction of a TATA-element into the U2 promoter converts it from being a promoter transcribed by $RNAP_{II}$ into one transcribed

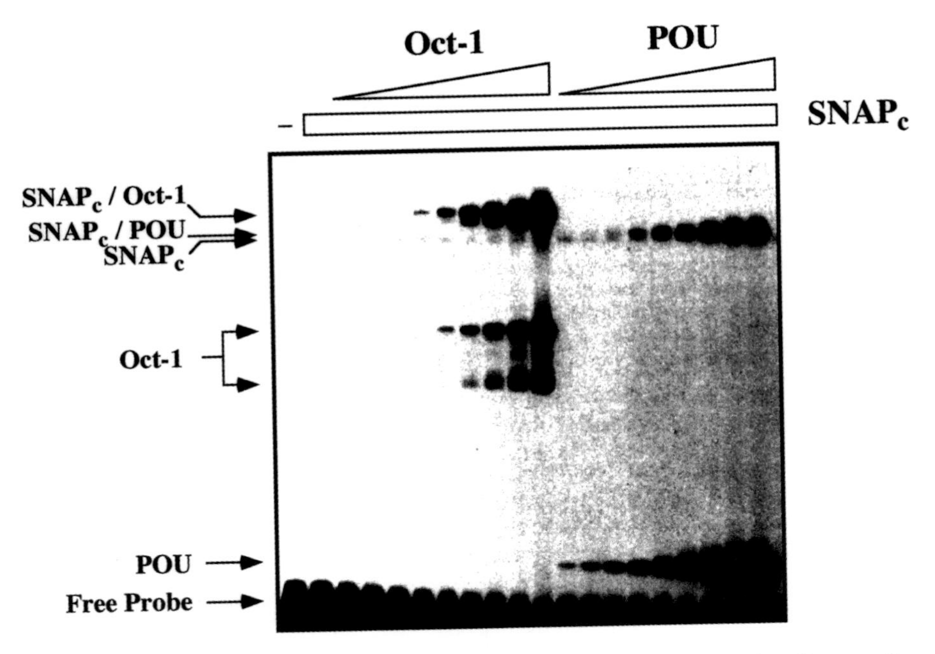

Figure 6.16. The Oct-1 DNA-Binding Domain (POU) is Sufficient for Cooperative Interactions With SNAP$_c$ on the Human U6 snRNA PSE.
An electrophoretic mobility shift assay of a DNA fragment containing the human U6 PSE with various concentrations of complete Oct-1, or the Oct-1 POU-domain, reveals that both interact effectively with the DNA-bound SNAP$_c$ complex. This experiment illustrates that the main interactions between SNAPc and Oct-1 occur through protein-protein contacts with the POU DNA-binding, rather than the RNAP$_{II}$-specific activation domains that are also present in Oct-1. From Ford and Hernandez (1997).

by RNAP$_{III}$ (see Figure 6.10; Lobo and Hernandez, 1989). It is therefore clear that a very subtle change (insertion/deletion of a 7 nucleotide TATA-element) causes a complete switch in the class-specific transcription patterns of these snRNA genes. Experiments with *in vitro* reconstituted extract fractions indicate that binding of recombinant (presumably 'uncomplexed') TBP to the U6 TATA-element is quite sufficient to direct at least basal levels of transcription of the U6 promoter by the RNAP$_{III}$ system (Lobo *et al.*, 1992, Sadowski *et al.*, 1993). It is likely, however, that in a typical *in vivo* situation only very little, if any,

TBP present in cells is not associated with TAFs. Although the general research literature is curiously silent on this issue, we will tentatively assume that the TBP required for binding to the TATA-element is probably part of the general $RNAP_{III}$ basal transcription factor TFIIIB (Figure 6.14; see also Whitehall *et al.*, 1995). The recruitment of TFIIIB to the TATA-element of the U6 gene would provide a relatively straightforward explanation why this gene is transcribed by $RNAP_{III}$ because we have already seen from the previous examples of other $RNAP_{III}$ promoters (type 1 and type 2) that this factor is directly involved in establishing specific protein-protein contacts with $RNAP_{III}$ subunits.

It is substantially less clear, however, in what manner the Oct-1/$SNAP_c$ protein complex recruits the $RNAP_{II}$ transcriptional machinery to the promoters of the U1 and U2 snRNA genes. Although it can not be excluded at this stage that $RNAP_{II}$ is directly contacted through specific contacts with certain $SNAP_c$ subunits, a much more likely scenario suggests that the DNA-bound $SNAP_c$-Oct 1 complex is recognized by $RNAP_{II}$ basal transcription factors. Once these basal factors are assembled around $SNAP_c$, $RNAP_{II}$ might then get recruited directly into this pre-initiation complex via the previously described mechanisms (Chapter 2).

Some of the counter-intuitive results obtained from studying the expression of snRNA genes, i.e. the fact that the existence of a TATA-element can actually favour transcription by the $RNAP_{III}$- rather than the $RNAP_{II}$ system, and the discovery of two distinct TBP-containing transcription factors bound to the same promoter, have already shattered several previously-held ‘texbook concepts’ that have dominated our views of eukaryotic transcription for a long time. One further surprise has come from investigating the functional domains of TBP required for transcription of the U6 snRNA gene. Sequence alignment of TBPs from species ranging from archaea to man have consistently revealed a highly conserved C-terminal domain present in all known TBPs. In higher eukaryotes this C-terminal domain is preceeded by an N-terminal domain that varies considerably in size and is not particulary well-conserved on the primary amino acid sequence level between different species (see Chapter 2; Figure 2.8). Since most TAFs and other transcription factors seem to exclusively interact with the C-terminal domain, the *raison d'etre* of this N-terminal expansion has been unclear for a long time. Studies described by Mittal and Hernandez (1997) provide, however, evidence for a functional significance of the N-terminal domain in enhancing the transcription of the U6 snRNA gene by several different processes, including mediating the binding of $SNAP_c$ to the PSE. Although it is quite possible that additional functional roles of the N-terminal TBP domains will be discovered in the future, this finding correlates very well with the evolution of the exclusive use of extragenic promoter elements of the snRNA genes in vertebrates (see below), and the specialized role of TBP in these promoters may have thus provided a strong evolutionary incentive for the expansion of the N-terminal TBP domain.

Extragenic Control Elements in $RNAP_{III}$-Type 3 Genes May Have Only Arisen During Evolution of Higher Eukaryotes

The presence of internal and extragenic promoter elements in $RNAP_{III}$-transcribed genes raises a number of fascinating questions. At the moment it is far from clear how actively transcribing $RNAPs_{III}$ deal with the situation of having to read through such protein-DNA complexes. Furthermore, from an evolutionary perspective, it would be interesting to find out whether such internal promoter elements are a 'primitive' feature that has been retained during eukaryotic evolution, or whether they are the outcome of an extreme degree of specialization that arose relatively late during evolution. Although we can not provide a definitive answer to this question at this point in time, there is some preliminary data that illuminates this particular issue. Investigations of $RNAP_{III}$-transcribed genes in yeast have revealed that all of the $RNAP_{III}$-type3 genes found in higher eukaryotes are organized as $RNAP_{III}$-type 2 genes in yeast, i.e. they lack the extragenic control elements (e.g. Hannon *et al.*, 1991; Bordonne and Guthrie, 1992). This result strongly supports the interpretation that the type-3 gene organization, with its apparently exclusive reliance on extragenic control elements, is therefore a recent evolutionary development that may have only occurred after the divergence of the metazoan and fungal lineages. It is possible that such an arrangement allows a greater dynamic range of transcription and an increased opportunity for control (Willis, 1993).

The overall functional similarities between the $RNAP_{II}$ and $RNAP_{III}$ transcriptional systems in terms of promoter structure and transcription factors is very clearly illustrated by the snRNA genes. This suggest that during the evolution of the eukaryotic transcription systems $RNAP_{II}$ and $RNAP_{III}$ have diverged later from each other than both of them have from the $RNAP_I$ system. This view is supported by other studies indicating that the type-specific $RNAP_{II}$ and $RNAP_{III}$ subunits are more closely related to each other on the primary sequence level, than either of them to $RNAP_I$ (Memet *et al.*, 1988). $RNAP_{II}$ and $RNAP_{III}$ also display a similar degree of sensitivity to α-amanitin and *Pseudomonas* tagetitoxin (Steinberg *et al.*, 1990), and there is some preliminary evidence suggesting that the basal factor TFIIA, long considered a standard component of the $RNAP_{II}$ system, may also play a stimulatory role in the transcription of $RNAP_{III}$ transcribed genes (Meissner *et al.*, 1993).

6.5. Control of $RNAP_I$ and $RNAP_{III}$ Transcriptional Machineries

Investigating the transcriptional control mechanisms regulating the activity of the $RNAP_I$ and $RNAP_{III}$ systems is not merely an academic exercise but has numerous important consequences for understanding normal cell functions, and how these are modified during the pathological conditions occuring during oncogenic and viral transformation processes. Interestingly, many of the controls in both transcription systems are similarly affected supporting the view that there are many common regulatory links.

Control of $RNAP_I$ Transcription through Controlling rRNA Gene Copy Number

The rate of rRNA transcription is generally considered to be one of the most important rate-limiting factors governing the overall rate of cell growth and division (Sollner-Webb and Tower, 1986; Sommerville, 1986; Jacob, 1995). Fast-growing cancer cells often have noticably enlarged nucleoli due to the substantially increased rate of rRNA gene transcription by $RNAP_I$ (Derenzini and Trere, 1991) and *vice versa*, cells often substantially reduce transcription of their rRNAs genes after undergoing final differentiation (Cavanaugh *et al.*, 1995). Genetic experiments in *Drosophila* with altered copy number of rRNA genes have shown that a reduction below 130 copies/haploid genome causes severe retardation in growth, and reduction of the copy number to 20 is lethal (Tartof and Hawley, 1992). Another naturally occuring phenomenon, the somatic amplification of rRNA genes in developing amphibian oocytes, also underscores the importance for cells to have a sufficiently high number of rRNA genes to sustain high growth rates (Bird, 1978).

Simian Virus 40 Large T-Antigen Stimulates Transcription by Direct Interactions with TBP-Containing Complexes

The variation of rRNA gene copy number is restricted to rather specialized circumstances and is not a universally used mechanism. All other forms of regulating the levels of rRNA exert themselves through a variety of regulatory mechanisms that target the $RNAP_I$ and $RNAP_{III}$ transcriptional machinery directly. Some of the best evidence for direct control comes form virus-infected cells. In a process that is dependent on functional viral large T antigen (Learned *et al.*, 1983), mammalian cells dramatically up-regulate their rRNA synthesis after infection with the SV40 tumor virus (up to 5-fold; May *et al.*, 1976; Pockl and Wintersburger, 1980). Large T-antigen is a versatile transcription factor that influences cell growth through targeting many transcriptional control mechanisms in all three nuclear RNAP systems (see Figure 6.17; Gruda *et al.*, 1993; Nevins, 1994; Berger *et al.*, 1996; Damania and Alwine, 1996; Johnston *et al.,* 1996; Zhai *et al.*, 1997; Damania *et al.*, 1998).

A recent study showed that large T-antigen binds to the SL1 component of the $RNAP_I$ transcription complexes through specific contacts with TBP and TAF_Is (Zhai *et al.*, 1997). This recruitment step explains the increased rate of rRNA gene transcription because the stimulatory effect can be reproduced in an *in vitro* transcription system containing highly-purified large T-antigen, UBF, SL1 and $RNAP_I$. A similar effect can also be observed to act on the $RNAP_{III}$ TBP-containing complexes, TFIIIB and $SNAP_c$. Large T-antigen appears to bind specifically to the BRF component of TFIIIB and at least to two $SNAP_c$ subunits (Damania *et al.*, 1998). In addition to the effects on TFIIIB and $SNAP_c$, increased levels of TFIIIC and a change in its phosphorylation state may account for the stimulation of transcription from type 1 and 2 $RNAP_{III}$ promoters. (White *et al.*, 1990; Sinn *et al.*, 1995).

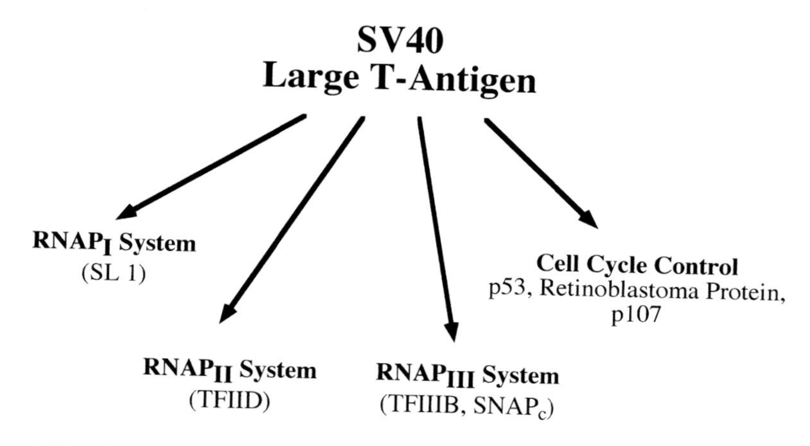

Figure 6.17. Transcriptional Control Targets of SV40 Large T Antigen.
The SV40 Large T-antigen affects important regulatory components present in all three nuclear transcriptional machineries and cell cycle control circuits.

Cell-cycle Specific Regulation of $RNAP_I$ and $RNAP_{III}$ Transcriptional Machineries by Retinoblastoma Protein

In contrast to the stimulatory effects of viral proteins, several cellular proteins with a known role in tumor suppression, such as retinoblastoma protein ('Rb') and p53, serve as general repressors of the $RNAP_I$ and $RNAP_{III}$ systems. Rb is one of the most important proteins involved in connecting cell cycle-specific events to transcriptional control mechanisms. The gene encoding Rb is often found to be mutationally inactivated in cancer cells, which leads to uncontrolled cell proliferation and cancer (reviewed in Weinberg, 1995). Rb exerts a direct inhibitory effect on the $RNAP_I$ transcription system through specific protein:protein contacts with UBF, which presumably interferes with the stimulatory role of UBF on $RNAP_I$ transcription (Shan *et al.*, 1992; Cavanaugh *et al.*, 1995). Furthermore, it has been postulated that SV 40 large T-antigen directly inhibits Rb activity, so that the increased level of $RNAP_I$-mediated transcription in virus-infected cells may ultimately be due to the combination of direct stimulation of the various TBP-containing complexes, coupled with a relieve of Rb-mediated transcriptional repression. It is thus likely that the presence of inactive Rb protein in cancer cells generally allows (among other functional consequences) the high level production of ribosomal RNAs necessary for sustaining the protein synthesis capacity required for the maintenance of the high degree of metabolic activity in such cells. Rb also displays a similar inhibitory effect on the transcription of genes encoding tRNA, 5S and other small nuclear RNAs by the $RNAP_{III}$ transcription system, hinting at a common mechanism responsible for the coordinated control of the $RNAP_{I}$ and $RNAP_{III}$ transcription systems. $RNAP_{III}$ transcription fluctuates at different stages of the cell cycle in normal cells, and is often substantially increased in cancer cells to support the increased growth rates in these cells (Singh *et al.*, 1985). *In vitro* transcription assays and nuclear run-on experiments demonstrate that retinoblastoma protein specifically represses the transcription of a number of genes transcribed by $RNAP_{III}$ through a direct interaction with TFIIIB. Rb protein contains regions that may be structurally similar to TBP and thus might effectively compete for protein-interaction sites in $RNAP_{III}$ transcription initiation complexes (White *et al.*, 1996; Larminie *et al.*, 1997). p53 has a similar repressive effect: overexpression of p53 reduces transcription of Alu- and snRNA genes under *in vitro* and *in vivo* conditions (Chesnokov *et al.*, 1996). Recent results obtained by Cairns and White (1998) suggest that p53 causes inhibition of TFIIIB activity in a manner reminscent of the effects of Rb described above.

6.6. Conclusions

The $RNAP_{I}$ and $RNAP_{III}$ transcription systems systems share many similarities with the $RNAP_{II}$ system, such as the use of of basal factors and gene-specific transcription factors to assemble specific preinitiation complexes on promoters. The presence of TBP as a subunit in various $RNAP_{I/III}$ basal factors, and the sharing of five communal RNAP subunits between the three different transcriptional machineries, strongly suggests a common evolutionary origin of all three transcription systems and may reflect a need to coordinate the synthesis of all types of RNA in the cell. The $RNAP_{I}$ transcription system consists of $RNAP_{I}$ and at least two basal factors, UBF and SL1. Both of these basal factors are required for the specific transcription of the promoters for the 5.7S/18S/28S precursor transcript from the tandemly repeated rRNA genes.

The promoters transcribed by the $RNAP_{III}$ system are highly variable in the arrangement of regulatory elements and their requirements for basal and gene-specific transcription factors. Transcription of tRNA genes depends on the presence of TFIIIC and the TBP-containing basal factor TFIIIB. TFIIIA is required in addition to TFIIIB and TFIIIC as a gene-specific transcription factor for the expression of 5S rRNA genes. The additional TBP-containing complex $SNAP_c$ directs the expression of the small nuclear RNA genes transcribed by the $RNAP_{III}$ system in conjunction with TFIIIB, TFIIIC and Oct-1. TFIIIB and SL1 are the main regulatory targets during various stages of the cell cycle and during viral infection because the expression of rRNAs represents a major growth rate-limiting factor during these processes.

References

Baker, R.E., Camier, S., Sentenac, A., and Hall, B.D. (1987). Gene size differentially affects the binding of yeast transcription factor tau to two intragenic regions. *Proc. Natl. Acad. Sci. USA* 84, 8768–8772

Bartholomew, B., Kassavetis, G.A., and Geiduschek, E.P. (1991). Two components of *Saccharomyces cerevisiae* transcription factor IIIB (TFIIIB) are stereospecifically located upstream of a tRNA gene and interact with the second-largest subunit of TFIIIC. *Mol. Cell. Biol.* 11, 5181–5189.

Bartholomew, B., Durkovich, D., Kassavetis, G.A., and Geiduschek, E.P. (1993). Orientation and topography of RNA polymerase III in transcription complexes. *Mol. Cell. Biol.* 13, 942–952.

Bateman, E., and Paule, M.R. (1986). Regulation of eukaryotic ribosomal RNA transcription by RNA polymerase modification. *Cell* 47, 445–450.

Bazett-Jones, D.P., LeBlanc, B., Herfort, M., and Moss, T. (1994). Short range DNA looping by the *Xenopus* HMG-box transcription factor, xUBF. *Science* 264, 1134–1136.

Bell, S.P., Learned, R.M., Jantzen, H.-M., and Tjian, R. (1988). Functional cooperativity between transcription factors UBF1 and SL1 mediates human ribosomal RNA synthesis. *Science* 241, 1192–1197.

Berger, L., Smith, D.B., Davidson, I., Hwang, J.-J., Fanning, E., and Wildeman, A.G. (1996). Interaction between T antigen and TEA domain of the factor TEF-1 derepresses simian virus 40 late promoter *in vitro*: identification of T-antigen domains important for transcriptional control. *J. Virol.* 70, 1203–1212.

Bird, A.P. (1978). A study of early events in ribosomal gene amplification. *Cold Spring Harbor Symp. Quant. Biol.* 42, 1179–1183.

Bond, V.C., and Wold, B. (1993). Nucleolar localization of myc transcripts. *Mol. Cell. Biol.* 13, 3221–3230.

Bordonne, R., and Guthrie, C. (1992). Human and human-yeast chimeric U6 snRNA genes identify structural elements required for expression in yeast. *Nucl. Acids Res.* 20, 479–485.

Buratowski, S., and Zhou, H. (1992). A suppressor of TBP mutations encodes an RNA polymerase III transcription factor with homology to TFIIB. *Cell* 71, 221–230.

Cairns, C., and McStay, B. (1995). HMG box 4 is the principal determinant of species-specificity in the RNA polymerase I transcription factor UBF. *Nucl. Acids Res.* 23, 4583–4590.

Cairns, C.A., and White, R.J. (1998). p53 is a general repressor of RNA polymerase III transcription. *EMBO J.* 17, 3112–3123.

Camier, S., Gabrielsen, O., Baker, R., and Sentenac, A. (1985). A split binding site for transcription factor tau on the tRNA3^{glu} gene. *EMBO J.* 4, 491–500.

Cavanaugh, A.H., Hempel, W.M., Taylor, L.J., Rogalsky, V., Todorov, G., and Rothblum, L.I. (1995). Activity of RNA polymerase I transcription factor UBF blocked by *Rb* gene product. *Nature* 374, 177–180.

Chesnokov, I., Chu, W.-M., Botchan, M.R. and Schmid, C.W. (1996). p53 inhibits RNA polymerase III-directed transcription in a promoter-dependent manner. *Mol. Cell. Biol.* 16, 7084–7088.

Clemens, K.R., Liao, X., Wolf, V., Wright, P.E., amd Gottesfeld, J.M. (1992). Definition of the binding sites of individual zinc fingers in the transcription factor IIIA-5S RNA gene complex. *Proc. Natl. Acad. Sci. USA* 89, 10822–10826.

Clemens, K.R., Wolf, V., McBryant, S.J., Zhang, P., Liao, X., Wright, P.E., and Gottesfeld, J.M. (1993). Molecular basis for specific recognition of both RNA and DNA by a zinc finger. *Science* 260, 530–533.

Clos, J., Buttgereit, D., and Grummt, I. (1986). A purified transcription factor (TIF-1B) binds to essential sequences of the mouse rDNA promoter. *Proc. Natl. Acad. Sci. USA* 83, 604–608.

Colbert, T., and Hahn, S. (1992). A yeast TFIIB-related factor involved in RNA polymerase III transcription. *Genes Dev.* 6, 1940–1949.

Colbert, T., Lee, S., Schimmack, G., and Hahn, S. (1998). Architecture of protein and DNA contacts within the TFIIIB-DNA complex. *Mol. Cell. Biol.* 18, 1682–1691.

Comai, L., Tanese, N., and Tjian, R. (1992). The TATA-binding protein and associated factors are integral components of the RNA polymerase I transcription factor, SL1. *Cell* 68, 965–976.

Comai, L., Zomerdijk, J.C., Beckmann, H., Zhou, S., Admon, A., and Tjian, R. (1994). Reconstitution of transcription factor SL1: exclusive binding of TBP by SL1 or TFIID subunits. *Science* 266, 1966–1972.

Damania, B., and Alwine, J.C. (1996). TAF-like function of S40 large T-antigen. *Semin. Virol.* 5, 349–356.

Damania, B., Mital., R., and Alwine, J.C. (1998). Simian virus 40 large T antigen interacts with human TFIIB-related factor and small nuclear RNA-activating protein complex for transcriptional activation of TATA-containing polymerase III promoters. *Mol. Cell. Biol.* 18, 1331–1338.

Das, G., Hinkley, C.S., and Herr, W. (1995). Basal promoter elements as a selective determinant of transcriptional activator function. *Nature* 374, 657–660.

Derenzini, M., and Trere, D. (1991). Importance of interphase nucleolar organizer regions in tumor pathology. *Virchows Arch. [B]* 61, 1–8.

Dover, G.A., and Flavell, R.B. (1984). Molecular co-evolution: rDNA divergence and maintenance of function. *Cell* 38, 622–623

Eberhard, D., Tora, L., Egly, J.M., and Grummt, I. (1993). A TBP-containing multiprotein complex (TIF-IB) mediates transcription specificity of murine RNA polymerase I. *Nucl. Acids Res.* 21,4180–4186.

Ford, E., and Hernandez, N. (1997). Characterization of a trimeric complex containing Oct-1, $Snap_c$, and DNA. *J. Biol. Chem.* 272, 16048–16055.

Geiduschek, E.P., and Kassavetis, G.A. (1995). Comparing transcriptional inititation by RNA polymerase I and III. *Curr. Opin. Cell Biol.* 7, 344–351.

Geiduschek, E.P., and Tochini-Valentini, G.P. (1988). Transcription by RNA polymerase III. *Ann. Rev. Biochem.* 57, 873–914.

Gerber, J.K.,Gogel, E., Berger, C., Wallisch, M., Muller, F., Grummt, I., and Grummt, F. (1997). Termination of mammalian rDNA replication: polar arrest of replication fork movement by transcription termination factot TTF-1. *Cell* 90, 559–567.

Gruda, M.C., Zablolotny, J.M., Xiao, J.H., Davidson, I., and Alwine, J.C. (1993). Transcriptional activation by simian virus 40 large T antigen: interaction with multiple components of the transcription complex. *Mol. Cell. Biol.* 13, 961–969.

Haltiner, M.M., Smale, S.T., and Tjian, R. (1986). Two distinct promoter elements in the human rRNA gene identified by linker scanning mutagenesis. *Mol. Cell. Biol.* 6, 227–235.

Hannon, G.J., Chubb, A., Maroney, P.A., Hannon, G., Altman, S., and Nielsen, T.W. (1991). Multiple *cis*-acting elements required for RNA polymerase III transcription of the gene encoding H1 RNA, the RNA component of RNAase P. *J. Biol. Chem.* 266, 22796–22799.

Henry, R.W., Sadowski, C.L., Kobayashi, R., and Hernandez, N. (1995). A TBP-TAF complex required for transcription of human snRNA genes by RNA polymerases II and III. *Nature* 374, 653–657.

Hernandez, N. (1992). Cold Spring Harbor Monographs Ser. 281–313. Cold Spring Harbor Laboratory Press, New York.

Hozak, P., Cook, P.R., Schofer, C., Mosgoller, W., and Wachtler, F. (1994). Site of transcription of ribosomal RNA and intranucleolar structure in HeLa cells. *J. Cell Sci.* 107, 639–648.

Hu, C.H., McStay, B., Jeong, S.-Y., and Reeder, R.H. (1994). xUBF, an RNA polymerase I transcription factor, binds crossover DNA with low sequence-specificity. *Mol. Cell. Biol.* 14, 2871–2882.

Huibregtse, J.M., and Engelke, D.R. (1989). Genomic footprinting of a yeast tRNA gene reveals stable complexes over the 5'-flanking region. *Mol. Cell. Biol.* 9, 3244–3252.

Jacob, S.T. (1995). Regulation of ribosomal gene transcription. *Biochem. J.* 306, 617–626.

Jackson, D.A., Hassan, A.B., Errington, R.J., and Cook, P.R. (1993). Visualization of focal sites of transcription within human nuclei. *EMBO J.* 12, 1059–1065.

Jantzen, H.-M., Admon, A., Bell, S.P., and Tjian, R. (1990). Nuclear transcription factor hUBF contains a DNA-binding motif with homology to HMG proteins. *Nature* 344, 830–836.

Jantzen, H.-M., Chow, A.M., King, D.S., and Tjian, R. (1992). Multiple domains of the RNA polymerase I activator hUBF interact with the TATA-binding protein complex hSL1 to mediate transcription. *Genes Dev.* 6, 1950–1963.

Johnston, S.D., Yu, X.-M., and Mertz, J.E. (1996). The major transcriptional transactivation domain of simian virus 40 large T antigen associates nonconcurrently with multiple components of the transcriptional preinitiation complex. *J. Virol.* 70, 1191–1202.

Karpen, G.H., Schaefer, J.E., and Laird, C.D. (1988). A *Drosophila* rRNA gene located in euchromatin is active in transcription and nucleolus formation. *Genes Dev.* 2, 1745–1763.

Kassavetis, G.A., Riggs, D.L., Negri, R., Nguyen, L.H., and Geiduschek, E.P. (1989). Transcription factor IIIB generates extended DNA interactions in RNA polymerase III transcription complexes on tRNA genes. *Mol. Cell. Biol.* 9, 2551–2566.

Kassavetis, G.A., Braun, B.R., Nguyen, L.H., and Geiduschek, E.P. (1990). *S.cerevisiae* TFIIIB is the transcription initiation factor proper of RNA polymerase III, while TFIIIA and TFIIIC are assembly factors. *Cell* 60, 235–245.

Kassavetis, G.A., Nguyen, S.T., Kobayashi, R., Kumar, A., Geiduschek, E.P., and Pisano, M. (1995). Cloning, expression, and function of TFC5, the gene encoding the B″ component of the *Saccharomyces cerevisiae* RNA polymerase III transcription factor TFIIIB. *Proc. Natl. Acad. Sci. USA* 92, 9786–9790.

Kassavetis, G., Bardeleben, C., Kumar, A., Ramirez, E., and Geiduschek, E.P. (1997). Domains of the Brf component of RNA polymerase III transcription factor IIIB (TFIIIB): functions in assembly of TFIIIB-DNA complexes and recruitment of RNA polymerase to the promoter. *Mol. Cell. Biol.* 17, 5299–5306.

Kooh, B., Brophy, B., and Jackson, S.P. (1994). Conserved functional domains of the RNA polymerase III general transcription factor BRF. *Genes Dev.* 8, 2879–2890.

Kownin, P., Iida, C.T., Brown-Shimer, S., and Paule, M.R. (1985). The ribosomal RNA promoter of *Acanthamoeba castellanii* determined by transcription in a cell-free system. *Nucl. Acids Res.* 13, 6237–6248.

L'Etoile, N.D., Fahnestock, M.L., Shen, Y., Aebersold, R., and Berk, A.J. (1994). Human transcription factor IIIC box B binding subunit. *Proc. Natl. Acad. Sci. USA* 91, 1652–1656.

Lagna, G., Kovelman, R., Sukegawa, J., and Roeder, R.G. (1994). Cloning and characterization of an evolutionarily divergent DNA-binding subunit of mammalian TFIIIC. *Mol. Cell. Biol.* 14, 3053–3064.

Langst, G., Becker, P.B., and Grummt, I. (1998). TTF-1 determines the chromatin architecture of the active rDNA promoter. *EMBO J.* 17, 3135–3145.

Larminie, C.G.C., Cairns, C.A., Mital, R., Martin, K., Kouzarides, T., Jackson, S.P., and White, R.J. (1997). Mechanistic analysis of RNA polymerase III regulation by the retinoblastoma protein. *EMBO J.* 16, 2061–2071.

Learned, R.M., Smale, S.T., Haltiner, M.M., and Tjian, R. (1983). Regulation of human ribosomal RNA transcription. *Proc. Natl. Acad. Sci. USA* 80, 3558–3562.

LeBlanc, B., Read, C., and Moss, T. (1993). Recognition of the *Xenopus* ribosomal core promoter by the transcription factor xUBF interdomain interaction. *EMBO J.* 12, 513–525.

Lee, M.S.L., Gippert, G.P., Soman, K.V., Case, D.A., and Wright, P.E. (1989). Three-dimensional solution structure of a single zinc finger DNA-binding domain. *Science* 245, 635–637

Leveillard, T., Kassavetis, G.A., and Geiduschek, E.P. (1991). *Saccharomyces cerevisiae* transcription factors IIIB and IIIC bend the DNA of a tRNA (Gln) gene. *J. Biol. Chem.* 266, 5162–5168.

Lobo, S.M., and Hernandez, N. (1989). A 7 bp mutation converts a human RNA polymerase II snRNA promoter into an RNA polymerase III promoter. *Cell* 58, 55–67.

Lobo, S.M., Tanaka, M., Sullivan, M.L., and Hernandez, N. (1992). A TBP-complex essential for transcription from TATA-less but not TATA-containing RNA polymerase III promoters is part of the TFIIIB fraction. *Cell* 71, 1029–1040.

Long, E.O., and Dawid, I.B. (1980). Repeated genes in eukaryotes. *Annu. Rev. Biochem.* 49, 727–764.

Lopez-de-Leon, A., Librizzi, M., Tuglia, K., and Willis, I. (1992). PCF4 encodes an RNA polymerase III transcription factor with homology to TFIIB. *Cell* 71, 211–220.

Maden, B.E.H., and Hughes, J.M.X. (1997). Eukaryotic ribosomal RNA: the recent excitement in the nucleotide modification problem. *Chromosoma* 105, 391–400.

Margottin, F., Dujardin, G., Gerard, M., Egly, J.-M., Huet, J., and Sentenac, A. (1991). Participation of the TATA factor in transcription of the yeast U6 gene by RNA polymerase C. *Science* 251, 424–426.

May, P., May, E., and Borde, J. (1976). Stimulation of cellular RNA synthesis in mouse kidney cell cultures infected with SV40 virus. *Exp. Cell Res.* 100, 433–439.

Meissner, W., Holland, R., Waldschmidt, R., and Seifert, K.H. (1993). Transcription factor IIA stimulates the expression of classical polIII genes. *Nucl. Acids Res.* 21, 1013–1018.

Memet, S., Saurin, W., and Sentenac, A. (1988). RNA polymerases B and C are more closely related to each other than to RNA polymerase A. *J. Biol. Chem.* 263, 10048–10051.

Miller, J., McLachlan, A.D., and Klug, A. (1985a). Repetitive zinc-binding domains in the protein transcription factor TFIIIA from *Xenopus* oocyctes. *EMBO J.* 4, 1609–1614.

Miller, K.G., Tower, J., and Sollner-Webb, B. (1985b). A complex control region of the mouse rRNA gene directs accurate initiation by RNA polymerase I. *Mol. Cell. Biol.* 5, 554–562.

Mittal., V., Cleary, M.A., Herr, W., and Hernandez, N. (1996). The Oct-1 POU-specific domain can stimulate small nuclear RNA gene transcription by stabilizing the basal transcription complex $SNAP_c$. *Mol. Cell. Biol.* 16, 1955–1965.

Mittal, V., and Hernandez, N. (1997). Role of the amino-terminal region of human TBP in U6 snRNA transcription. *Science* 275, 1136–1140.

Moss, T., and Stefanovsky, V.Y. (1995). Promotion and regulation of ribosomal transcription transcription in eukaryotes by RNA polymerase I. *Progr. Nucleic Acids Res. Mol. Biol.* 50, 25–66.

Murphy, S., Yoon, J.-B., Gerster, T., and Roeder, R.G. (1992). Oct-1 and Oct-2 potentiate functional interactions of a transcription factor with the proximal sequence element of small nuclear RNA genes. *Mol. Cell. Biol.* 12, 3247–3261.

Nevins, J.R. (1994). Cell cycle targets of the DNA tumor viruses. *Curr. Opin. Genet. Dev.* 4, 130–134.

Nikolov, D.B., Chen, H., Halay, E.D., Usheva, A.A., Hisatake, K., Lee, D.K., Roeder, R.G., and Burley, S.K. (1995). Crystal structure of a TFIIB-TBP-TATA-element ternary complex. *Nature* 377, 119–128.

Paule, M.R. (1994). Transcription of ribosomal RNA by eukaryotic RNA polymerase I. In Conaway, R.C. and Conaway, J.W. (Eds) 'Transcription: mechanisms and regulation'. Raven Press, New York.

Pavletich, N.P., and Pabo, C.O. (1991). Zinc finger-DNA recognition: crystal structure of a Zif268-DNA complex at 2.1Å. *Science* 252, 809–817.

Pieler, T., and Theunissen, O. (1993). TFIIIA: nine fingers - three hands? *Trends Biochem. Sci.* 18, 226–230.

Pockl, E., and Wintersburger, E. (1980). Increased rate of RNA synthesis: early reaction of primary mouse kidney cells to infection with polyoma virus or simian virus 40. *J. Virol.* 35, 8–19.

Radebaugh, C.A., Matthews, J.L., Geiss, G.K., Liu, F., Wong, J., Bateman, E., Camier, S., Sentenac, A., and Paule, M.R. (1994). TATA box-binding protein (TBP) is a constituent of the polymerase I-specific transcription initiation factor TIF-IB (SL1) bound to the rRNA promoter and shows differential sensitivity to TBP-directed reagents in polymerase I, II, and III transcription factors. *Mol. Cell. Biol.* 14, 597–605

Read, C., Larose, A.M., LeBlanc, B., Bannister, A.J., Firek, S., Smith, D.R., and Moss, T. (1992). High resolution studies of the *Xenopus laevis* ribosomal gene promoter *in vivo* and *in vitro*. *J. Biol. Chem.* 267, 10961–10967.

Read, C.M., Cary, P.D., Crane-Robinsonm C., Driscoll, P.C., and Norman, D.G. (1993). Solution structure of a DNA-binding domain from HMG1. *Nucl. Acids Res.* 21, 3427–3437.

Roussel P. , Andre, C., Comai, L., and Hernandez-Verdun, D. (1996). The rDNA transcription machinery is assembled during mitosis in active NORs and absent in inactive NORs. *J. Cell Biol.* 133, 235–246.

Rudloff, U., Eberhard, D., Tora, U., Stunnenberg, H., and Grummt, I. (1994). TBP-associated factors interact with DNA and govern species-specificity of RNA polymerase I transcription. *EMBO J.* 13, 2611–2616.

Sadowski, C.L., Henry, R.W., Lobo, S., and Hernandez, N. (1993). Targeting TBP to a non-TATA box *cis*-regulatory element: a TBP-containing complex activates transcription from snRNA promoters through PSE. *Genes Dev.* 7, 1535–1548.

Shan, B., et al. (1992). Molecular cloning of cellular genes encoding retinoblastoma-associated proteins: identification of a gene with properties of the transcription factors E2F. *Mol. Cell. Biol.* 12, 5620–5631.

Scheer, U., and Weisenberger, D. (1994). The nucleolus. *Curr. Opin. Cell Biol.* 6, 354–359.

Schnapp, A., and Grummt, I. (1991). Transcription complex formation at the mouse rDNA promoter involves the stepwise association of four transcription factors and RNA polymerase I. *J. Biol. Chem.* 266, 24588–24595.

Schnapp, A., Schnapp, G., Erny, B., and Grummt, I. (1993). Function of the growth-regulated transcription initiation factor TIF-1A in initiation complex formation at the murine ribosomal gene promoter. *Mol. Cell. Biol.* 13, 6723–6732.

Schnapp, G., Schnapp, A., Rosenbauer, H. , and Grummt, I. (1994). TIF-IC, a factor involved in both transcription in:tiation and elongation of RNA polymerase I. *EMBO J.* 13, 4028–4035.

Schultz, P., Marzouki, N., Marck, C., Ruet, A., Oudet, P., and Sentenac, A. (1989). The two DNA-binding domains of yeast transcription factor τ as observed by scanning transmission electron microscopy. *EMBO J.* 8, 3815–3824.

Segall, J., Matsui, T., and Roeder, R.G. (1980). Multiple factors are required for the accurate transcription of purified genes by RNA polymerase III. *J. Biol. Chem.* 255, 11986–11991.

Shen, Y., Kassavetis, G.A., Bryant, G.O., and Berk, A.J. (1998). Polymerase (Pol) III TATA box-binding protein (TBP)-associated factor Brf binds to a surface on TBP also required for activated pol II transcription. *Mol. Cell. Biol.* 18, 1692–1700.

Singh, K., Carey, M., Saragosti, S., and Botchan, M. (1985). Expression of enhanced levels of small RNA polymerase III transcripts encoded by the *B2* repeats in simian virus 40-transformed mouse cells. *Nature* 314, 553–556.

Sinn, E., Wang, Z., Kovelmann, R., and Roeder, R.G. (1995). Cloning and characterization of a TFIIIC2 subunit (TFIIICβ) whose presence correlates with activation of RNA polymerase III-mediated transcription by adenovirus E1A expression and serum factors. *Genes Dev.* 9, 675–685.

Sollner-Webb, B., and Tower, J. (1986). Transcription of cloned eukaryotic ribosomal RNA genes. *Annu. Rev. Biochem.* 55, 801–830.

Sommerville, J. (1986). Nucleolar structure and ribosome biogenesis. *Trends Biochem. Sci.* 11, 438–442.

Steinberg, T.H., Mathews, D.E., Durbin, R.D., and Burgess, R.R. (1990). Targetitoxin: a new inhibitor of eukaryotic transcription by RNA polymerase III. *J. Biol. Chem.* 265, 499–505.

Tartoff, K.D., and Hawley, R.S. (1992). In Lindsey, D.l., and Zimm, G.G. (Eds.) 'The Genome of *Drosophila melanogaster*'. Academic Press, London.

Teichmann, M., Dieci, G., Huet, J., Ruth, J., Sentenac, A., and Seifart, K.H. (1997). Functional interchangability of TFIIIB components from yeast and human cells *in vitro*. *EMBO J.* 16, 4708–4716.

Tower, J., and Sollner-Webb, B. (1987). Transcription of mouse rDNA is regulated by an activated subform of RNA polymerase I. *Cell* 50, 873–883.

Tyler, B.M., and Giles, B.H. (1985). Structure of a *Neurospora* RNA polymerase I promoter defined by transcription *in vitro* with homologus extracts. *Nucl. Acids Res.* 13, 4311–4332.

Wallace, H., and Birnstiel, M.L. (1966). Ribosomal cistrons and the nucleolar organizer. *Biochem. Biophys. Acta* 114, 296–310.

Wang, Z., and Roeder, R.G. (1995). Structure and function of a human transcription factor TFIIIB subunit that is evolutionarily conserved and contains both TFIIB and high-mobility-group protein 2-related domains. *Proc. Natl. Acad. Sci. USA* 92, 7026–7030.

Weinberg, R.A. (1995). The retinoblastoma protein and cell cycle control. *Cell* 81, 323–330.

Weir, H.M., Kraulis, P.J., Hill, C.S., Raine, A.R.C., Laue, E.D., and Thomas, J.O. (1993). Structure of the HMG box motif in the B domain of HMG1. *EMBO J.* 12, 1311–1319.

Werner, M., Chaussivert, N., Willis, I.M., and Sentenac, A. (1993). Interaction between a complex of RNA polymerase III subunits and the 70 kDa component of TFIIIB. *J. Biol. Chem.* 268, 20721–20724.

White, R., Stott, D., and Rigby, P.J. (1990). Regulation of RNA polymerase III transcription in response to simian virus 40 transformation. *EMBO J.* 9, 3713–3721.

White, R.J., Khoo, B.C-E., Inostroza, J.A., Reinberg, D., and Jackson, S.P. (1994). Differential regulation of RNA polymerase I, II, and III by the TBP-binding repressor Dr1. *Science* 266, 448–450.

White, R.J., Trouche, D., Martin, K., Jackson, S.P., and Kouzarides, T. (1996). Repression of RNA polymerase III transcription by the retinoblastoma protein. *Nature* 382, 88–90.

Whitehall, S.K., Kassavetis, G.A., and Geiduschek, E.P. (1995). The symmetry of the yeast U6 RNA gene's TATA box and the orientation of the TATA-binding protein in yeast TFIIIB. *Genes Dev.* 9, 2974–2985.

Willis, I.M. (1993). RNA Polymerase III. Genes, factors and transcriptional specificity. *Eur. J. Biochem.* 212, 1–11.

Windle, J.B., and Sollner-Webb, B. (1986). Upstream domains of the *Xenopus laevis* rDNA promoter are revealed in microinjected oocytes. *Mol. Cell. Biol.* 6, 1228–1234.

Wong, M.W., Henry, R.W., Ma, B., Kobayashi, R., Klages, N., Matthias, P., Strubin, M., and Hernandez, N. (1998). The large subunit of basal transcription factor $SNAP_c$ is a Myb domain protein that interacts with Oct-1. *Mol. Cell. Biol.* 18, 368–377.

Yamamoto, R.T., Nogi, Y., Dodd, J.A., and Nomura, M. (1996). *RRN3* gene of *Saccharomyces cerevisiae* encodes an essential RNA polymerase I transcription factor which interacts with the polymerase independently of DNA template. *EMBO J.* 15, 3964–3973.

Yoon, J.-B., Murphy, S., Bai, L., Wang, Z., and Roeder, R.G. (1995). Proximal sequence element-binding transcription factor (PTF) is a multisubunit complex required for transcription of both RNA polymerase II- and RNA polymerase III-dependent small nuclear RNA genes. *Mol. Cell. Biol.* 15, 2019–2027.

Zhai, W., Tuan, J., and Comai, L. (1997). SV40 large T antigen binds to the TBP-TAF_I complex SL1 and coactivates ribosomal RNA transcription. *Genes Dev.* 11, 1605–1617.

Zomerdijk, J.C., Beckmann, H., Comai, L., and Tjian, R. (1994). Assembly of transcriptionally active RNA polymerase I initiation factor SL1 from recombinant subunits. *Science* 266, 2015–2018.

Zomerdijk, J.C.B.M., and Tjian, R. (1998). Structure and assembly of human selectivity factors SL1. In 'Transcription of Eukaryotic Ribosomal RNA Genes by RNA Polymerase I' (Ed. Paule, M.R).R.G. Landes & Co, Austin, Texas.

Chapter 7

Chromatin

Up to this stage we have mainly considered the function of gene-specific and basal transcription factors as if DNA was freely accesssible to them in the nucleus. The DNA of eukaryotic and archaeal cells is, however, packaged into chromatin structures that have a substantial influence on the expression of individual genes and even whole chromosomes. Chromatin is the natural 'habitat' of transcription factors, and many aspects of activator-, coactivator- and basal transcription factor functions can only be appreciated in context of nucleosomal- and higher order structures. The struggle between keeping the long DNA molecules packaged tightly, but also leaving active regions accessible to the enzymatic machinery of the cell (for transcription, replication, recombination and repair processes) necessitates a tight regulatory link between the expression of individual genes and overall chromatin structure. Here we will explore several fundamental aspects of this dynamic process, such as the structure and function of nucleosomes, the factors involved in establishing active chromatin environments, and the maintenance of stable chromatin configurations on the chromosomal level. We will then continue this line of investigation in Chapter 8 to explore the effects of higher order chromatin arrangement in the context of eukaryotic nuclear architecture. This chapter was primarily designed to provide an overview of some of the most salient features that help us to understand gene expression mechanisms more clearly. There are two recently published treaties on chromatin structure and function to which the interested reader is referred for more detailed information (Wolf, 1995; van Driel and Otte, 1997).

7.1. The Evolutionary Origin of Chromatin

DNA Packaging in Bacteria and Archaea

Histones and chromatin have historically always been regarded as a typical eukaryotic feature. Although there is evidence for some form of DNA packaging and organization in bacteria, it is clearly not based on regular nucleosome-like packaging (Zillig *et al.*, 1988; Pettijohn, 1988). Two of the proteins that have long been considered to be most 'histone-like' in their function, HU and H-NS, share no significant primary sequence homology with eukaryotic histones (Drlica and Rouviere-Yaniv, 1987; Schmidt, 1990), and recent structural studies have confirmed the lack of homology on the protein fold level (Figure 7.1; Shindo *et al.*, 1995; Vis *et al.*, 1995). Furthermore, the HU and H-NS proteins are present in relatively low quantities in bacterial cells and would therefore only be sufficient for packing a relatively small percentage of a typical bacterial genomes into higher order structures. Recent research

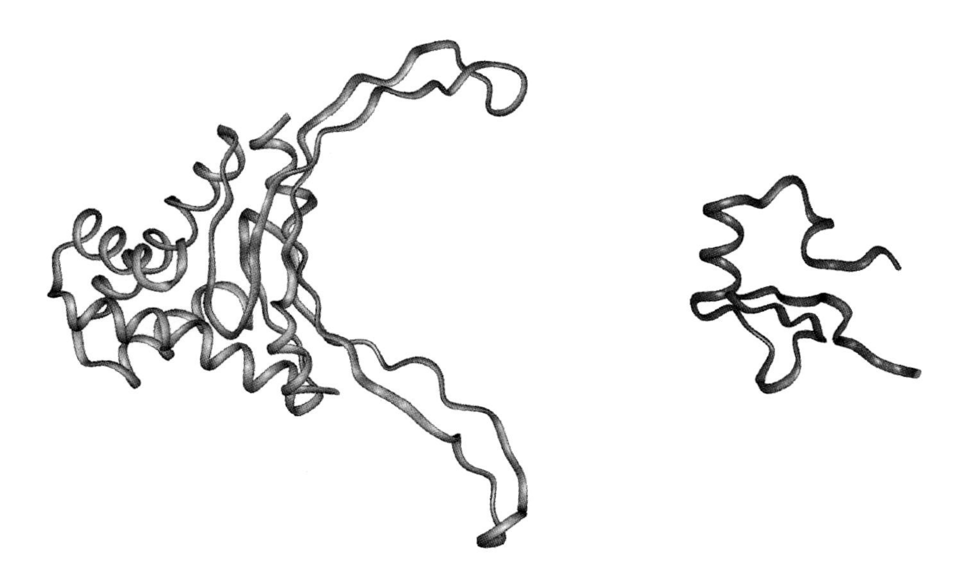

Figure 7.1. Bacterial DNA Packaging Proteins HU and H-NS are Structurally Distinct from Histones.
The solution structures HU (left; PDB Access No. 1HUE), and the DNA-binding motif of H-NS (right; PDB Access No. 1HNR) are shown. Neither of them contain a histone fold.

indicates that these proteins are likely to be involved in setting up the architecture of a subset of bacterial promoters by inducing DNA-bending at specific posititons. Clearly, HU and H-NS do not carry out any functions that are directly comparable to eukaryotic histones.

It came therefore as a considerable surprise when evidence for chromatin-like packaging of DNA emerged from studies of the other prokaryotic life domain, the archaea (see Chapters 1 and 2 for a more detailed discussion of the evolutionary significance of archaea and their transcriptional machineries). Figure 7.2. shows an electron micrograph of chromatin extracted from the extremely halophilic ('salt loving') archaeon *Halobacterium salinarum,* illustrating the presence of protease-sensitive 'beads-on-a-string' structures that are strongly reminiscent of the electron microscopic appearance of nucleosomally-packaged DNA found in eukaryotic chromatin (Takayanagi *et al.,* 1992). Other independent biochemical studies led to the isolation and characterization of basic histone-like proteins from several additional archaeal species (Sandman *et al.*, 1990 and 1994; Tabassum *et al.*, 1992; Darcy *et al.*,

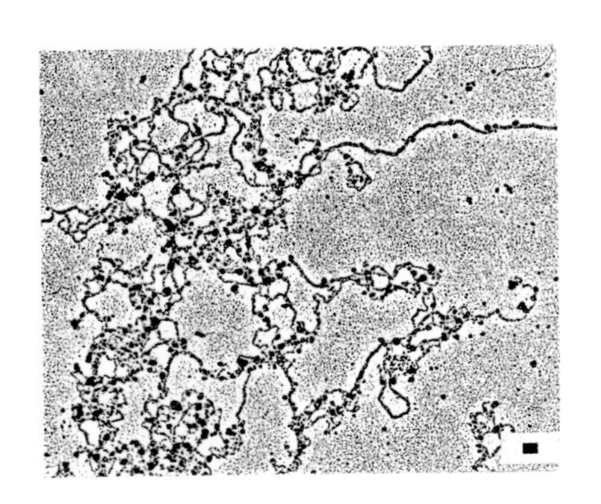

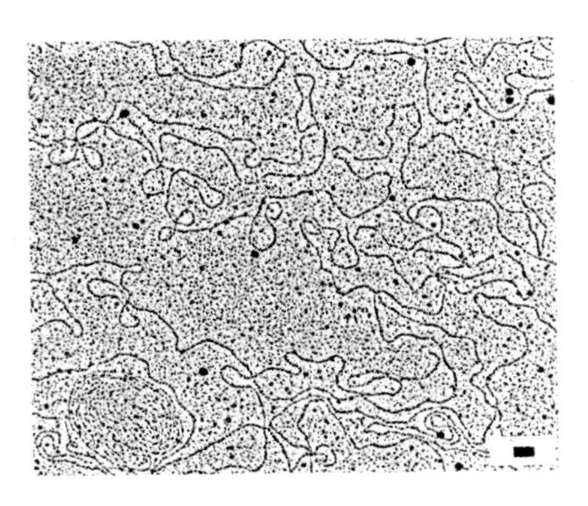

- Proteinase K **+ Proteinase K**

Figure 7.2. Archaeal Genomic DNA is Organized in Chromatin-like Structures. Chromosomal preparations from the archaeon *Halobacterium salinarium* in the late exponential growth phase were examined by electron microscopy. The observed fibres (left) are composed of DNA and associated proteins. Treatment with proteinase K leaves only the DNA strands intact (right). Data from Takayanagi *et al.*, 1992.

1995), and structural studies have convincingly demonstrated that archaeal histones do indeed contain a *bona fide* histone folding motif (Figure 7.3; Starich *et al.*, 1996; see also below). From the complete genomic sequence of *Methanococcus jannaschii* we now know that there are no less than five (!) distinct histone-like genes in this particular archaeon (Bult *et al.*, 1996). The existence of *bona fide* histones and chromatin structures in archaea indicates that the evolutionary origin of ancestral histones almost certainly predates the split between archaeal and eukaryotic lineages, which is commonly thought to have occured approximately 2–3 billion years ago. The symmetrical nature of the canonical histone fold allows the further suggestion that primitive histones arose from a gene duplication of a simple helix-strand-helix motif (Arents and Moudrianakis, 1995).

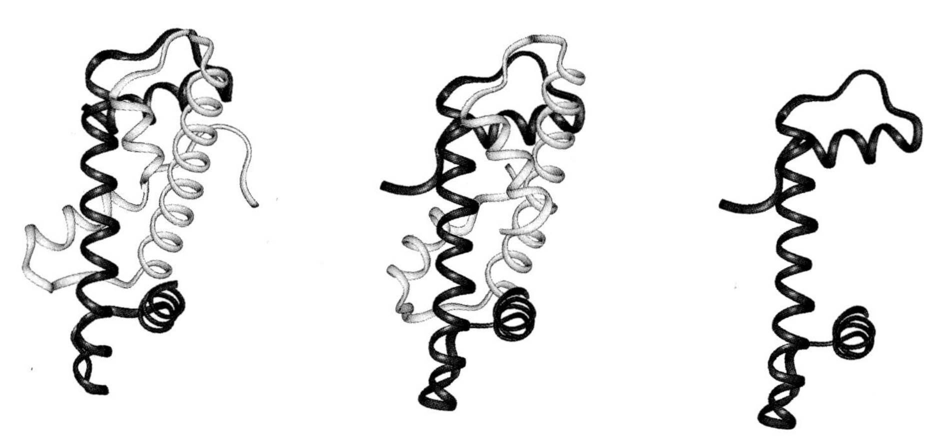

Figure 7.3. 'Histone' Fold.
The histone folds present in the *Drosophila* TFIID complex subunits $TAF_{II}40$ (left) and $TAF_{II}60$ (middle) are shown and compared to an archaeal histone (right). Each of these proteins contains a bona fide histone fold. PDB Access Nos. 1BFM (archaeal histone) and 1TAF ($TAF_{II}40$ and $TAF_{II}60$).

Thermal Protection of DNA as *the* Major Driving Force in the Evolution of Chromatin?

From the universal presence of histones in eukaryotes and archaea it is reasonable to conclude that histone-like proteins were already present in common ancestral cells living several billion years ago, i.e. close to the origin of life on earth. At present, many archaeal species live in specialized environmental niches such as hot sulphur springs and vulcanic thermal vents on the ocean floors. These conditions are thought to be similar to the ones prevailing on Earth 4 billion years ago. It is therefore tempting to conclude from such observations that one of the major driving forces leading to the evolution of histones and nucleosomes might, at least initially, have been the necessity to protect the DNA from such inhospitable conditions. Hyperthermophilic archaea, such as *Methanotermus*, grow optimally at 83°C in the laboratory. Their DNA is rich in A/T nucleotides (67%), and is therefore easily heat-denatured at the temperatures at which these archaea normally grow best. So how do these cells manage to prevent such a disastrous event from occuring? The double-stranded state of *Methanotermus* DNA is at least partially stabilized *in vivo* by the high concentration of intracellular monovalent ions (approximately 950 mM K^+!) and by the presence of high concentrations of metabolites, such as cyclic diphosphoglycerate (Hensel and Konig, 1988; Stroup and Reeve, 1992). In addition, it can be demonstrated that the histones from *Methanothermus* play a major role in preventing heat-denaturation of double-stranded DNA. Under *in vitro* conditions the melting temperature of linear DNA in low ionic-strength buffer rises from 60.5°C to 84.6°C when the DNA is complexed stoichiometrically with archaeal histones (Sandman *et al.*, 1990).

Thus protection of DNA from environmental conditions may have been the primary driving force for the evolution of of histones. This was probably an important event in the evolutionary history of life on earth, which may have subsequently allowed the formation and propagation of stable DNA genomes and, eventually, underpinned the huge expansion eukaryotic genome sizes. Obviously most eukaryotes nowadays live in environments where thermal denaturation of their genomic DNA is no further an issue. Like many other known instances in evolution, it looks as if a feature developed for a particular reason was eventually subverted for an entirely different purpose. Changes in chromatin structure and conformation are now important elements directing the controlled expression of genes in eukaryotic cells and chromatin allowed the evolution of complex genomes through its ability to compact long DNA molecules.

7.2. Nucleosome Structure and Function — The Basics

The Nucleosome Core is the Fundamental Building Block of Chromatin

The discovery of nucleosomes as a fundamental structural element for packaging DNA into ordered arrays was a historic breakthrough that initiated much of the contemporary research aimed at defining the structure and function of chromatin (Kornberg and Thomas, 1974; Kornberg, 1974). In all eukaryotes nucleosomes are composed of four different 'core' histones, H2A, H2B, H3 and H4. Histones are unusual proteins containing many basic amino acids (thought to be required for ionic interactions with the acidic phosphate-pentose backbone of DNA), and flexible N-terminal domains that lack a defined three-dimensional structure in solution. The core histone folding motif is conserved across the archaeal/eukaryotic boundary (see above), and the overall amino acid sequences are highly similar across the entire eukaryotic range indicating the functional significance of such a structural features (Figure 7.4). This high degree of structural conservation is also reflected on the functional level: yeast and human core nucleosomes bind exactly the same amount of double-stranded DNA (146 base pairs).

The elucidation of the arrangement of the four different histone subunits within the nucleosome particle, and of the path that DNA takes through each nucleosome, has been subject of a continuous research effort ever since the existence of nucleosomes became known. The investigations started off with low resolution structures obtained by electron microscopy (Klug *et al.*, 1980)

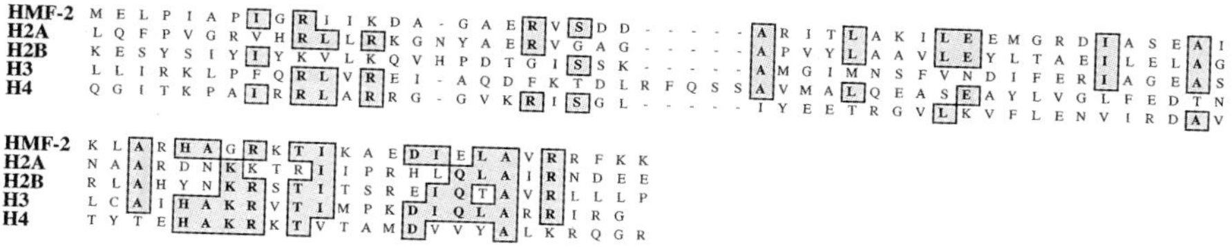

```
HMF-2  M E L P I A P I G R I I K D A - G A E R V S D D - - - - - A R I T L A K I L E E M G R D I A S E A I
H2A    L Q F P V G R V H R L L R K G N Y A E R V G A G - - - - - A P V Y L A A V L E Y L T A E I L E L A G
H2B    K E S Y S I Y I Y K V L K Q V H P D T G I S S K - - - - - A M G I M N S F V N D I F E R I A G E A S
H3     L L I R K L P F Q R L V R E I - A Q D F K T D L R F Q S S A V M A L Q E A S E A Y L V G L F E D T N
H4     Q G I T K P A I R R L A R R G - G V K R I S G L - - - - - I Y E E T R G V L K V F L E N V I R D A V

HMF-2  K L A R H A G R K T I K A E D I E L A V R R F K K
H2A    N A A R D N K K T R I I P R H L Q L A I R N D E E
H2B    R L A H Y N K R S T I T S R E I Q T A V R L L L P
H3     L C A I H A K R V T I M P K D I Q L A R R I R G
H4     T Y T E H A K R K T V T A M D V V Y A L K R Q G R
```

Figure 7.4. Primary Amino Acid Sequence Conservation of Archaeal Histone and the Four Eukaryotic Core Histones.

The sequence alignment of an archaeal histone (HMF-2) with the central domains of the four eukaryotic histones H2A, H2B, H3 and H4 illustrates the conservation of key amino acid residues, especially towards the C-terminal end.

and recently culminated in successively higher resolution structures obtained by X-ray crystallographic studies (Richmond *et al.*, 1984; Arents *et al.*, 1991; Luger *et al.*, 1997). The structures have yielded a wealth of detailed information that help us to understand many fundamental aspects of nucleosome function at the atomic level (Figure 7.5).

From biochemical *in vitro* studies with purified histones it became evident that assembly of each nucleosome particle starts with the formation of two

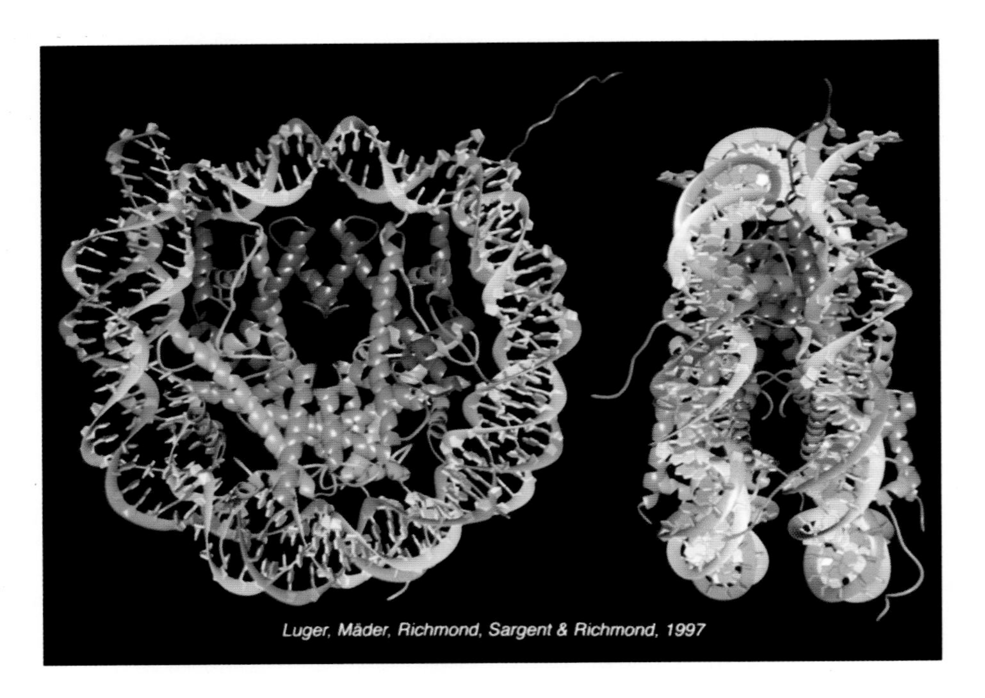

Figure 7.5. High-Resolution Structure of Eukaryotic Nucleosome Core.
Frontal and side view of a nucleosomal core particle with asociated DNA. The four different types of histones present in the nucleosomal core structure are color-coded (blue = histone H3; green = histone H4; red = histone H2B; yellow = histone H2A). Note the presence of the distinct histone folds and the similarity to those shown in Figure 7.3. The helical pitch of DNA varies considerably at different positions of the nucleosome. The majority of the highly flexible N-terminal domains of the various histones protruding from different positions of the nucleosome particle are not 'visible' in this X-ray crystallographic structure because they do not take up a defined tertiary conformation. From Luger *et al.*, 1997.

copies of a H3-H4 complex that bind approximately 120 bp of DNA. This $DNA_{120bp}/(H3\text{-}H4)_2$ complex then recruits two copies of the pre-assembled H2A-H2B dimer, which allows a further extension of specific contacts between the nucleosome particle and DNA to increase the amount of packaged DNA to 146 base pairs. We thus end up with a DNA/protein complex with an overall stoichiometry of: $DNA_{146bp}/(H3\text{-}H4)_2/(H2A\text{-}H2B)_2$. The DNA is wound around each nucleosome core particle for approximately 1.65 turns in a pathway that is under the direct guídance of individual histone pairs at the various locations. The most recent high-resolution X-ray crystallographic studies by Luger *et al.* (1997) identified five distinct types of DNA-protein interactions that anchor DNA specifically to the nucleosome surface:

- Spatial fixation of single phosphate groups from the DNA phosphate-sugar backbone through a positive charge created by histone helix dipoles.
- Specific hydrogen-bonding of phosphates through main chain amide nitrogen atoms of amino acids located at or near the turns present in the histone fold.
- Intercalation of arginine side chains into the minor groove of DNA to contact phosphates across the groove.
- Nonpolar contacts between the nucleosome surface and deoxyribose sugar moieties.
- Various additional hydrogen bonds and ionic interactions between nucleosome surface and the DNA phosphate groups from both major and minor grooves.

The overall path of DNA around the nucleosome core is far from even (Figure 7.5). Many of the energetically strong DNA:histone interactions cause a substantial distortion of the standard double-helical DNA conformation, including formation of numerous bulges and buckles in the phosphate-sugar backbone, and a variation in the overall twist between 9.4 and 10.9 base pairs per turn (the normal twist for 'free' DNA is 10.5). These features have important consequences because they help us to understand several important aspects of the effects of nucleosomes on the regulation of gene expression (Luger *et al.*, 1997). First, the distortions that nucleosome-bound double-helical DNA undergoes can be either facilitated or made substantially more difficult by the different nucleotide sequences present. This point is best exemplified by the observation that the penetration of arginine side chains into the minor groove of double helical DNA at regular 10 nucleotide distances (see above) is structurally greatly facilitated by having A-T basepairs in these positions. The regular occurance of A-T base pairs at ten nucleotide intervals in a DNA sequence greatly stabilizes the winding of such a DNA molecule around a nucleosome and thus lead to a preferred placement of a nucleosome onto a predetermined position along a stretch of DNA. It is therefore possible for a gene to evolve specific nucleosome 'phasing' sequences to ensure that nucleosomes are placed in precise positions relative to functionally important regulatory sequence motifs or coding regions. Further below we will see that this strategy is used in eukaryotic organisms, and positioned nucleosomes are known to play a significant role in controlling transcription on several promoters. Conversely, certain other nucleotide sequences can be so unfavourable for packaging into nucleosomes that they are likely to remain as freely exposed DNA in chromatin. Second, the distortions that nucleosome-bound DNA undergoes can completely prevent gene-specific transcription factors from recognizing their target seqence, even if it is present in an exposed form on the nucleosome surface. As is evident from several of the examples of specific protein-DNA complexes shown in Chapter 3, the contacts between individual amino acid side chains present in DNA-binding domains of gene-specific transcription factors are spatially precise and act over very short distances. The presence of nucleosome-induced bulges and kinks can sterically prevent many of the specific contacts between nucleotides and DNA-binding domains from occurring. Finally, the large number of specific contacts of the core histones along the bound DNA molecule makes spontanous dissociation of a nucleosome from DNA an extremely unlikely event due to cooperativity effects. It is therefore in most instances energetically unfavourable for a transcription factor to destabilize nucleosome-packaged DNA sufficiently to create a nucleosome-free 'patch' on a promoter for setting up high levels of gene-specific transcription. We will see below that this problem is overcome by the presence of chromatin-remodelling complexes that either actively remove nucleosome from DNA templates, or mobilize them so that they shift their positions relatively to the underlying DNA sequences. These chromatin-remodelling complexes need to use energy generated from ATP hydrolysis to accomplish this task.

Nucleosome Core Properties are Altered by Post-Translational Modifications

Apart from the histone fold-containing central domains which allow the formation of a stable nucleosome particle, all four histones contain N- and C-terminal 'tails' that are mostly freely exposed, unstructured, and located on the periphery of the nucleosome particle (and usually not 'visible' in X-ray crystallographic studies of nucleosome core particles because of their flexibility). Genetic studies carried out in yeast show that different genes have different requirements for the histone N-termini. The presence of an intact H2A N-terminus is essential for the correct transcription of the *SUC2* gene (Hirschhorn *et al.*, 1995), and intact H3/H4 N-termini are required for the regulated expression of several other genes (including *GAL1*, *GAL7*, *GAL10*, *PHO5*, and *HM*; Grunstein *et al.*, 1995), and nucleosome assembly (Ling *et al.*, 1996). The gene-specific phenotypes that mutations in particular histone N-termini impart on yeast cells thus provide a strong hint that histones affect the expression of genes in different manners. Many, if not all, of these effects are exerted by differential post-translational modifications of specific N-terminal amino acid residues by acetylation, phosphorylation, methylation, poly ADP-ribosylation and ubiquitinilation (Figure 7.6; reviewed in Hansen, 1997). We do not currently understand the functional significance of many of these modifications, except the effects of histone acetylation. The use of antibodies that specifically bind to acetylated histones, but not to unmodified ones, show that there is a positive correlation between the presence of acetylated nucleosomes and actively transcribed genes in chromatin. In the majority of cases (but not all!) acetyled nucleosomes are highly indicative of transcribed regions of chromatin, whereas transcriptionally-silent chromatin frequently contains histones with only low levels of acetylation (Pogo *et al.*, 1966; Braunstein *et al.*, 1993; Clark *et al.*, 1993; Fletcher and Hansen, 1996). We have already seen in Chapter 4 that several of the coactivators required for gene-specific activators to exert their stimulatory functions contain acetyl-transferase enzymatic activities with a high degree of substrate-specificity for histones (summarized in Figure 7.7; reviewed in Wade *et al.*, 1997). The yeast GCN5 histone acetyltransferase, and its homolog from *Tetrahymena*, provided the first clues for a molecular link between acetylated histones, histone

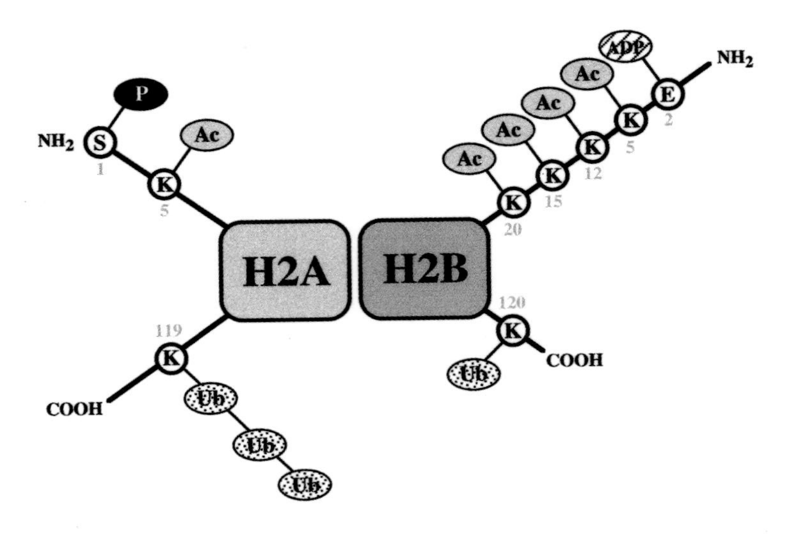

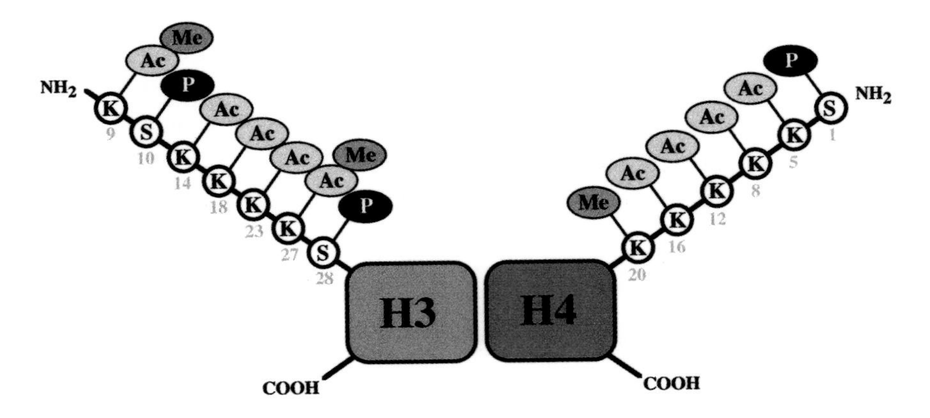

Figure 7.6. Post-Translational Modifications of Histone 'Tails'.
All four types of core histones contain flexible N-termini that are mostly outside the nucleosome core structure and are available for post-translational modification at specific residues by acetylation ('Ac'), phosphorylation ('P'), methylation ('Me') or ADP-ribosylation ('ADP'). Two lysines residues (K) located near the C-termini of histones H2A and H2B are also subject to ubiquitinilation ('Ub'). The modifications are thought to have a substantial influence on nucleosome/chromatin structure and function.

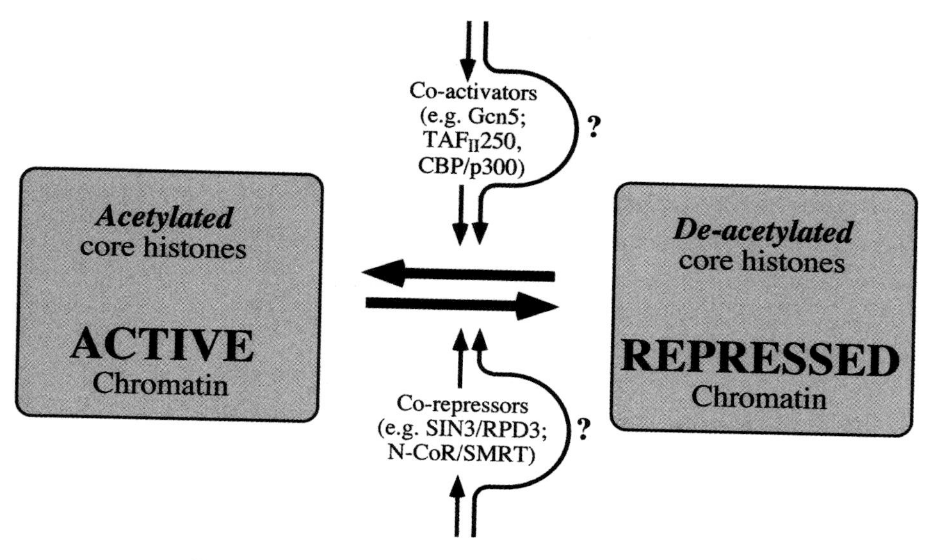

Figure 7.7. Specific Acetylation/De-acetylation of Core Histones in Response to Gene-Specific Transcription Factors.
Several coactivators contain histone acetylase activities that may be used to acetylate nucleosomes during the creation of transcriptionally active promoters. Co-repressors probably exert part of their functions through reversing these processes.

acetyltransferase type A ('HAT-A') activity, and transcriptional activation (Brownell *et al.*, 1996). GCN5 is one of the components of the yeast ADA2-ADA3 coactivator complex that is required for transcriptional stimulation by certain gene-specific activators containing 'acidic' activation domains (Berger *et al.*, 1992). Mutations in GCN5 that eliminate its enzymatic activities interfere with transcriptional activation *in vivo*, and thus provide supporting evidence for the importance of the link between acetylation and gene expression (Candau *et al.*, 1997; Wang *et al.*, 1997; Zhang *et al.*, 1998). Similar acetyltransferase activities have been found in other types of transcriptional coactivators, such as the TFIID subunit $TAF_{II}250$ (Mizzen *et al.*, 1996), p300/CBP (Bannister and Kouzarides, 1996; Ogryzko *et al.*, 1996) and the p300/CBP-associated factor P/CAF (Yang *et al.*, 1996). P/CAF is the human homolog of the yeast GCN5 acetylases we encountered above. In addition to these general coactivators, two more highly-specialized coactivators, SRC1 and ACTR, that interact with ligand-bound nuclear steroid receptors also contain HAT activity. SRC1 and ACTR are known to interact with P/CAF and CBP/p300, suggesting that some gene-specific transcription activators may have the ability to recruit a diverse range of distinct acetylase-containing coactivators to their target promoters (Chen *et al.*, 1997; Spencer *et al.*, 1997).

A simple model based on these findings proposes that any direct (or possibly indirect) interaction of gene-specific activators with these acetyltransferase-containing coactivators leads to the localized acetylation of nearby histones, 'High-Mobility-Group' ('HMG') proteins (Sterner *et al.*, 1981; Elton *et al.*, 1986), and possibly also of the basal transcription factors TFIIE and TFIIF (Imhof *et al.*, 1997). What is the overall effect of acetylation? The transfer of acetyl groups to lysine residues on the N-terminal tails of histones (especially H4) leads to a reduction in the overall positive charge density of the modified nucleosome, which may account at least partially for the increased ability of gene-specific transcription factors to bind to nucleosomal DNA (Turner, 1991; Wolffe and Pruss, 1996). In many cases the resulting conformational changes are probably relatively minor, but extremely high levels of acetylation have been reported to cause the complete disintegration of the nucleosomal core (Oliva *et al.*, 1990), suggesting that a partial destabilization of the nucleosomal core structure may also occur at more moderate acetylation levels. It is likely that acetylation allows the formation of loosely-packed chromatin structures by influencing the path of DNA between individual nucleosomes (Bauer *et al.*, 1994), and by affecting their interaction with non-histone chromatin components (Garcia-Ramirez *et al.*, 1995; Edmondson *et al.*, 1996). Taken together, it can be easily seen that coactivator-mediated acetylation processes could certainly play an important role in establishing an 'active' promoter architecture. Acetylated nucleosome-containing promoters are more accessible to many transcription factors that are normally prevented from binding to chromatin-packaged DNA. Such promoters would also be more effective in supporting high levels by $RNAP_{II}$ due to destabilization of the interactions between the modified nucleosomes and DNA. This model explains (at least in principle) the previously observed link between histone acetylation and actively-transcribed genes because it implies that the acetyltransferase activities mediated by coactivators are only locally stimulated on promoters that are switched on by gene-specific activators. Not all observed histone acetylation modifications *in vivo* are, however, the consequence of activator/coactivator-mediated acetylation. A cytoplasmically-located histone acetyltransferase ('HAT-B') specifically modifies two lysine residues (positions 5 and 12; Figure 7.6.) in histone H4 molecules before they get incorporated into complete nucleosomes. This may play an important role in controlling the rate of assembly and deposition of newly synthesized nucleosomes (Chang *et al.*, 1997).

Since the expression of individual genes is generally a dynamic process, the degree of histone acetylation has to be reversible, so that transcription can be switched off under appropriate circumstances. This view correlates well with the discovery that the activities of several histone deacetylases ('HDACs') responsible for removing acetyl groups from modified histones are associated with known transcriptional repressors (summarized in Figure 7.7). Many of the HDAC activities seem to be organized as large complexes around the conserved SIN3 corepressor and its associated RPD3-type deacetylase (e.g. Kadosh and Struhl, 1997; Karsten *et al.*, 1997; Yang *et al.*, 1997; reviewed in Wolffe, 1997). Non-liganded nuclear receptors mediate repression through recruitment of the SIN3/RPD complex via the N-CoR and SMRT corepressors (Figure 7.7; Heinzel *et al.*, 1997; Nagy *et al.*, 1997). From the generally stimulatory effects of acetylated chromatin on the transcriptional machinery it follows that the activities of the various deacetylase complexes mainly reverse the 'opening up' effects and eventually lead to the formation of denser deacetylated chromatin structures, that are less accessible to the various gene-specific transcription factors and components of the basal transcriptional machinery. The ultimate acetylation state of the chromatin environment of individual genes can therefore be seen as predominantly resulting from the interplay between histone acetylation and deacetylation enzymes, whose local activity is controlled through specific recruitment by gene-specific transcriptional activators and repressors, respectively. Repressed chromatin states are also further consolidated through the increased binding of transcriptional repressors, such as the yeast gene-specific transcriptional repressor Tup1, to the underacetylated N-termini of histones H3 and H4 (Edmondson *et al.*, 1996).

Apart from genetic approaches in yeast, the effects of histone acetylation can also be experimentally studied through the action of specific drugs, such as sodium butyrate. Sodium butyrate is a simple, but effective inhibitor of the enzymes that are responsible for removing acetyl groups from modified histones. After addition of sodium butyrate to cells grown in culture it is often possible to see, in accordance with the observations described earlier, an increased sensitivity of chromatin to nucleases and an overall loss of higher order chromatin structures (Perry and Chalkley, 1981; Perry and Annunciato, 1989; Ridsdale *et al.* 1990). Amazingly enough, butyrate-based treatments are used in clinical situations to successfully activate dormant fetal globin gene expression in β-thalassemia patients, and it is possible that changes in overall chromatin configuration are involved in this process (see Chapter 10 for more details).

'Linker' Histones Organize Core Nucleosomes into Higher Order Chromatin Structures

Most nucleosomes are spaced at an interval of 200 ± 40 nucleotides along the DNA to form a linear 'bead-on-a-string' structure in native chromatin. Nucleosome core particles bind, however, only around 146 nucleotides in approximately 1.7 left-handed superhelical turns. What happens with the remaining 60 nucleotides that are present in each nucleosomal repeat? The binding of these additional DNA sequences, and the organization of nucleosomes into the higher order 'solenoid' or 30 nanometer (nm) filament, is the responsibility of the 'linker histones', such as histone H1 (Figure 7.8; Finch and Klug, 1976; Thoma *et al.*, 1979). Neutron scattering studies indicate that histone H1 is actually located in the interior of the 30 nm filament, where it is in direct contact with DNA entering and exiting individual nucleosomes (Figure 7.9; Graziano *et al.*, 1994). A similar compaction of nucleosome arrays can be achieved *in vitro* in the absence of H1 by a simple variation of cation concentrations. This suggests that counteracting the electrostatic repulsion between the negatively charged phosphodiester DNA backbones wound around neighbouring nucleosomes is one of the major force that needs to be neutralized

Figure 7.8. The 'Solenoid' Model of Higher Order Nucleosome Packaging.
In the presence of linker histones, or under particular ionic conditions, nucleosomes form higher order structures (the 30 nm 'solenoid' fibres). See 'http://www.average.org/~pruss/nucleosome.html' for this and further examples of nucleosomal structure models.

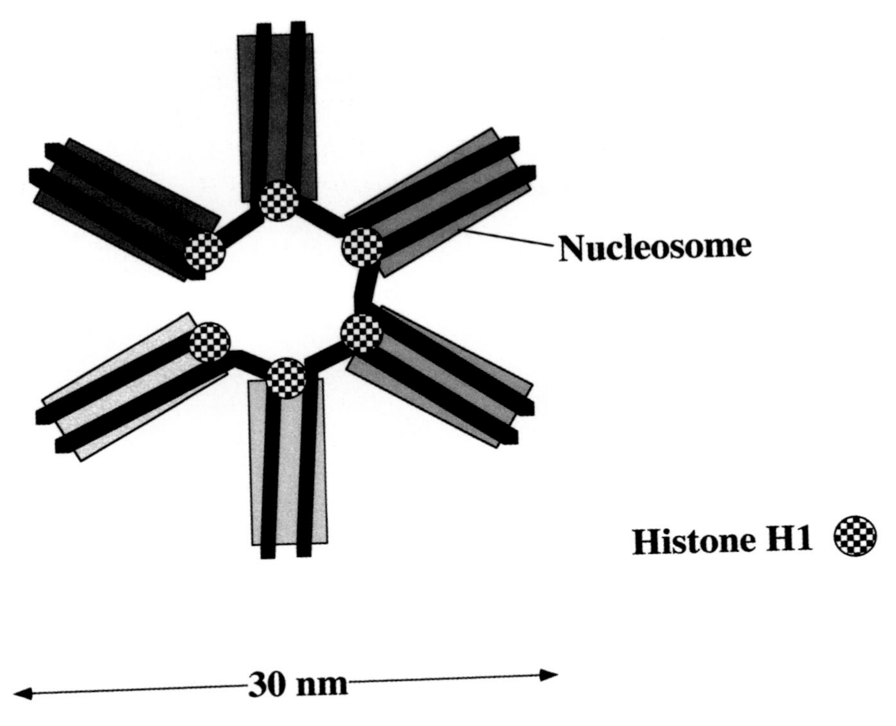

Figure 7.9. Central Location of Linker Histones.
This diagram shows a schematic cross-section of the 30 nm filament. The individual nucleosomes (seen from the direction of the filamental axis) are arranged into a helical structure with an approximate periodicity of approximately 6 nucleosomes per turn. The DNA molecules wound around them are shown in black. Note the central position of histone H1 positioned over the entry and exit sites of DNA from each nucleosome. After Galiano *et al.* (1994).

during formation of higher order chromatin filaments (e.g. Schwarz and Hansen, 1994). Linker histones, like H1, play a role in such a charge-shielding operation by binding specifically to the free DNA between individual nucleosomes through their extensive N- and C-terminal domains (Clark *et al.*, 1988; Clark and Kimura, 1990). Differential post-translational phosphorylation of H1 at various stages of the cell cycle may also contribute to a certain extent to the condensation/decondensation of chromosomes by regulating the DNA-binding ability of H1 (Roth and Allis, 1992). Nevertheless, there is a logical dilemma

1995; Blank and Becker, 1996). This 'positioning information' encoded in DNA sequences plays an important role, and can even override species-specific nucleosomal packing patterns if DNA is transferred between different organisms (Chavez *et al.*, 1995). Certain DNA sequences, such as long homopolymeric runs of single nucleotides, actively 'discourage' nucleosome binding and may thus be used to keep key genomic regions, such as promoters and replication origins, free of nucleosomes (Prunell, 1982). Although the DNA sequence can be used in this way to dictate nucleosome positions to a certain extent, there is clearly also considerable room for flexibility, which is exploited by the various mechanisms involved in transcriptional regulation. The binding of gene-specific transcription factors to DNA target sites often has a substantial influence the overall nucleosomal positioning pattern on a gene. Roth *et al.* (1990) demonstrated that the sequence-specific binding of a transcription factor complex containing the yeast transcriptional repressor α2 to a stretch of DNA with disordered nucleosomal positions leads to the formation of a highly ordered array of precisely positioned nucleosomes. Similar results obtained with several other transcription factors (including GAL4, NFκ-B, GAGA- and heat shock factor) confirm that α2 is not exceptional in this respect. Binding of gene-specific transcription factors to DNA target sites can thus influence the subsequent binding of other transcription factors by either making DNA target motifs available for access, or by masking them with positioned nucleosomes. Such results hint at the possibility that the transcriptional regulation of individual promoters in a chromatin environment may depend strongly (substantially more that under *in vitro* conditions) on the precise temporal sequence of individual gene-specific transcription factors binding to their target sites. Such a temporal dependency on the recruitment of gene-specific transcription factor may provide the basis for an important regulatory control strategy in activating transcription of specific genes during embryonic development and cellular differentiation. *In vitro* experiments using a *lac* repressor variant to induce specific nucleosome positioning suggests that nucleosome positions become rapidly disordered once gene-specific transcription factors dissociate from their target motifs (Pazin *et al.*, 1997). Nucleosome repositioning is thus fully reversible and needs probably to be continously maintained by the presence of stable transcription factor complexes on promoters.

Highly Regulated Genes Often Contain Precisely Positioned Nucleosomes in Their Promoter Region

The use of specifically-positioned nucleosomes to control physical access of transcriptional activators to promoter sequences has been convincingly documented in several dozen cases (e.g. Simpson and Stafford, 1983; Venter *et al.*, 1994; Li and Wrange, 1995; Fragoso *et al.*, 1995). This access-control mechanism can either facilitate (Vettese-Dadey *et al.*, 1994), or prevent binding of transcription factors to specific sequence motifs (e.g. Venter *et al.*, 1994). The available evidence suggests that most positioned nucleosomes play an inhibitory role by preventing gene-specific transcription factors from recognizing their target sequences. One of the most dramatic illustrations of this fact comes from genetic studies in yeast. Unlike higher eukaryotic genomes, that usually contain several hundred histone-encoding genes per haploid genome, the yeast genome contains only two copies of each of the genes encoding the four different core histones (H2A, H2B, H3 and H4). It is therefore relatively straightforward to mutate/delete one or more of the histone genes in order to study the phenotypical consequences of such an action. Although the deletion of both gene copies encoding any of the four histones causes a lethal phenotype, it is possible to observe the effects of histone depletion by genetic methods involving conditional complementation. As we have seen earlier, H4 is a crucial component of the nucleosome core structure and function (e.g. Dong and van Holde, 1991), and the number of functional nucleosomes is reduced by around 50% if the synthesis of H4 is stopped. This degree of nucleosome loss results in the spontanous activation of a number of genes that are normally repressed and require specific inducers to activate transcription (such as *CUP1*, *HIS3* and, to a certain extent, *PHO5*; Figure 7.12; Han and Grunstein, 1988; Han *et al.*, 1988; Durrin *et al.*, 1992). The result clearly indicates that chromatin plays a major role in keeping many inducible genes transcriptionally silent in the absence of an inducer through a general repressive effect. The severe loss of nucleosomes allows uncontrolled access of the transcriptional machinery to the freely exposed promoters, which then become available for transcription.

The loss of nucleosomes, such as in the experimental system described above, is obviously an extreme situation that would not be used under normal

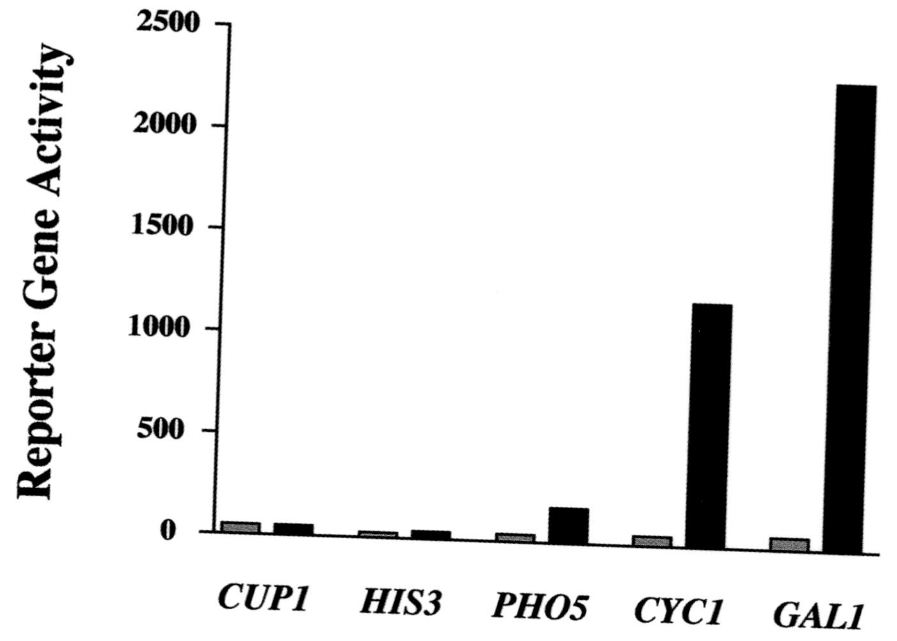

Figure 7.12. Repression Release of Certain Promoters in the Absence of Nucleosomes. The levels of reporter gene expression due to nucleosome loss are shown for five different promoters (grey columns). These levels are compared with the values of reporter gene activity obtainable by specific induction of each of the five promoters. Note that the expression levels of the CUP1 and HIS3 genes resulting from either nucleosome loss, or specific induction, are essentially identical. The specifically-induced expression levels obtained from the PHO5, CYC1 and GAL1 promoters substantially exceed the transcriptional activity observed after nucleosome loss. The data suggests that nucleosome loss results in a generally low and comparable level of transcription from many different promoters, and that the high levels of transcription from strongly-inducible promoters (such as GAL1) must involve additional activation mechanisms. Redrawn from Durrin *et al.* (1992).

in vivo circumstances to activate inducible genes in response to specific signals. The removal of positioned nucleosomes interfering with transcription of inducible promoters occurs under physiologically conditions through the concerted action of gene-specific activators and chromatin remodelling complexes. One of the best-understood model systems, the yeast *PHO5* promoter, contains six precisely positioned nucleosomes that play a distinct role in the transcriptional control mechanism of this gene (Figure 7.13; Almer and Horz, 1986; Gaudreau *et al.*, 1997). One of the target sites for a transcriptional activator (PHO4) that controls PHO5 expression is located in a short nucleosome-free region of the promoter and is freely accessible. The binding of PHO4 to this site is necessary to start the induction process but is not sufficient by itself. Another PHO4 binding site (which is much more important for stimulating PHO5 transcription) is located more closely to the transcription start site, but is at this stage still shielded by a positioned nucleosome nearby. The binding of PHO4 to the accessible site leads to a

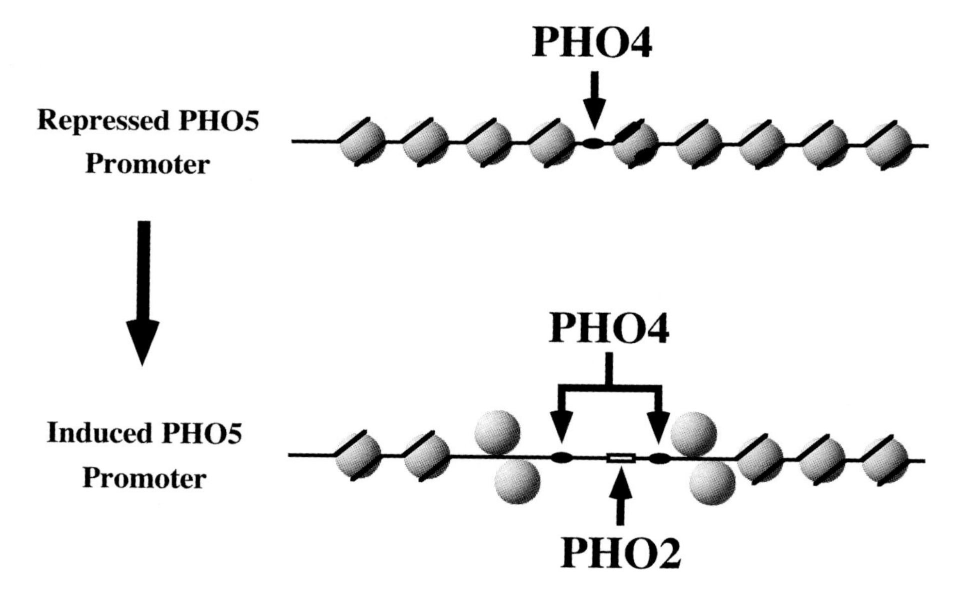

Figure 7.13. Positioned Nucleosomes in the PHO5 Promoter.
The arrangement of positioned nucleomes on the PHO5 promoter allows access of the transcriptional activator PHO4 exclusively to a single DNA target site when the promoter is in the repressed state. Binding of PHO4 to this site causes a rearrangement of the surrounding nucleosomes and exposes additional promoter sequences. Specific binding of the transcription factor, PHO2, and an additonal PHO4 molecule to the newly accessible DNA target sites results in a transcriptionally active, induced PHO5 promoter.

gradual displacement of the inhibitory nucleosome and allows the crucial promoter-proximal PHO4 site to become available for binding of another PHO4 molecule (Venter *et al.*, 1994). Eventually, all of the remaining positioned nucleosomes are removed which allows increased access of further transcription factors to the promoter (Almer *et al.*, 1986). The repressive effect of positioned nucleosomes is probably due to several reasons. The most obvious one is that nucleosomes are a physical obstacle that prevents other transcription factors from binding to promoter sequences by physical obstruction. Depending on where the DNA binding motifs are located relative to the positioned nucleosome, it is possible that the distortion of the helical pitch of DNA during nucleosome packing (see Figure 7.5.!) sufficiently distorts the transcription factor target site so that it can no longer be recognized, even if it faces away from the nucleosome surface and is thus, in principle, physically accessible to a sequence-specific DNA-binding protein. A further possibility is that the flexible histone N-terminal tails emanating from the nucleosome core (Figure 7.6.) interfere with the binding of gene-specific transcription factors at nearby sites (Lee *et al.*, 1993; Juan *et al.*, 1994; Vettese-Dadey *et al.*, 1994).

Chromatin is a Tough Working Environment for RNA Polymerases: Elongation Control

We have seen earlier that the acetylation of the N-termini of histones present in each nucleosome core particle 'opens' chromatin structures and makes them easier to transcribe. It is nevertheless clear that even when promoters become physically accessible after such chromatin modifications, the transcription of a DNA template covered with nucleosomes is still not necessarily a straightforward task for eukaryotic RNA polymerases (Morse, 1989; Izban and Luse, 1991; Adams and Workman, 1993; Studitsky *et al.*, 1997). In many cases, actively transcribed genes retain their nucleosomal packing during transcription (De Bernardin *et al.*, 1986; Nacheva *et al.*, 1989; Ericsson *et al.*, 1990), although there may be changes due to nucleosome movement on the DNA template (e.g. Pennings *et al.*, 1991; Villeponteau *et al.*, 1992). It is likely that the nucleosome array in highly transcribed genes, with a continuous stream of elongating RNAPs moving through the coding region, becomes increasingly disorganized and eventually may become substantially depleted of nucleosomes (Cavalli and Thoma, 1993; Tsukiyama *et al.*, 1994). This process may by itself solve some of the problems that $RNAP_{II}$ encounters in chromatin templates by decreasing the density of nucleosomes that present an obstacle to high levels of RNA synthesis. Several recent discoveries suggest, however, that there are other solutions. The heterodimeric elongation factor FACT ('*Fa*cilitates *C*hromatin *T*ranscription') enables increased levels of *in vitro* transcription of nucleosome-packaged DNA templates in the presence of only purified basal factors and $RNAP_{II}$ (Orphanides *et al.*, 1998). Unlike some of the other known chromatin remodelling complexes (see below), FACT does not require ATP hydrolysis and does not cause nucleosome loss or repositioning. The transcriptional elongation rate through chromatin is also greatly facilitated by the presence of the 'non-histone' protein HMG-14 (Ding *et al.*, 1994). HMG-14 binds directly to nucleosome core particles, but is present in substoichiometric quantities so that only a subset of nucleosomes can associate with it (Mardian *et al.*, 1980; Sandeen *et al.*, 1980). Chemical cross-linking studies indicate that HMG-14 interacts with DNA at the entry/exit point of nucleosomes and thus influences, similar to the effects of histone H1, the higher order structure of chromatin (Alfonso *et al.*, 1994)

The existence of specific chromatin-dependent elongation factors (such as FACT and HMG-14) is highly reminiscent of the tight control of the the elongation properties of $RNAP_{II}$ under the control of gene-specific transcriptional activators and elongation factors on nucleosome-free DNA templates (Chapter 5), and suggests that transcription elongation is regulated at multiple levels all the way from controlling $RNAP_{II}$ processivity to the ability of $RNAP_{II}$ to read through nucleosomally-packed DNA.

7.4. Chromatin Remodelling of Promoters Packed into Repressive Chromatin Complexes

Replication-Dependent and -Independent Access to Chromatin-Packaged DNA

We have seen above that positioned nucleosomes exert in most cases a negative effect on the transcription of individual genes containing them. In the case of PHO5, the initial binding of a transcriptional activator initiates a cascade of nucleosome displacement which eventually results in the binding of another copy of the activator to a crucial promoter site. This is, however, not the only way of converting the repressive chromatin configuration of a promoter into one that is compatible with high levels of activated transcription. In this section we will investigate additional mechanisms that accomplish in a similar outcome, although via a number of different routes.

One of the most straightforward way to allow transcription factors unimpeded access to DNA target sequences is during the DNA replication process. As the replication machinery moves through chromatin and unwinds double-helical DNA, the nucleosomes dissociate from the 'old' and newly-formed DNA strands to leave promoters at least temporarily free of nucleosomes. This process gives transcription factors a brief window of opportunity to establish active transcription complexes, and is often referred to as 'replication-dependent' chromatin activation. Although it is not clear how widespread such a mechanism is *in vivo*, it has been demonstrated to occur in in reconstituted synthetic nuclei (Barton and Emerson, 1995). Most other known chromatin activation processes are replication-independent, and thus do not have to rely on cell-division to occur before the gene expression pattern can be changed. This allows genes to be switched on regardless of the stage of the division cycle a cell is currently in.

Replication-Independent Chromatin Remodelling Complexes: SWI/SNF, NURF, CHRAC

Our understanding of how DNA packaged into nucleosomes can be reconfigured into a transcriptionally active templates has made a big leap forward with the discovery of a number of chromatin-remodelling activities present in various eukaryotic cells. One of the best understood activities is the SWI/SNF complex, originally discovered through a detailed genetic dissection of the the regulatory features of certain yeast genes (reviewed in Peterson and Tamkun, 1995). The components of the SWI/SNF complex are required for the high level expression of many inducible genes (such as ADH1, ADH2, GAL1 and GAL10; Peterson and Herskowitz, 1992), where they play a supporting role for gene-specific transcription factors. The biochemical characterization of SWI/SNF revealed it to be organized as a large complex (> 2 megadaltons) consisting of approximately 10 different subunits (Cote *et al.*, 1994; Peterson *et al.*, 1994). Homologs of many of these subunits have been discovered in higher eukaryotes, supporting the notion that SWI/SNF-like chromatin remodelling complexes are a common feature of all eukaryotic organisms (Tamkun *et al.*, 1992; Kalpana *et al.*, 1994; Kwon *et al.*, 1994; Wang *et al.*, 1996). One of the most intriguing discoveries was the revelation that the SWI2 subunit contained a number of sequence motifs that had been previously found to be characteristic of DNA-stimulated ATPases (Henikoff, 1993; Laurent *et al.*, 1993). This was an important finding because this homology immediately suggested a possible mechanism of action: SWI/SNF might use ATP-hydrolysis to undergo conformational changes for disrupting various protein-protein and protein-DNA interactions during its chromatin remodelling function. Experimental observation confirmed that purified SWI/SNF greatly enhances the binding of gene-specific and basal transcription factors to nucleosomal DNA (Cote *et al.*, 1994; Kwon *et al.*, 1994, Imbalzano *et al.*, 1994), and alters the linking number of circular minichromosomes in an ATP-dependent manner (Kwon *et al.*, 1994). Currently we still are far from understanding the precise functional significance of these activities (and the function of the many other SWI/SNF complex subunits with no recognizable motifs!). Initial studies reporting the specific association and copurification of SWI/SNF components with the yeast $RNAP_{II}$ holoenzyme raised the exciting possibility that the recruitment of the holoenzyme to specific promoters would lead to localized chromatin remodelling activity (Wilson *et al.*, 1996), but subsequent investigations failed to support such a conclusion (Cairns *et al.*, 1996).

In addition to SWI-SNF, there are several other chromatin remodelling activities that have been characterized through various genetical and biochemical approaches. Using an *in vitro* chromatin remodelling system on the *Drosophila* hsp70 heat-shock promoter, Wu and coworkers identified a new nucleosome disruption activity, NURF ('nucleosome remodelling factor'), that depends on the presence of the gene-specific 'GAGA' factor for activity (Tsukiyama *et al.*, 1994 and 1995). A variety of other transcriptional activators (such as Sp1 and NFκ-B, Pazin *et al.*, 1996) also have the ability to recruit NURF, suggesting that all these transcription factors have the ability to remodel chromatin configurations on the promoters they are bound to, and thus assist in setting up actively transcribed promoters (Mizuguchi *et al.*, 1997).

Another recently characterized chromatin remodelling factor, CHRAC ('chromatin-accessibility complex') contains enzymatic activities that converts irregular nucleosomal arrays into regular chromatin with evenly spaced nucleosomes and affects overall DNA topology. Interestingly, NURF and CHRAC have opposite effects (nucleosome perturbance and nucleosome ordering, respectively), but they share a common subunit with ATPase activity (ISWI; Varga-Weisz *et al.*, 1997). The presence of ISWI in both NURF and CHRAC suggests that many of the chromatin remodelling complexes may be formed by different combinations of discrete functional modules, and may be recruited to different promoters via different transcription factors.

7.5. Heritable Maintenance of Gene Expression States through Regulated Heterochromatinization

Fixation of Developmental Decisions

Once particular chromatin states have been set up through mechanisms involving nucleosome displacement and differential post-translational histone modifications, the need arises to maintain this 'activated' chromatin state in a stable configuration which can be maintained during disruptive events, such as cell division. During the replication process DNA is (at least temporarily) in a state where it is more accessible to the transcriptional machinery, and thus 'vulnerable' to changes in gene expresssion states (see Section 7.4. on 'Replication-Dependent Chromatin Activation'). Since many decisions during embryonic development and cellular differentiation are taken in an hierarchical manner, which requires a high degree of irreversibility during the early stages, it is clear that mechanisms must exist that stabilize chromatin configurations once a developmental decision has been reached in a particular cell. Some of the most clear-cut examples of the functional decoupling of mechanisms reponsible for establishment of chromatin states, and separate mechanisms reponsible for subsequently maintaining them, have been discovered by genetic investigations in yeast and *Drosophila*.

Members of *Polycomb* and *trithorax* Groups Maintain Regional Differential Gene Expression during *Drosophila* Development

During the first few hours of fruitfly embryonic development, many of the regional differences are specified through the spatially-controlled expression of a group of homeotic genes. The expression of individual homeotic genes is initially set up in an approximate manner through morphogen gradients (e.g. *bicoid*), and subsequently refined by 'gap-' and other segmentation genes (see Chapter 4). Once the expression of a particular homeotic gene is transiently specified in a particular location of the embryo, the proteins encoded by several dozen genes, including the '*Polycomb* group' ('PcG') and '*trithorax* group' ('trxG'), are required to 'freeze' the spatially-restricted expression pattern. Mutants in the genes encoding the various members of the PcG/trxG families show normal expression of homeotic genes during early stages of embryogenesis, but fail to maintain the specific expression pattern during the later stages, indicating that they are responsible for maintaining the gene expression pattern once it is established through a network of other genes (Figure 7.14; McKeon and Brock, 1991; Kennison, 1995). Interestingly, the proteins encoded by the two families seem to act in a mutually antagonistic manner: members of the PcG prevent improper activation of inactive chromatin configurations, wheras members of the trxG are responsible for keeping open chromatin configuration transcriptionally active. It is likely that such an arrangement is beneficial for keeping actively transcribed genes switched on and keeping inactive genes permanently switched off, which is of course ideal for propagating developmental decisions in a stable and essentially irreversible manner from one cell generation to the next.

The molecular analysis of representatives of the PcG has demonstrated that the encoded proteins represent a diverse set of nuclear proteins that are probably arranged into multisubunit complexes (Franke *et al.*, 1992; Rastelli *et al.*, 1993; Lonie *et al.*, 1994). None of the characterized PcG proteins displays any ability by itself to interact sequence-specifically with DNA and it is likely that they

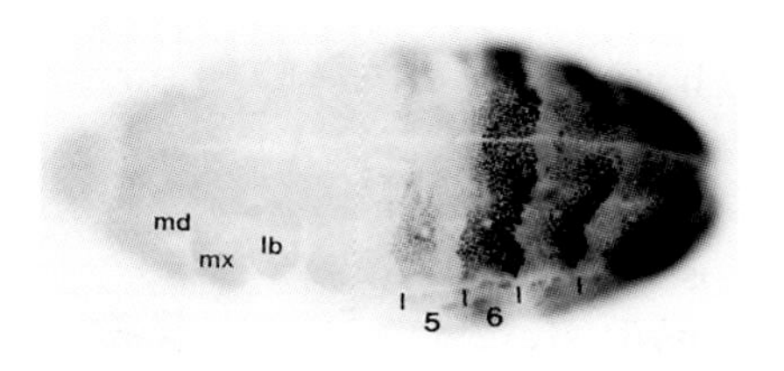

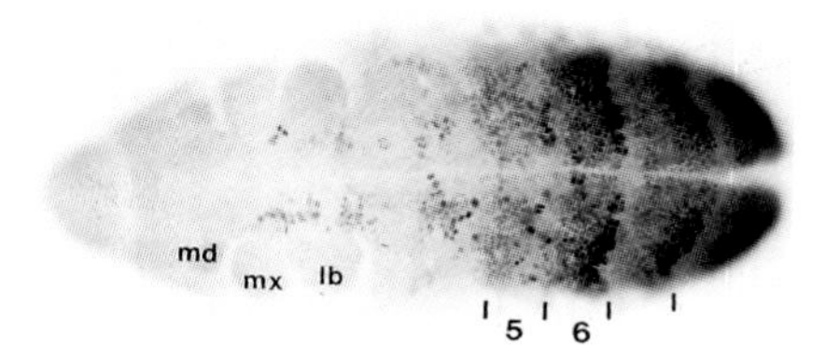

Figure 7.14. Abnormal Maintenance of Homeotic Gene Expression Patterns in *Polycomb* Mutants.

The expression pattern of the *Ultrabithorax* (*Ubx*) protein is shown in a wildtype embryo and a *Polycomb* mutant. Both embryos have been stained with the same anti-*Ubx* antibody and are at a comparable developmental stage. The *Ubx* gene encodes homeodomain-containing transcription factors that control the regional differentiation of parts of the thoracic and abdominal regions during embryonic development. *Left*: In wildtype embryos *Ubx* is typically expressed at its highest level in parasegment 6 and in the abdominal regions, whereas the expression levels are relatively low level in parasegment 5. *Right*: In *Polycomb* mutants the *Ubx* expression pattern is less clearly spatially defined and extends substantially towards the head segements (md, mx, and lb). The presence of ectopically expressed *Ubx* protein in these segments indicates a failure to maintain the initially defined expression state of the *Ubx* gene due to the lack of *Polycomb* protein in the mutant embryos. Figure from Gould *et al.* (1990).

exert their functions through direct protein: protein interactions with yet to be identified chromatin components. Certain characteristic protein motifs, such as the 'chromo' domain present in *Polycomb* and other nuclear proteins, have been postulated to play a crucial role in such recognition events (Messmer *et al.*, 1992; Koonin *et al.*, 1995; Platero *et al.*, 1995). Detailed investigations of the localization of endogenous *Polycomb* protein on polytene chromosomes have revealed that these proteins are not uniformly distributed, but are clustered at a number of discrete sites, the PcG response elements ('PREs'). This result suggests that PcG proteins may carry out their role of maintaining transcriptionally passive chromatin configurations by shielding potentially accessible target sites from transcriptional activators and the transcriptional machinery (Simon *et al.*, 1993). This view has been substantially strenghtened by the high-resolution mapping of PcG-containing chromatin complexes to discrete regions surrounding promoter and regulatory elements of homeotic genes (Figure 7.15; Orlando and Paro, 1993).

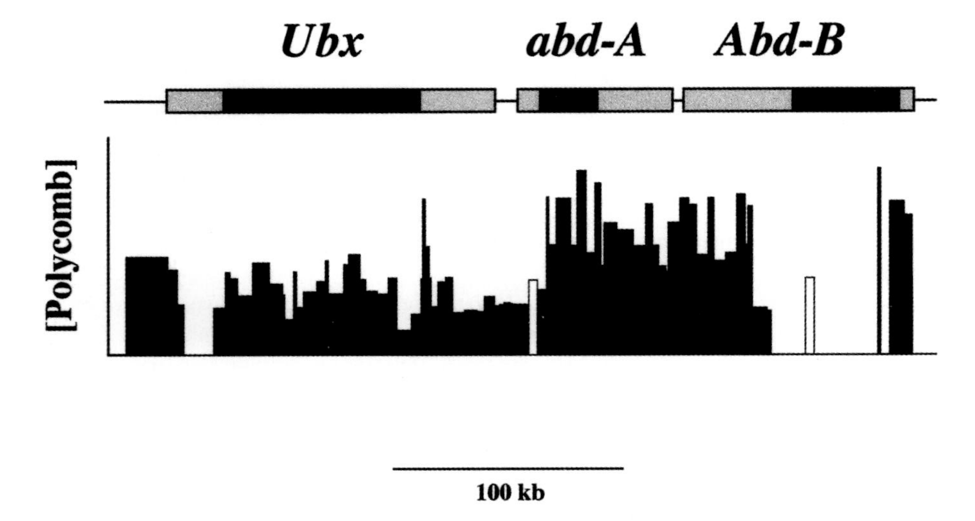

Figure 7.15. Distribution of Polycomb Containing Protein Complexes On Bithorax Complex.
Polycomb-containing chromatin fragments were immunopurified, the DNA extracted, cleaved with a restriction enzyme, and prepared for Southern blotting. Probing such a blot with bacteriophage clones spanning various parts of the bithorax complex reveals the amount of *Polycomb* protein present at various position (bottom histogram). The relative position of the three homeotic genes present in the bithorax complex (*Ultrabithorax* [*Ubx*], *abdominal-A* [*abd-A*] and *Abdominal-B* [*Abd-B*] is shown schematically and in alignment with the *Polycomb* distribution data. The two bars shown in white in the histogram contain repetitive sequences and may thus not represent the amount of *Polycomb* located on them accurately. Redrawn from Orlando and Paro (1993).

7.6. Constitutive Heterochromatin: A Transcriptional Graveyard For Most Genes!

What is Constitutive Heterochromatin?

Most eukaryotic genomes contain a large proportion of repetitive sequences that are very low in sequence complexity. For example, in the fruitfly *Drosophila virilis* up to half of the genome consists of long tandem repeats of individual motifs that are only seven nucleotides in length ('simple sequences'; see Table 7.1). In mammalian species most of the repeats are multimerized

Table 7.1. Simple Sequences Present in the Genome of *Drosophila virilis*. More than 40% of the genome consists of tandem repeats of the three types of repeat units in the proportions indicated.

Type	Sequence of Repeat Unit	Proportion of Genome
I	$(ACAAACT)_n$	25%
II	$(ATAAACT)_n$	8%
III	$(ACAAATT)_n$	8%

derivatives of a fundamental nine nucleotide motif with the consensus GAAAAA$^A/_T{}^G/_C$T. Such sequences do not have any obvious potential for encoding functional proteins and display significant variations between closely related species, indicating that they have little or no phenotypic impact on the organism containing them. Apart from highly repetitive sequences, many genomes also contain 'middle repetitive' sequences that are mostly made up of functionally inactive remnants of transposable and retroviral elements. Middle repetitive sequences constitute up to ten percent of the genome of *Drosophila melanogaster* (Pimpinelli *et al.*, 1995).

The presence of repetitive sequences in eukaryotic genomes can be correlated with histological observations that revealed the presence of intensely staining, dense particles ('heterochromatin') in eukaryotic nuclei (Heitz, 1928). From recent cytogenetic studies in *Drosophila melanogaster*, it has become clear that most of the middle- and highly repetitive sequence elements are arranged into blocks of 50–900 kb that are packaged in a heterochromatic configuration, usually in the vicinity of the centromeric regions of eukaryotic chromosomes (e.g. Lohe *et al.*, 1993; Le *et al.*, 1995). Genetic screens (see below) and biochemical studies have up to now identified a small number of proteins that are preferentially associated with heterochromatin and may be responsible for the tight packaging of nucleosomes in such structures. Although we are still very far from understanding the structural nature of heterochromatic packaging, some explanations are slowly emerging. The preferential presence of specific proteins in heterochromatin, such as 'Heterochromatin Protein 1'

('HP1'; James and Elgin, 1986; Eissenberg *et al.*, 1990; Saunders *et al.*, 1993), and 'GAGA' factor (Raff *et al.*, 1994) suggests that these proteins may have a distinct structural or enzymatic functions enabling the initial formation and subsequent maintenance of high-density nucleosomal arrays (e.g. Paro and Hogness, 1991). Heterochromatically-packaged regions of the genome are also specifically located within nuclei *in vivo*, which can substantially affect the transcription and replication of genes located in or near heterochromatin (Csink and Henikoff, 1996; see also Chapter 8).

What is the Function of Constitutive Heterochromatin?

It is clear that packaging of functionally inert DNA is one of the major function of heterochromatin, and this may effectively neutralize any damaging effect of 'junk' DNA by shielding it from access by the eukaryotic transcriptional machineries. While this may explain the function of the bulk of heterochromatin in cells, it has also become clear that some functionally active genes are actually embedded in an heterochromatin environment. Genetic and molecular biological studies in *Drosophila* have revealed more than 40 actively expressed genes that are surrounded by middle-repetitive sequences and appear to be packaged into a genuine heterochromatic environment (Gatti and Pimpinelli, 1992; Eberl *et al.*, 1993; Howe *et al.*, 1995). One of the most surprising outcomes of such studies has been the realization that these genes need the heterochromatic environment to be properly expressed, and translocation to a normal euchromatic environment actually interferes with their transcription through an unknown mechanism (Eberl *et al.*, 1993; Howe *et al.*, 1995). Conversely, the expression of the vast majority of genes is negatively affected when they are placed into (or even just near) a heterochromatically packed region of a chromosome. This process has been most effectively studied in the *Drosophila* model system and is usually referred to as 'Positional Effect Variegation' ('PEV').

As we have seen earlier, most of the heterochromatin in *Drosophila* occurs near the centromer and coincides with the presence of middle- and highly repetitive sequence elements located there. Since the majority of the unique sequences occur in the substantially more open euchromatin configuration, a distinct 'boundary' has to be established between the two chromatin configurations. Genetic experiments utilizing mutants, in which parts of the euchromatic portions of chromosomes are brought into the vicinity of the euchromatin-heterochromatin boundary, have been very informative for studying the transcriptional silencing effects of heterochromatin. One of the best-studied examples of PEV utilizes a chromosomal inversion of an eye-pigment gene in *Drosophila*, *white* ^{m4} (Tartof *et al.*, 1984). In this mutant the wildtype *white* gene (which actually is responsible for encoding the red eye pigment in wild type fruit flies) is brought within 20 kb of the euchromatic-heterochromatic boundary, resulting in the appearance of a red/white mottled eye (Figure 7.16). This phenotype is due to the random silencing effect of the expression of the *white* genes by fluctuations of the precise position of the heterochromatin/euchromatin borders in various eye-precursor cells, resulting in the stochastic inactivation of *white* gene expression in some cells, but not in others. Moving the *white* gene by genetic means further into the euchromatic portion of the chromosome substantially reduces the effect of PEV (increase in red eye color), whereas translocation of *white* closer to the boundary substantially increases the frequency of transcriptional silencing of the gene (increase in white eye color).

Position Effect Variegation-Based Genetic Screens Allow Identification of Genes Affecting Heterochromatin Function

The easily detectable phenotype of PEV in combination with certain genetic markers (such as *white*, see above) provides a convenient assay for detecting genes encoding heterochromatin components and proteins involved in modifying heterochromatin function. Mutants in such genes enhance PEV if they result in the increased stabilization of heterochromatin, and *vice versa*. Up to now more than 50 distinct loci have been detected that either enhance or supress PEV(Wustmann *et al.*, 1989; Sinclair *et al.*, 1992). The molecular analysis of the mutated genes has revealed numerous proteins involved in heterochromatin structure and function, including histones, heterochromatin protein 1, protein phosphatase 1 and a variety of proteins involved in DNA replication (reviewed in Weiler and Wakimoto, 1995). It seems that many of the components identified play rather general roles in overall cell function and may influence PEV only indirectly. In future, a more biochemically-oriented approach may be necessary to identify heterochromatin components (cf. Csink *et al.*, 1997).

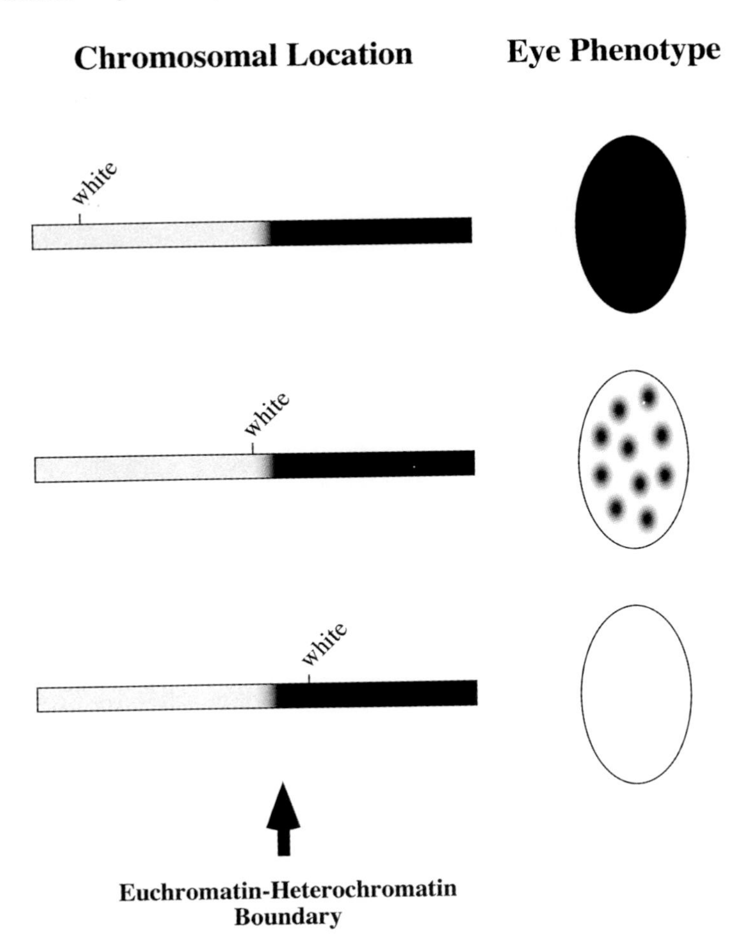

Figure 7.16. Schematic Diagram of 'Position Effect Variegation' in *Drosophila*. The relationship between eye coloration (depicted schematically on the right hand side) and the position of the *white* gene relative to a heterochromatic boundary (left hand side; euchromatin is shown in light grey and heterochromatin in black) are depicted. *Top*: If the white gene is present in its normal euchromatic position, the eye colour of the fly is red. *Middle*: If the *white* gene is moved (with genetic techniques) closer to the euchromatin/heterochromatin boundary, the eye colour becomes patchy because the white gene is actively expressed in some cells, but not in others due to slight fluctuations in the precise position of the euchromatin/heterochromatin border in different cells. *Bottom*: If the *white* gene is entirely surrounded by heterochromatin it becomes transcriptionally inactive, resulting in a white-eyed fly.

7.7. Conclusions

Histones, the fundamental building blocks of the nucleosomes, are ancient proteins that almost certainly already existed in the common ancestor cell predating the evolution of the archaeal and eukaryotic domains. Although thermal protection may have been the major driving force leading the adaptation of histones for DNA packing in such cells, the subsequent evolution of more elaborate chromatin structures rapidly lead to their integration into a variety of gene expression control mechanisms. Core nucleosomes interact with DNA through a variety of specific protein-DNA contacts which confer a certain degree of selectivity for the binding of nucleosomes to positions defined by the underlying DNA sequence. Gene-specific transcription factors and a variety of chromatin-remodelling complexes play, however, a decisive role in the repositioning of nucleosomes to create access routes for the transcriptional machineries to individual promoters. The selective modification of amino acid residues in the flexible N-termini of histones, especially by acetylation, is one of the major control mechanisms through which various coactivators create transcriptionally active chromatin environments. Special chromatin elongation factors are required for efficient elongation by $RNAP_{II}$ through nucleosomal DNA templates. Once established, active and inactive chromatin environments need to be maintained through the activity of the PcG/trx group of transcription factors to 'freeze' the various decisions of differentiation- and developmental programs. Large parts of the genome, mainly consisting of repetitive sequences, are permanently maintained in a densely-packed heterochromatic state. Translocation of the majority of genes into heterochromatin (or into the vicinity of the heterochromatic border) represses their transcription.

References

Internet Sites of Interest

'http://rampages.onramp.net/~jrbone/chrom.html'
General information about chromatin and links to other related sites.

'http://www.mol.biol.ethz.ch/richmond/'
Website based on nucleosome core particle structure and related projects.

'http://www.average.org/~pruss/nucleosome.html'
Images of nucleosomes and higher order chromatin structures.

Research Literature

Adams, C.C., and Workman, J.L. (1993). Nucleosome displacement in transcription. *Cell* 72, 305–308.

Alevizopoulos, A., Duserre, Y., Tsai-Pflugfelder, M., Von der Weid, T., Whali, W., and Mermod, N. (1995). A proline-rich TGF-β-responsive transcriptional activator interacts with histone H3. *Genes Dev.* 9, 3051–3066.

Alfonso, P.J., Crippa, M.J., Hayes, J.J., and Bustin, M. (1994). The footprint of chromosomal proteins HMG-14 and HMG-17 on chromatin subunits. *J. Mol. Biol.* 236, 189–198.

Almer, A., Rudolph, H., Hinnen, A., and Horz, W. (1986). Removal of positioned nucleosomes from the yeast *PHO5* promoter upon *PHO5* induction releases additional upstream activating DNA elements. *EMBO J.* 5, 2689–2696.

Arents, G., and Moudrianakis, E. (1995). The histone fold: a ubiquitous architectural motif utilized in DNA compaction and protein dimerization. *Proc. Natl. Acad. Sci. USA* 92, 11170–11174.

Arents, G., Burlingame, R.W., Wang, B.-C., Love, W.E., Moudrianakis, E.N. (1991). The nucleosomal core histone octamer at 3.1Å resolution: a tripartite protein assembly and a left-handed superhelix. *Proc. Natl. Acad. Sci. USA* 88, 10148–10152.

Barton, M.C., and Emerson, B.M. (1995). Regulated expression of the β-globin gene locus in synthetic nuclei. *Genes Dev.* 8, 2453–2465.

Bannister, A.J., and Kouzarides, T. (1996). The CBP coactivator is a histone acetyltransferase. *Nature* 384, 641–643.

Bauer, W.R., Hayes, J.J., White, J.H., and Wolffe, A.P. (1994). Nucleosome structural changes due to acetylation. *J. Mol. Biol.* 236, 685–690.

Berger, S.L., Pina, B., Silverman, N., Marcus, G.A., Apapite, J., Regier, J.L., Triezenberg, S.J., and Guarente, L. (1992). Genetic isolation of ADA2: a potential transcriptional adaptor required for function of certain acidic activiation domains. *Cell* 70, 251–265.

Blank, T.A., and Becker, P.B. (1996). The effect of nucleosome phasing sequences and DNA topology on nucleosome phasing. *J. Mol. Biol.* 250, 1–8.

Braunstein, M., Rose, A.B., Holmes, S.G., Allis, C.D., and Broach, J.R. (1993). Transcriptional silencing in yeast is associated with reduced histone acetylation. *Genes Dev.* 7, 592–604.

Brownell, J.E., Zhou, J., Ranalli, T., Kobayashi, R., Edmondson, D.G., Roth, S.Y., and Allis, C.D. (1996). *Tetrahymena* acetyltransferase A: a homolog to yeast Gcn5p linking histone acetylation to gene activation. *Cell* 84, 843–851.

Bult, C.J., *et al.* (1996). Complete genome sequence of the methanogenic archaeon, *Methanococcus jannaschii*. *Science* 273, 1058–1073.

Cairns, B.R., Kim, Y.-J., Sayre, M.H., Laurent, B.C., and Kornberg, R.D. (1994). A mutisubunit complex containing the SWI1/ADR6, SWI2/SNF2, SWI3, SNF5, and SNF6 gene products isolayed from yeast. *Proc. Natl. Acad. Sci. USA* 91, 1950–1954.

Cairns, B.R. *et al.* (1996). RSC, an essential, abundant chromatin remodelling complex. *Cell* 87, 1249–1260.

Candau, R., Zhou, J., Allis, C.D., and Berger, S.L. (1997). Histone acetyltransferase activity and interaction with ADA2 are critical for GCN5 function *in vivo*. *EMBO J.* 16, 555–565.

Chang, L., Loranger, S.S., Mizzen, C., Ernst, S.G., Allis, C.D., and Annunziato, A.T. (1997). Histones in transit: cytosolic histone complexes and diacetylation of H4 during nucleosome assembly in human cells. *Biochemistry* 36, 469–480.

Chen, H., Lin, R.J., Schiltz, R.L., Chakravrti, D., Nash, A., Nagy, L., Privalsky, M.L., Nakatani, Y. and Evans, R.M. (1997). Nuclear receptor coactivator ACTR is a novel histone acetyltransferase and forms a multimeric activation complex with P/CAF and CBP/p300. *Cell* 90, 569–580.

Clark, D.J., and Kimura, T. (1990). Electrostatic mechanism of chromatin folding. *J. Mol. Biol.* 211, 883–896.

Clark, D.J., Hill, C.S., Martin, S.R., and Thomas, J.O. (1988). α-helix in the carboxy-terminal domains of histones H1 and H5. *EMBO J.* 7, 69–75.

Clark, D.J., O'Neil, L.P., and Turner, B.M. (1993). Selective use of H4 acetylation sites in the yeast *Saccharomyces cerevisiae*. *Biochem J.* 294, 557–561.

Cote, J., Quinn, J., Workman, J.L, and Peterson, C.L. (1994). Stimulation of GAL4 derivative binding to nucleosomal DNA by yeast SWI/SNF complex. *Science* 265, 53–60.

Csink, A.K., and Henikoff, S. (1996). Genetic modification of heterochromatin association and nuclear organization in *Drosophila*. *Nature* 381, 529–531.

Csink, A.K., Sass, G.L., and Henikoff, S. (1997). *Drosophila* heterochromatin: retreats for repeats. In 'Nuclear organization: chromatin structure and gene expression' (Eds. Van Driel, R., and Otte, A.P.). Oxford University Press.

Davey, C., Pennings, S., Meersseman, G., Wess, T.J., and Allen, J. (1995). Periodicity of strong nucleosome positioning sites around the chicken adult beta-globin gene may encode regularly spaced chromatin. *Proc. Natl. Acad. Sci. USA* 92, 11210–11214.

Dimitri, P. (1991). Cytogenetic analysis of the second chromosome heterochromatin of *Drosophila melanogaster*. *Genetics* 127, 553–564.

Dong, F., and van Holde, K.E.(1991). Nucleosome positioning is determined by the $(H3\text{-}H4)_2$ tetramer. *Proc. Natl. Acad. Sci. USA* 88, 10596–10600.

Drlica, K., and Rouviere-Yaniv, J. (1987). Histone-like proteins in bacteria. *Microbiol. Rev.* 51, 301–319.

Durrin, L.K., Mann, R.K., and Grunstein, M. (1992). Nucleosome loss activates *CUP1* and *HIS3* promoters to fully induced levels in the yeast *Saccharomyces cerevisiae*. *Mol. Cell. Biol.* 12, 1621–1629.

Eberl, D.F., Duyf, B.J., and Hilliker, A.J. (1993). The role of heterochromatin in the expression of a heterochromatic gene, the *rolled* locus of *Drosophila melanogaster*. *Genetics* 134, 277–292.

Edmondson, D.G., Smith, M.M., and Roth, S.Y. (1996). Repression domain of the yeast global repressor Tup1 interacts directly with histones H3 and H4. *Genes Dev.* 10, 1247–1259.

Eissenberg, J.C., James, T.C., Foster-Hartnett, D.M., Ngan, V., and Elgin, S. (1990). Mutation in a heterochromatin-specific chromosomal protein is associated with suppression of position-effect variegation in *Drosophila melanogaster*. *Proc. Natl. Acad. Sci. USA* 87, 9923–9927.

Elton, T.S., and Reeves, R. (1986). Purification and postsynthetic modifications of Friend erythroleukemic cell high mobility group protein HMG-1. *Anal. Biochem.* 157, 53–62.

Finch, J.T., and Klug, A. (1976). Solenoidal model for superstructure in chromatin. *Proc. Natl. Acad. Sci. USA* 73, 1897–1901.

Fletcher, T.M., and Hanse, J.C. (1996). The nucleosome array: structure/function relationships. *Crit. Rev. Eukaryot. Gene Expr.* 6, 149–188.

Franke, A., DeCamilis, M., Zink, D., Cheng, N., Brock, H.W., and Paro, R. (1992). *Polycomb* and *Polyhomeotic* are constituents of a multimeric protein complex in chromatin of *Drosophila melanogaster*. *EMBO J.* 11, 2941–2950.

Fritton, H.P., Igo-Kemenes, T., Nowock, J., Strech-Jurk, U., Theisen, M., and Sippel, A.E. (1984). Alternative sets of DNAase I-hypersensitive sites characterize the various functional states of the chicken lysozyme gene. *Nature* 311, 163–165.

Garcia-Ramirez, M., Rocchini, C. and Ausio, J. (1995). Modulation of chromatin folding by histone acetylation. *J. Biol. Chem.* 270, 17923–17928,

Gatti, M., and Pimpinelli, S. (1992). Functional elements in *Drosophila melanogaster* heterochromatin. *Ann. Rev. Genetics* 26, 239–275.

Gatti, M., Pimpinelli, S., and Santini, G. (1976). Characterization of *Drosophila* heterochromatin. I. Staining and decondensation with Hoechst 33258 and quinacrine. *Chromosoma* 75, 351–357.

Gaudreau, L., Schmid, A., Blaschke, D., Ptashne, M. and Horz, W. (1997). RNA polymerase II holoenzyme recruitment is sufficient to remodel chromatin at the yeast *PHO5* promoter. *Cell* 89, 55–62.

Gould, A.P., Lai, R.Y.K., Green, M.J., and White, R.A.H. (1990). Blocking cell division does not remove the requirement for *Polycomb* function in *Drosophila* embryogenesis. *Development* 110, 1319–1325.

Graziano, V., Gerchman, S.E., Schneider, D.K., and Ramakrishnan, V. (1994). Histone H1 is located in the interior of the chromatin 30 nm filament. *Nature* 368, 351–354.

Grunstein, M., hecht, A., Fisher-Adams, G., Wan, J., Mann, R.K., Strahl-Bolsinger, S., Laroche, T., and Gasser, S. (1995). The regulation of euchromatin and heterochromatin by histones in yeast. *J. Cell Sci.* 19 (suppl.), 29–36.

Han, M., and Grunstein, M. (1988). Nucleosome loss activates yeast downstream promoter *in vivo*. *Cell* 55, 1137–1145.

Han, M., Kim, U.-J., Kayne, P., and Grunstein, M. (1988). Depletion of histone H4 and nucleosomes activates the *PHO5* gene in *Saccharomyces cerevisiae*. *EMBO J.* 7, 2221–2228.

Hansen, J.C. (1997). The core histone amino-termini: combinatorial interaction domains that link chromatin structure with function. *Chem. Tracts: Biochem. Mol. Biol.* 10, 56–59.

Hecht, A., Strahl-Bolsinger, S., and Grunstein, M. (1996). Spreading of transcriptional repressor SIR3 from telomeric heterochromatin. *Nature* 383, 92–96.

Heinzel, T., Lavinsky, R.M., Mullen, T.M., Soderstrom, M., Laherty, C.D., Torchia, J., Yangm W.M., Brard, G., Ngo, S.D., Davie, J.R. *et al.* (1997). A complex containing N-CoR, mSin3A and histone deacetylase mediates transcriptional repression. *Nature* 387, 43–48.

Heitz, E. (1928). Das Heterochromatin der Moose. *Jahrbucher fur Wissenschaftliche Botanik* 69, 762–818.

Henikoff, S. (1993). Transcriptional activator components and poxvirus DNA-dependent ATPases comprise a single family. *Trends Biochem. Sci.* 18, 291–292.

Hensel, R., and Konig, H. (1988). Thermoadaptation of methanogenic bacteria by intracellular ion concentration. *FEMS Microbiol. Lett.* 49, 75–79.

Hirschhorn, J.N., Bortvin, A.L., Ricupero-Hovasse, S.L., and Winston, F. (1995). A new class of histone H2 mutants in *Saccharomyces cerevisiae* causes specific transcriptional defects *in vitro*. *Mol. Cell Biol.* 15, 1999–2009.

Howe, M., Dimitri, P., Berloco, M., and Wakimoto, B.T. (1995). *Cis*-effects of heterochromatin on heterochromatic and euchromatic gene activity in *Drosophila melanogaster*. *Genetics* 140, 1033–1045.

Huber, M.C., Graf, T., Sippel, A.E., and Bonifer, C. (1995). Dynamic changes in the chromatin of the chicken lysozyme gene domain during differentiation of multipotent progenitors to macrophages. *DNA Cell Biol.* 14, 397–402.

Imhof, A., Yang, X.J., Ogryzko, V.V., Nakatani, Y., Wolffe, A.P., Ge, H. (1997). Acetylation of general transcription factors by histone acetyltransferases. *Curr. Biol.* 7, 689–692.

Izban, M.G., and Luse, D.S. (1991). Transcription on nucleosomal templates by RNA polymerase II *in vitro*: inhibition of elongation with enhancement of sequence-specific pausing. *Genes Dev.* 5, 683–696.

James, T.C., and Elgin, S.C.R. (1986). Identification of a nonhistone chromosomal protein associated with heterochromatin in *Drosophila melanogaster* and its gene. *Mol. Cell. Biol.* 6, 3862–3872.

Juan, L.-J., Utley, R.T., Adams, C.C., Vettese-Dadey, M., and Workman, J.L. (1994). Differential repression of transcription factor binding by histone H1 is regulated by the core histone amino termini. *EMBO J.* 13, 6031–6040.

Kadosh, D., and Struhl, K. (1997). Repression by Ume6 involves recruitment of a complex containing Sin3 corepressor and Rpd3 histone deacetylase to target promoters. *Cell* 89, 365–371.

Kalpana, G.V., Marmon, S., Wang, W., Crabtree, G.R., and Goff, S.P. (1994). Binding and stimulation of HIV-1 integrase by a human homolog of yeast transcription factor SNF5. *Science* 266, 2002–2006.

Kamaka, R.T. (1997). Silencers and locus control regions: opposite sides of the same coin. *Trends Biochem. Sci.* 22, 124–128.

Karsten, M.M., Dorland, S., and Stillman, D.J. (1997). A large protein complex containing the yeast Sin3p and Rpd3 transcriptional regulators. *Mol. Cell. Biol.* 17, 4852–4858.

Kennison, J.A. (1995). The *Polycomb* and *trithorax* group proteins of *Drosophila*: transregulators of homeotic gene function. *Ann. Rev. Genet.* 29, 289–303.

Klug, A., Rhodes, D., Smith, J., Finch, J.T., and Thomas, J.O. (1980). A low resolution structure for the histone core of the nucleosome. *Nature* 287, 509–516.

Koonin, E.V., Zhou, S.B., and Lucchesi, J.C. (1995). The chromo superfamily: new members, duplication of the chromo domain and possible role in delivering transcriptional regulators to chromatin. *Nucl. Acids Res.* 23, 4229–4233.

Kornberg, R.D. (1974). Chromatin structure: a repeating unit of histones and DNA. *Science* 184, 868–871.

Kornberg, R.D., and Thomas, J.O. (1974). Chromatin structure: oligomers of the histones. *Science* 184, 865–868.

Kwon, H., Imbalzano, A.N., Khavari, P.A., Kingston, R.E., and Green, M.R. (1994). Nucleosome disruption and enhancement of activator binding by a human SWI/SNF complex. *Nature* 370, 477–481.

Landsman, D. (1996). Histone H1 in *Saccharomyces cerevisiae*: a double mystery solved? *Trends Biochem. Sci.* 21, 287–288.

Laurent, B.C., Treich, I., and Carlson, M. (1993). The yeast SWI/SNF protein has DNA-stimulated ATPase activity required for transcriptional activation. *Genes Dev.* 7, 583–591.

Laybourn, P.J., and Kadonaga, J.T. (1991). Role of nucleosomal cores and histone H1 in regulation of transcription by RNA polymerase II. *Science* 254, 238–245.

Le, M.-H., Duricka, D., and Karpen, G.H. (1995). Islands of complex DNA are widespread in *Drosophila* centric heterochromatin. *Genetics* 141, 283–303.

Lee, D.Y., Hayes, J.J., Pruss, D., and Wolfe, A.P. (1993). A positive role for histone acetylation in transcription factor access to nucleosomal DNA. *Cell* 72, 73–84.

Ling, Y., Harkness, T.A., Schultz, M.C., Fisher-Adams, G., and Grunstein, M. (1996). Yeast histone H3 and H4 amino termini are important for nucleosome assembly *in vivo* and *in vitro*: redundant and position-independent functions in assembly but not in gene regulation. *Genes Dev.* 10, 686–699.

Lohe, A.R., Hilliker, A.J., and Roberts, P.A. (1993). Mapping simple repeated DNA sequences in heterochromatin of *Drosophila melanogaster*. *Genetics* 134, 1149–1174.

Lonie, A., D'Andrea, R., Paro, R., and Saint, R. (1994). Molecular characterization of the *Polycomblike* gene of *Drosophila melanogaster*, a trans-acting negative regulator of homeotic gene expression. *Development* 120, 2629–2636.

Luger, K., Mader, A.W., Richmond, R.K., Sargent, D.F., and Richmon, T.J. (1997). Crystal structure of the nucleosome core particle at 2.8Å resolution. *Nature* 389, 251–260.

Mardian, J.K.W., Paton, A.E., Bunick, G.K., and Olins, D.E. (1980). Nucleosome cores have two specific binding sites for nonhistone chromosomal proteins HMG14 and HMG17. *Science* 209, 1534–1536.

McKeon, J., and Brock, H.W. (1991). Interaction of the *Polycomb* group genes with homeotic loci of *Drosophila. Roux's Arch. Dev. Biol.* 199, 387–396.

Messmer, S., Franke, A., and Paro, R. (1992). Analysis of the functional role of the *Polycomb* chromo domain in *Drosophila melanogaster. Genes Dev.* 6, 1241–1254.

Mizuguchi, G., Tsukiyama, T., Wisniewsky, J., and Wu, C. (1997). Role of nucleosome remodeling factor NURF in transcriptional activation of chromatin. *Mol. Cell* 1, 141–150.

Mizzen, C.A., Yang, X.J., Kokubo, T. Brownell, J.E., Bannister A.J., Owen-Hughes, T., Workman, J., Wang, L., Berger, S.L., Kouzarides *et al.* (1996). The $TAF_{II}250$ subunit of TFIID has histone acetyltransferase activity. *Cell* 87, 1261–1270.

Morse, R.H. (1989). Nuclosomes inhibit both transcriptional initiation and elongation by RNA polymerase III *in vitro. EMBO J.* 8, 2343–2351.

Nagy, L., Kao, H.-Y., Chakravarti, D., Lin, R.J., Hassig, C.A., Ayer, D.E., Schreiber, S.L., and Evans, R.M. (1997). Nuclear receptor repression mediated by a complex containing SMRT, mSin3A, and histone deacetylase. *Cell* 89, 373–380.

Ogryzko, V.V., Schiltz, R.L., Russanova, V., Howard, B.H., and Nakatani, Y. (1996). The transcriptional coactivators p300 and CBP are histone acetyltransferases. *Cell* 87, 953–959.

Oliva, R., Bazett-Jones, D.P., Locklear, L., and Dixon, G.G. (1990). Histone hyperacetylation can induce unfolding of the nucleosomal core particle. *Nucl. Acids Res.* 18, 2739–2747.

Orlando, V., and Paro, R. (1993). Mapping *Polycomb*-repressed domains in the *bithorax* complex using *in vivo* formaldehyde cross-linked chromatin. *Cell* 75, 1187–1198.

Orphanides, G., LeRoy, G., Change, C.-H., Luse, D.S., and Reinberg, D. (1998). FACT, a factor that facilitates transcript elongation through nucleosomes. *Cell* 92, 105–116.

Paro, R., and Hogness, D.S. (1991). The Polycomb protein shares a homologous domain with a heterochromatin-associated protein of *Drosophila. Proc. Natl. Acad. Sci. USA* 88, 263–267.

Pazin, M.J., Sheridan, P.L., Cannon, K., Cao, Z., Keck, J.G., Kadonaga, J.T., and Jones, K.A. (1996). NF-κB mediated chromatin reconfiguration and transcriptional activation of the HIV-1 enhancer *in vitro. Genes Dev.* 10, 37–49.

Pazin, M.J., Bhargava, P., Geiduschek, E.P., and Kadonaga, M. (1997). Nucleosome mobility and the maintenance of nucleosome positioning. *Science* 276, 809–812.

Perry, C.A., and Annunciato, A.T. (1989). Influence of histone acetylation on the solubility, H1 content and DNAase I sensitivity of newly assembled chromatin. *Nucl. Acids Res.* 17, 4275–4291.

Perry, M., and Chalkley, R. (1981). The effect of histone hyperacetylation on the nuclease sensitivity and solubility of chromatin. *J. Biol. Chem.* 256, 3313–3318.

Peterson, C.L., and Herskowitz, I. (1992). Characterization of the yeast *SWI1*, *SWI2*, and *SWI3* genes, which encode a global activator of transcription. *Cell* 68, 573–583.

Peterson, C.L., and Tamkun, J.W. (1995). The SWI-SNF complex: a chromatin remodelling machine? *Trends Biochem. Sci.* 20, 143–146.

Peterson, C.L., Dingwall, A., and Scott, M.P. (1994). Five SWI/SNF gene products are components of a large multisubunit complex required for transcriptional enhancement. *Proc. Natl. Acad. Sci. USA* 91, 2905–2908.

Pettijohn, D.E. (1988). Histone-like proteins and bacterial chromosome structure. *J. Biol. Chem.* 263, 12793–12796.

Pimpinelli, S., Berloco, M., Fanti, L., Dimitri, P., Bonaccorsi, S., Marchetti, E., *et al.* (1995). Transposable elements are stable structural components of *Drosophila melanogaster* heterochromatin. *Proc. Natl. Acad. Sci. USA,* 92, 3804–3808.

Platero, J.S., Hartnett, T., and Eissenberg, J.C. (1995). Functional analysis of the chromo domain of HP1. *EMBO J.* 14, 3977–3986.

Pogo, B.G., Allfrey, V.G., and Mirsky, A.E. (1966). RNA synthesis and histone acetylation during the course of gene activation in lymphocytes. *Proc. Natl. Acad. Sci. USA* 55, 805–812.

Raff, J.W., Kellum, R., and Alberts, B. (1994). The *Drosophila* GAGA transcription factor is associated with specific regions of heterochromatin throughout the cell cycle. *EMBO J.* 13, 5977–5983.

Rastelli, L., Chan, C.S., and Pirotta, V. (1993). Related chromosome binding sites for *Zeste*, *Supressor of zeste* and *Polycomb* group proteins in *Drosophila* and their dependence on *Enhancer of zeste* function. *EMBO J.* 12, 1513–1522.

Richmond, T.J., Finch, J.T., Rushton, B., Rhodes, D., and Klug, A. (1984). Structure of the nucleosome core particle at 7Å resolution. *Nature* 311, 532–537.

Ridsdale, J.A., Henzdel, M.F., Decluve, G.P., and Davie, J.R. (1990). Histone acetylation alters the capacity of the H1 histones to condense transcriptionally active/competent chromatin. *J. Biol. Chem.* 265, 5150–5156.

Roth, S.Y., and Allis, C.D. (1992). H1 phosphorylation and chromatin condensation: exceptions define the rule? *Trends Biochem. Sci.* 17, 93–98.

Roth, S.Y., Dean, A., and Simpson, R.T. (1990). Yeast α2 repressor positions nucleosomes in TRP1/ARS1 chromatin. *Mol. Cell. Biol.* 10, 2247–2260.

Sandeen, G., Wood, W.I., and Felsenfeld, G. (1980). The interaction of high mobility group proteins HMG14 and 17 with nucleosomes. *Nucl. Acids Res.* 8, 3757–3778.

Sandman, K., Krzycki, J.A., Dobrinski, B., Lurz, R., and Reeve, J.N. (1990). HMf, a DNA-binding protein isolated from the hyperthermophilic archaeon *Methanothermus fervidus*, is most closely related to histones. *Proc. Natl. Acad. Sci. USA* 97, 5788–5791.

Sandman, K., Grayling, R.A., Dobrinski, B., Lurz, R., and Reeve, J.N. (1994). Growth-phase dependent synthesis of histones in the archaeon *Methanothermus fervidus*. *Proc. Natl. Acad. Sci. USA* 91, 12624–12628.

Saragosti, S., Moyne, G., and Yaniv, M. (1980). Absence of nucleosomes in a fraction of SV40 chromatin between the origin of replication and the region coding for the late leader RNA. *Cell* 20, 65–73.

Saunders, W.S., Chue, C., Goebl, M., Craig, C., Clark, R.F., Powers, A.J., *et al.* (1993). Molecular cloning of a human homologue of *Drosophila* heterochromatin protein HP1 using anti-centromere autoantibodies with anti-chromo specificity. *J. Cell Sci.* 104, 573–582.

Schmidt, M.B. (1990). More than just 'histone-like' proteins. *Cell* 53, 451–453.

Schwarz, P.M., and Hansen, J.C. (1994). Formation and stability of higher order chromatin structures. *J. Biochem.* 269, 16284–16298.

Shen, X., Yu, L., Weirm J.W., and Gorovsky, M.A. (1995). Linker histones are not essential and affect chromatin condensation *in vivo*. *Cell* 82, 47–56.

Shindo, H., Iwaki, T., Ieda, R., Kurumizaka, H., Ueguchi, C., Mizuno, T., Morikawa, S., Nakamura, H., and Kuboniwa, H. (1995). Solution structure of the DNA binding domain of a nucleoid-assoicated protein, H-NS, from *Escherichia coli*. *FEBS Lett.* 360, 125–131.

Simon, J. Chiang, A., Bender, W., Shimell, M.J., and O'Connor, M. (1993). Elements of the *Drosophila bithorax* complex that mediate repression by *Polycomb* group products. *Dev. Biol.* 158, 131–144.

Simpson, R.T., and Stafford, D.W. (1983). Structural features of a phased nucleosome core particle. *Proc. Natl. Acad. Sci. USA* 80, 51–55.

Sinclair, D.A., Ruddell, A.A., Brock, J.K., Clegg, N.J., Lloyd, V.K., and Grigliatti, T.A. (1992). A cytogenetic and genetic characterization of a group of closely linked second chromosome mutations that suppress position-effect variegation in *Drosophila melanogaster*. *Genetics* 130, 333–344.

Sirotkin, A.M., Edelman, W., Cheng, G., Klein-Szanto, A., Kucherlapati, R., Skoultchi, A.I. (1995). Mice develop normally without the H1^{0} linker histone. *Proc. Natl. Acad. Sci. USA* 92, 6434–6438.

Spencer, T.E., Jenster, G., Burcin, M.M., Allis, C.D., Zhou, J., Mizzen, C.A., McKenna, N.J., Onate, S.A., Tsai, S.Y., Tsay, M.J. *et al.* (1997). Steroid receptor coactivator-1 is a histone acetyltransferase. *Nature* 389, 194–198.

Starich, M.R., Sandman, K., Reeve, J.N., and Summers, M.F. (1996). NMR structure of HMfB from the hyperthermophile, *Methanothermus fervidus*, confirms that this archaeal protein is a histone. *J. Mol. Biol.* 255, 187–203.

Sterner, R., Vidali, G., and Allfrey, V.G. (1981). Studies of acetylation and deacetylation in high mobility group proteins: identification of the sites of acetylation in high mobility group proteins 14 and 17. *J. Biol. Chem.* 156, 8892–8895.

Strahl-Bolsinger, S., Hecht, A., luo, K., and Grunstein, M. (1997). Sir2 and Sir4 interactions differ in core and extended telomeric heterochromatin in yeast. *Genes Dev.* 11, 83–93.

Stroup, D., and Reeve, J.N. (1992). Histone HMf from the hyperthermophilic archaeon *Methanothermus fervidus* binds to DNA *in vitro* using physiological conditions. *FEMS Microbiol. Lett.* 91, 271–276.

Studitsky, V.M., Kassavetis, G.A., Geiduschek, E.P., and Felsenfeld, G. (1997). Mechanism of transcription through the nucleosome by eukaryotic RNA polymerase. *Science* 278, 1960–1963.

Tabassum, R., Sandman, K.M., and Reeve, J.N. (1992). HMt, a histone-related protein from *Methanobacterium thermoautotrophicum* strain DH. *J. Bacteriol.* 174, 7890–7895.

Takayanagi, S., Morimura, S., Kusaoke, H., Yokoyama, Y., Kano, K., and Shioda, M. (1992). Chromosomal structure of the halophilic Archaebacterium *Halobacterium salinarium. J. Bacteriol.* 174, 7207–7216.

Tamkun, J.W., Deuring, R., Scott, M.P., Kissinger, M., Pattatucci, A.M., Kaufman, T.C. *et al.* (1992). *brahma*: a regulator of *Drosophila* homeotic genes structurally related to the yeast transcriptional activator SNF2/SWI2. *Cell* 68, 561–572.

Tartof, K.D., Hobbs, C., and Jones, M. (1984). A structural basis for variegating position effects. *Cell* 37, 869–878.

Thoma, F., Losa, R., and Klug, A. (1979). Involvement of histone H1 in the organization of the nucleosome and of salt dependent superstructures of chromatin. *J. Cell Biol.* 83, 403–427.

Truss, M., Bartsch, J., Schelbert, A., Hache, R.G.G., and Beato, M. (1995). Hormone induces binding of receptors and transcription factors to a rearranged nucleosome on the MMTV promoter *in vivo. EMBO J.* 14, 1737–1751.

Tsukiyama, T., Becker, P.B., and Wu, C. (1994). ATP-dependent nucleosome disruption at a heat-shock promoter mediated by binding of GAGA-transcription factor. *Nature* 367, 525–532.

Tsukiyama, T., and Wu, C. (1995). Purification and properties of an ATP-dependent nucleosome remodeling factor. *Cell* 83, 1011–1020.

Turner, B.M. (1991). Histone acetylation and control of gene expression. *J. Cell. Sci.* 99, 13–20.

van Driel, R., and Otte, A.P. (1997). Nuclear organization, chromatin structure and gene expression. Oxford University Press, Oxford.

Varga-Weisz, P.D., Wilm, M., Bonte, E., Dumas, K., Mann, M., and Becker, P. (1997). Chromatin-remodelling factor CHRAC contains the ATPases ISWI and topoisomerase II. *Nature* 388, 598–602.

Venter, U., Svaren, J., Schmitz, J., Schmid, A., and Horz, W. (1994). A nucleosome precludes binding of the transcription factor Pho4 *in vivo* to a critical target site in the PHO5 promoter. *EMBO J.* 13, 4848–4855.

Vettese-Dadey, M., Walter, P., Chen, H., Juan, L.-J, and Workman, J.L. (1994). Role of the histone amino termini in facilitated binding of a transcription factor, GAL4-AH, to nucleosomal cores. *Mol. Cell. Biol.* 14, 970–981.

Vis, H., Mariani, M., Vorgias, C.E., Wilson, K.S., Kaptein, R., and Boelens, R. (1995). Solution structure of the HU protein from *Bacillus stearothermophilus*. *J. Mol. Biol.* 254, 692–703.

Wade, P.A., Pruss, D., and Wolffe, A.P. (1997). Histone acetylation: chromatin in action. *Trends Biochem. Sci.* 22, 128–132.

Wang, L., Mizzen, C., Ying, C., Candau, R., Barlev, N., Brownell, J., Allis, C.D., and Berger, S.L. (1997). Histone acetyltransferase activity is conserved between yeast and human GCN5 and is required for complementation of growth and transcriptional activation. *Mol. Cell. Biol.*17, 519–527.

Wang, W.D., *et al.* (1996). Diversity and specialization of mammalian SWI/SNF complexes. *Genes Dev.* 10, 2117–2130.

Weiler, K.S., and Wakimoto, B.T. (1995). Heterochromatin and gene expression in *Drosophila*. *Ann. Rev. Genetics* 9, 577–605.

Wilson, C.J., Chao, D.M., Imbalzano, A.N., Schnitzler, G.R., Kingstin, R.E. and Young, R.A. (1996). RNA polymerase II holoenzyme contains *SWI/SNF* regulators involved in chromatin remodeling. *Cell* 84, 235–244.

Wolffe, A.P. (1995). Chromatin: Structure and Function. Academic Press, London.

Wolffe, A.P. (1997). Sinful repression. *Nature* 387, 16–17.

Wolffe, A.P., and Pruss, D. (1996). Targeting chromatin disruption: transcription regulators that acetylate histones. *Cell* 84, 817–189.

Wu, C., Binham, P.M., Livak, K.J., Holmgren, R. and Elgin, S.C.R. (1979). The chromatin structure of specific genes: evidence for higher order domains of defined DNA sequences. *Cell* 16, 797–806.

Wu, R.S., Panusz, H.T., Hatch, C.L., and Bonner, W.M. (1984). Histones and their modifications. *CRC Crit. Rev. Biochem.* 20, 201.

Wustmann, G., Szidonya, J., Taubert, H., and Reuter, G. (1989). The genetics of position-effect variegation modifying loci in *Drosophila melanogaster*. *Mol. Gen. Genetics* 217, 520–527.

Yang, X.J., Ogryzko, V.V., Nishikawa, J., Howard, B.H., and Nakatani, Y. (1996). A p300/CBP-associated factor that competes with the adenoviral oncoprotein E1A. *Nature* 382, 319–324.

Yang, W.M., Inouye, C., Zeng, Y.Y., Bearss, D., and Seto, D. (1997). Transcriptional repression by YY1 is mediated by interaction with a mammalian homolog of the yeast global regulator RPD3. *Proc. Natl. Acad. Sci. USA* 93, 12845–12850.

Zhang, W., Bonem J.R., Edmondson, D.G., Turner, B., and Roth, S.Y. (1998). Essential and redundant functions of histone acetylation revealed by mutation of target lysines and loss of the Gcn5p acetyltransferase. *EMBO J.* 17, 3155–3167.

Zillig, W., Palm, P., Reiter, W., Gropp, F., Puhler, G., and Klenck, H. (1988). Comparative evaluation of gene expression in archaebacteria. *Eur. J. Biochem.* 173, 473–482.

Chapter 8

Nuclear Matrix, Chromosome Scaffolds and Transcriptional Factories

For a long time the fine structure of the eukaryotic nucleus was a rather neglected field of research. Apart from chromatin, nucleoli and nuclear pores there seemed to be little of interest for cell biologists, and many molecular biologists focused their efforts on recreating simplified *in vitro* transcription systems. During the last two decades, however, several laboratories have made discoveries that have revolutionized our views of nuclear ultrastructure. We now know that the eukaryotic nucleus has a distinct and highly dynamic structure that influences many aspects of transcriptional mechanisms (Lamond and Earnshaw, 1998).

In this chapter we will investigate a few of the most salient concepts that have emerged from this relatively new and fascinating interface between molecular biology and cell biology. Much remains to be explored, but some fundamental facts are starting to become clearly established. We will see that eukaryotic chromosomes, together with the various transcriptional- and post-transcriptional processing machineries attached to them, are arranged with a high degree of spatial order within the three-dimensional network formed by the nuclear matrix. This new insight into the complex morphology of nuclear architecture poses a new challenge for our understanding of the topological constraints affecting nuclear processes and their functional integration with each other. Many of these cell biological findings will therefore eventually have to be fully taken into account in the models describing the molecular mechanisms that underpin differential gene expression in eukaryotic cells.

8.1. What is Nuclear Matrix?

In 1942 biochemists already knew about the existence of an insoluble nuclear protein fraction that was left behind after extraction of purified nuclei with high ionic strength buffers (Mayer and Gulick, 1942). It then took another thirty years before further progress was made. Several independent discoveries transformed our view of the eukaryotic nucleus as 'a bag of chromatin' to an image of a highly structured organelle with sophisticated internal organization (Berezney and Coffey, 1975; Cook and Brazell, 1975). This paradigm shift was the consequence of detailed biochemical and electron microscopic studies that resulted in the discovery of the nuclear matrix.

So what is the nuclear matrix? Although the term was originally coined as a morphological term to describe the histologic features of the interior of nuclei (Fawcett, 1966), it is nowadays used mostly to describe a biochemically-defined entity: if isolated nuclei purified from a number of different eukaryotic organisms are subjected to DNAase I treatment, followed by differential washes in salt-containing buffers, one eventually ends up with a preparation that lacks most of the chromatin present in intact nuclei, but still retains much of the original nuclear morphology (Figure 8.1).

The nuclear matrix left behind after such treatment is therefore a 'non-chromatin' structure that consists of clearly recognisable nucleoli surrounded by a fibrogranular network of thin protein filaments. Inspection of intact nuclei reveals the presence of a similar network tightly intermixed with the euchromatic (transcriptionally active) portion of the decondensed chromosomes. The close association of euchromatin with the nuclear matrix immediately suggests an important role of the nuclear matrix in controlling the accessibility and spatial distribution of transcriptionally active DNA within the eukaryotic nucleus.

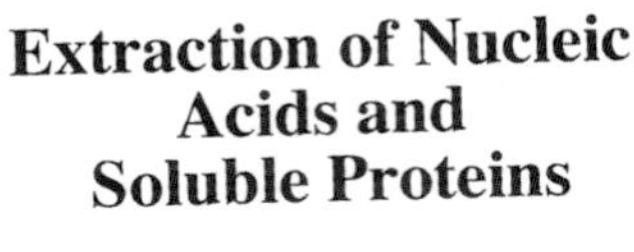

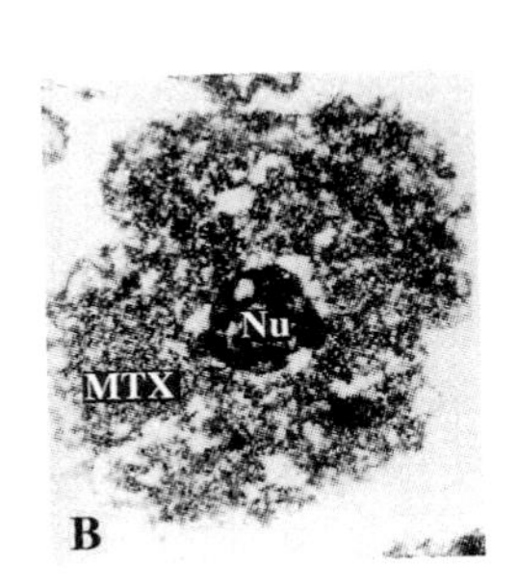

Figure 8.1. What is The Nuclear Matrix ?
Nuclease treatment and various salt extractions of purified nuclei leave a protein-meshwork (nuclear matrix; panel *B*) behind that retain most of the ultrastructural morphology of the original nuclei (panel *A*). *Nu*, nucleolus; *CC*, condensed chromatin; *MTX*, nuclear matrix. After Berezney *et al.*, 1995.

8.2. Biochemical Composition of the Nuclear Matrix

Methods for Nuclear Matrix Preparation

As mentioned earlier, the nuclear matrix is mainly defined by biochemical criteria. Different laboratories have therefore come up with different types of purification procedures and named the end product differently, including e.g. 'nucleoskeletons', 'nuclear cages' and 'nucleoids'.

Although these preparations may differ in their detailed morphology and behaviour in various *in vitro* assays, the overall consensus is that they are all (more or less!) representative examples of the same nuclear matrix present in intact nuclei *in vivo*. It has to be recognized, however, that small differences in the protocols for isolating the nuclear matrix (or the use of different chemicals) can have a substantial effect on the type of nuclear matrix isolated. Figure 8.2 illustrates the steps leading to an increasing depletion of various types of molecules from nuclei and the concomittant changes in morphology and molecular composition. More recent variations of the protocol involve the use of detergents such as LIS (Mirkovitch *et al.*, 1984), or electrophoresis applied to agarose-embedded cells to remove solubilized chromatin from the nuclear matrix (Jackson and Cook, 1985; Jackson *et al.*, 1988). These modifications of nuclear matrix preparation techniques were specifically developed to overcome

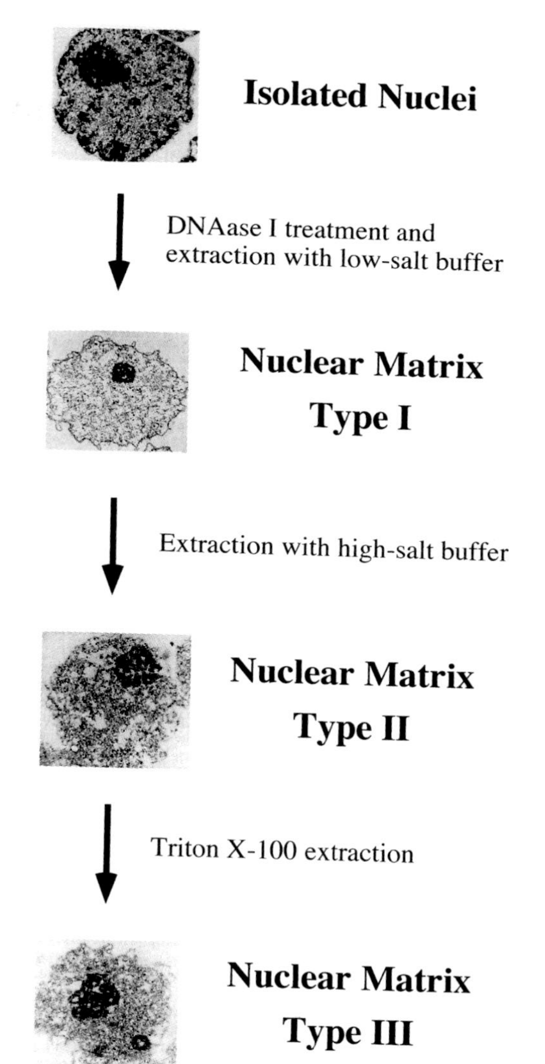

Figure 8.2. Biochemical Definition of Different Nuclear Matrix Types.
Sequential treatments extract different types of molecules from purified nuclei. Note the good morphological conservation up to the final stages of nuclear matrix purification despite the extraction of a major percentage of nucleic acids and nuclear soluble proteins. After Berezney *et al.*, 1995.

potential artefacts that might be caused by the nonphysiologically high salt extraction procedures used in the more traditional methods (Kirov *et al.*, 1984; Mirkovitch *et al.*, 1984; Cook, 1988; Jacks and Eggert, 1992). For example, a major concern was that the transcription activities found to be associated with the nuclear matrix (see below) were the result of non-specific precipitation of soluble transcription complexes onto nuclear matrix components during the high salt extraction steps. Ironically perhaps, comparison of nuclear matrix preparations obtained by LIS or electrophoretic extraction essentially confirmed, rather than contradicted, the validity of the results obtained with nuclear matrices purified by the high salt extraction procedures (Jackson and Cook, 1985; Jackson and Cook, 1986). In fact, these studies contributed to identifying low ionic strength buffers as a major cause leading to various artefacts by disrupting the anchoring of transcriptionally active chromatin to the nuclear matrix (Razin *et al.*, 1985). Another critical factor important for the preservation of good nuclear matrix morphology was found to be the necessity to include RNAase inhibitors in the extraction buffers, because RNA-containing protein complexes are now known to make a major contribution to the nuclear matrix architecture (Fey *et al.*, 1986; He *et al.*, 1990). Treatment of cells with transcription inhibitors destabilizes the nuclear matrix purified from them (Nickerson *et al.*, 1989).

Together the many different experimental approaches have thus led to a good and reproducible characterization of the components and structure of the nuclear matrix. It has to be realized, however, that even the best protocols may introduce unknown changes in the nuclear matrix during the extraction procedure and all results have to be interpreted with caution and will eventually need to be supplemented with direct *in vivo* observations.

Nuclear Matrix Proteins

The fact that up to 90% of the proteins present in isolated nuclei can be removed without significantly disturbing the gross morphology points immediately towards the major architectural role the nuclear matrix plays in determining and maintaining the overall nuclear structure. In this respect the nuclear matrix can be regarded as an extension of the cytoplasmic network of intermediate filaments that are crucially involved in controlling cell shape. An interesting, but largely unproven hypothesis exploring how such close coupling of the nuclear and cytoplasmic intermediate filament networks could be used to change gene expression patterns has been proposed by Ingber *et al.* (1994).

Examination of the protein composition of purified nuclear matrices from a number of eukaryotic sources by SDS-gel electrophoresis reveals the presence of three major proteins in addition to numerous other proteins present in substoichiometric amounts. These three major bands correspond mainly to the lamins A/B/C that form a dense network at the nuclear periphery (Aaronson and Blobel, 1975; Gerace and Blobel, 1980; McKeon *et al.*, 1986), and are also found internally during the G_1 phase of the cell cycle (Bridger, 1993). In addition to the lamins up to 200 different proteins can be detected by 2D-electrophoresis, including common and tissue-specific nuclear matrix components (Fey and Penman, 1988; Stuurman *et al.*, 1989; Stuurman *et al.*, 1990). Post-translational modifications, especially phosphorylation, also seem to play a major role in determining the structure of the nuclear matrix. The existence of numerous kinases in nuclear matrix preparations, including protein kinase C, are a strong indication that there may be signal transduction pathways leading directly from cell-surface membrane receptors all the way to the heart of the nucleus (Capitani *et al.,* 1987). Such mechanisms could establish links between control of nuclear architecture and large scale changes in the transcription patterns of numerous genes during normal cell differentiation and in cancerous malignancies. There is indeed a substantial body of evidence documenting detectable changes in nuclear matrix properties in cancer cells in support of such a view. (Getzenberg *et al.*, 1991; Khanuja *et al.*, 1993; Partin *et al.*, 1983; Pienta and Lehr, 1993; Bidwell *et al.*, 1994; Keesee *et al.*, 1994).

One of the major structural elements in purified nuclear matrices visible by high resolution electron microscopy are the 10nm core filaments organized into a highly branched three-dimensional network, that is revealed after high salt extraction (2M NaCl; He *et al.*, 1990). Unfortunately, despite systematic attempts to characterize nuclear matrix proteins, the core filaments have up to now eluded any attempt to be defined on the protein level. One candidate for involvement in core filament assembly is NuMA (nuclear mitotic apparatus protein), but attempts to use antibodies directed against NuMA to label core filaments have only been partially successful (Zeng *et al.*, 1994). It is therefore likely that NuMA only participates in a limited way in core filaments, and that the major components have yet to be identified.

Nuclear Matrix RNA

Nuclear matrix preparations that retain original nuclear morphology have to be prepared under conditions designed to protect RNA from degradation. Even under careful conditions, however, approximately 30% of nuclear RNA is usually lost (He *et al.*, 1990). The remaining 70% of RNAs are composed of roughly equal amounts of hnRNA and rRNA precursors. An interesting fact is that the majority of hnRNAs present in nuclei are not mRNA precursors and are therefore never exported into the cytoplasm to encode proteins (Harpold *et al.*, 1981; Salditt-Georgieff, *et al.*, 1981). The abundance of 'non-coding' hnRNA in nuclei, coupled with RNA-dependent stabilization of chromatin structures *in vitro* suggests a structural role of hnRNA in maintaining nuclear architecture. This correlates with the disruption of nuclear ultrastructure observed *in vivo* when cells are grown in presence of transcription inhibitors actinomycin D or 5,6-dichloro-1-β-ribofuranosylbenzimidazole, indicating that continuous RNA synthesis is required for the structural role of nuclear RNAs (Nickerson *et al.*, 1989).

8.3. The Nuclear Matrix Organizes Eukaryotic Chromatin into Higher Order Structures

Chromosomes are 'Parked' in Chromosomal Territories during Interphase

The only stage where all chromosomes are highly condensed and microscopically visible is during mitosis. In comparison, interphase nuclei do not contain many recognizable features and for a long time it was thought that decondensed chromosomes were randomly spread throughout the nucleus. It came therefore as a considerable surprise when *in situ* hybridization studies with chromosome-specific probes revealed distinct nuclear subdomains where particular chromosomes are reproducibly located. The nuclear subcompartment allocated to specific chromosomes is often referred to as 'chromosomal territories' and is conceptually equivalent to a reserved 'parking space' for specific chromosomes during interphase (Manuelides, 1985; Schardin *et al.*, 1985; Manuelides, 1990; Hadlaczky *et al.*, 1986; Cremer *et al.*, 1993).

With hindsight, the concept of chromosomal territories seems almost a logical necessity, since such an arrangement allows the mitotically condensed chromosomes to unfold in a controlled manner during interphase without becoming irreversibly entangled with each other. It also implies that interphase chromosomes are arranged with a high degree of spatial order within nuclei with potentially important consequences for the expression of genes located on them. Further down we shall see that transcription complexes are probably organized in discrete 'transcription factories' within restricted areas of the nuclear matrix. The three-dimensional structure of a chromosome therefore might determine the accessibility of such transcriptional factories to particular parts of the chromosome, and thus act as a genomic-scale regulatory mechanism for enhancing or limiting transcription of individual chromosomal domains. This view is supported by the observation that transcriptionally active genes are often located at the periphery of individual chromosomal territories (Kurtz *et al.*, 1996; Wansink *et al.*, 1996), and near the nuclear envelope (Marshall *et al.*, 1996; Ferreira *et al.*, 1997).

Structure of Interphase Chromosomes: Arrangement of Chromatin Fibres into Loops

Up to now we have focused most of our attention on the nuclear matrix left behind after extensive DNAase I digestion of purified nuclei. If we want to find out more about the arrangement of chromatin fibres within the nucleus, we need to study the molecular characteristics of the released and retained chromatin. This is a controversial subject area because chromatin structure and organization are highly dependent on ionic concentrations, and different estimates for the lengths of chromatin loops along the nuclear matrix have been obtained with different techniques. Probably the most careful studies carried out under near-physiological conditions (avoiding the use of hypotonic buffers, high concentrations of magnesium ions and NaCl) have utilized agarose-embedded cells. Measurements based on the quantity and size of chromatin remaining after endonuclease treatment yielded an *average* chromatin loop size of 86 kb in HeLa cells (Jackson *et al.*, 1990). This suggests that chromosomal DNA is organized into loops with nuclease-resistant attachment points spaced at fairly regular intervals (reviewed in Vogelstein *et al.*, 1980; Gasser and Laemmli, 1987; Zlatanova and van Holde, 1992; Razin *et al.*, 1993).

So what are these attachment sequences in DNA? Matrix attachment sequences ('MARs'; often also referred to as 'SARs' for scaffold attachment regions) have been found to be AT-rich sequences without strict consensus (Mirkovitch *et al.*, 1984), that are often found near transcriptional enhancers (Cockerill and Yuen, 1986; Jarman and Higgs, 1988; Gasser and Laemmli, 1986; Cockerill and Yuen, 1987), or map near chromatin boundaries (Phi-Van and Stratling, 1988; Bode and Maass, 1988; Levy-Wilson and Fortier, 1989) and even occur within genes (Kas and Chasin, 1987). Although most MAR sites are presumably universally recognized in different cell types, there is evidence for the existence of tissue-specific nuclear matrix components. MARs can have a substantial influence on the efficiency of transcription of genes located near them (see next section) and are therefore very obvious targets for gene expression control mechanisms. Most of the available evidence points tentatively towards gene-specific activators as crucial elements in tissue-specific MAR function.

Eukaryotic Transcription and Transcript Processing Occurs in Spatially-Controlled Nuclear 'Factories'

It has been known for a considerable time that actively transcribed genes, $RNAP_{II}$ and newly synthesized mRNAs are specifically retained in purified nuclei that have been depleted of bulk chromatin after DNAase I or restriction enzyme digestion (Herman *et al.*, 1978; Jackson and Cook, 1985). Similarly, a wide variety of gene-specific transcriptional activators (such as Sp1, Oct-1 and AP-1) have been found to be tightly associated with purified nuclear matrix preparations (van Wijnen *et al.*, 1993). These results are backed up by the *in vivo* distribution of transcriptionally active and inactive genes: matrix-associated DNA from chicken oviduct cells is enriched for actively transcribed ovalbumin genes, but not for β-globin genes that are switched off in this cell type (Robinson *et al.*, 1982; Ciejek *et al.*, 1983). Similar observations have been made with rRNA encoding genes (Pardoll and Vogelstein, 1980; Keppel, 1986).

Such observations have several implications, but most importantly suggest that actively transcribed DNA is in some form specifically anchored to the nuclear matrix and thus prevented from diffusing out of the nucleus after nuclease treatment. Since chromatin fibres are organized in loops with an average length of 86 kb in interphase nuclei (see above), it follows that actively transcribed genes must be located near the matrix attachment points to explain the preferential retention of transcribed genes in purified nuclear matrices. Any genes located near the tips of chromatin loops would presumably be easily accessible to DNAase I and efficiently extracted from the matrix. The *in vivo* association of actively transcribed genes with the nuclear matrix implies that the transcriptional machineries are preferentially located near these nuclear matrix anchoring points. It has indeed been shown that addition of MAR sequences increases the transcription of transgenic constructs and makes them less suceptible to inappropriate regulatory influences of the surrounding genes (positional effects; Stief *et al.*, 1989; Phi-Van *et al.*, 1990; Klehr *et al.*, 1991). On the molecular level, it can be imagined that the presence of strong MAR sites within a particular chromosomal region allows efficient anchorage of such sequences to the nuclear matrix, leading to the establishment of an independent chromatin loop and bringing enhancer sequences closer to the transcriptional machineries associated with the nuclear matrix. MAR sequences are also often associated with consensus sites recognized by topoisomerase II, and this raises the possibility that chromatin loops are topological units with individually adjustable degrees of supercoiling density.

How do such events on the supramolecular level influence the way we need to think about transcription on the molecular level? For a long time it was generally assumed that transcription factors were predominantly guided to their genomic targets via diffusion-controlled molecular interactions. Although specific protein-nucleic acid contacts do indeed play a major role *in vivo* (Chapter 3), they are almost certainly substantially enhanced and complemented in their specificity through a high degree of nuclear compartmentalization. In previous chapters we have already encountered persuasive evidence (e.g. from studies of eukaryotic RNAP holoenzymes) that many of the components involved in transcription are arranged as large complexes, rivalling other macromolecular assemblies (such as e.g. ribosomes) in size. The proposed functional linkage of transcription with RNA processing, DNA repair and chromatin modification systems in the holoenzyme model suggests that many activities, that previously were thought to be separate, occur in close physical proximity to each other with a high degree of spatial coordination under *in vivo* conditions. The concept of '$RNAP_{II}$ transcription factories' takes this view one step further by combining the idea of large transcription complexes with the existence of topologically distinct chromatin loops.

The distribution of $RNAP_{II}$ within the nucleus can be determined by immunolocalization techniques with antibodies directed against $RNAP_{II}$-specific epitopes (Bregman *et al.*, 1995; Wei *et al.*, 1995). Similarly, the sites of active mRNA synthesis can be visualized through the α-amanitin sensitive incorporation of labelled nucleotides (Jackson *et al.*, 1993; Wansink *et al.*, 1993). The results from such experiments have revealed many hundred discrete 'transcription foci,' rather than non-distinct overall nuclear labelling. Significantly, these foci are still detectable after most of the chromatin is removed (Xing and Lawrence, 1991; Jackson *et al.*, 1993). The simplest interpretation compatible with these observations is that $RNAP_{II}$-containing transcription complexes are highly localized in 'transcription factories' that are attached to the nuclear matrix. Since an average eukaryotic cell transcribes thousands of genes simultanously, it follows that each of the foci must be engaged in the transcription of several different genes.

Such a model has several startling implications. Firstly, the attachment of $RNAP_{II}$ to the nuclear matrix means that any DNA/chromatin template has to be physically fetched towards the transcription factory for transcription, because $RNAP_{II}$ would not be free to find promoters by diffusion throughout the nucleus. Secondly, transcription of a DNA template by immobilized $RNAP_{II}$ generates supercoils that must be effectively neutralized by topoisomerase activities to allow transcription to proceed (Cook, 1994). Finally, the generation of pre-mRNA transcripts from a limited number of distinct transcription factories creates the possibility of channelling these transcripts efficiently into the post-transcriptional processing machinery, including RNA splicing, transport towards the nuclear periphery, and export into the cytoplasm (reviewed in Jiminez-Garcia and Spector, 1993; Greenleaf, 1993). This seems indeed to be the case in intact nuclei *in vivo*. Fluorescent *in situ* hybridization techniques have shown that certain cellular and viral pre-mRNAs are distributed as curved linear tracks radiating from transcription factories out towards the nuclear periphery (Lawrence *et al.*, 1989; Huang and Spector, 1991; Raap *et al.*, 1991; Sui and Spector, 1991; Carter *et al.*, 1993; Xing *et al.*, 1993). Certain tracks cross 'speckled domains' that are known to contain components of the splicing machinery, whereas others do not. Other pre-mRNAs do not appear as tracks, but display an elongated (or punctuated) *in situ* hybridization pattern, suggesting the presence of alternative RNA transport, processing and storage systems within the nucleus. These observations are particularly significant in the light of recent discoveries showing that elongating $RNAP_{II}$s are associated with splicing- and polyadenylation factors (McCracken *et al.*, 1997).

Transcription During Mitosis: Inheritance of Gene Expression Patterns through Segregating Transcription Factories?

Once during every cell cycle the nucleus undergoes dramatic morphological changes that involve an extensive (albeit temporary!) breakdown of higher order nuclear structures, followed by their reconstruction after completion of mitosis. This is, without doubt, a rather traumatic event for transcriptional machineries that are either trying to initiate new transcripts during this stage of the cell cycle or are caught in the middle of their elongation phase. We have already seen earlier (Chapter 5) that the transcription of very large genes is probably so extensively disrupted at this stage that no mature mRNAs can be produced from such genes during periods of rapid nuclear division. This conclusion is backed up by biochemical studies, showing that incomplete transcripts are aborted during mitosis (Shermoen and O'Farrell, 1991). Such observations indicate that mitotically condensed chromatin is transcriptionally essentially silent (although many of the key components of the transcriptional machineries are still present and potentially active during this stage of the cell cycle [e.g. Johnson and Holland, 1965]). While a temporary stop of ongoing transcriptional activities may not pose any particular difficulties for cells on the whole, an interesting problem arises: if transcription ceases altogether, how can cells propagate the differential transcription status of the genome from one generation to the next? We have already seen previously (Chapter 7) that the transcription of many genes is highly dependent on the formation of localized 'active' chromatin domains that are set up by gene-specific transcription factors, coactivators, ATP-driven chromatin remodelling complexes and the transcriptional activities of elongating RNAPs. Due to the extensive compaction of chromosomal structures during mitosis it is likely that the structural integrity of such activated chromatin domains can not be easily maintained on mitotic chromosomes. The solution to this problem requires a 'molecular bookmarking' system to encode the various chromatin configurations that were present before compaction. One possible insight into the biochemical basis of a bookmarking system has emerged from the work of Michelotti *et al.* (1997), suggesting that regions of single-stranded DNA within active promoters could serve as molecular tags that earmark genes for rapid reactivation of transcription after completion of mitosis. The molecular encoding of gene expression states during the mitotic process could, however, also result directly from the supramolecular organization of transcription factories along the central chromosomal axis (Cook, 1994). This theory depends on the supposition that individual chromatin loops are anchored at their base onto transcriptional factories, which in turn are attached to the nuclear matrix. If these immobilized transcriptional factories are evenly divided up during the division of the chromosomal scaffold, the associated duplicated genomic loops could maintain their 'activated' state simply by remaining attached to the segregating transcriptional factories. Such a mechanism would, in principle, allow defined genomic expression patterns to be inherited over multiple cell division cycles with the help of a relatively straightforward topological mechanism. Although this theory is still essentially unproven, it constitutes one of the clearest examples of how key aspects of eukaryotic transcriptional control mechanisms could result directly from the structural order intrinsically present in chromosomes and nuclei.

It seems likely that further advances in our understanding of the molecular basis of transcriptional control mechanisms will in future depend more and more on gathering further insights into the structure and function of nuclear matrix components. With all the data before us, although clearly incomplete and possibly wrong in many important details, the potential role of nuclear organization in determining many aspects of the global gene expression programs of eukaryotic cells can no longer be easily ignored. In the next chapter

(Chapter 9) we will find out more about the large-scale transcription patterns that are the result of the gene expression mechanisms acting on such molecular and supramolecular scales.

8.4. Conclusions

The three-dimensional arrangement of eukaryotic genomes within the nucleus is of fundamental importance for understanding regulated gene expression in eukaryotes. Decondensed eukaryotic chromosomes occupy defined spaces within the nucleus during interphase. The nuclear matrix consists of a fibrogranular network of thin protein filaments and its biochemical properties depend extensively on the chosen purification method. Transcriptionally active genes are frequently found to be specifically attached to nuclear matrix components and may participate in the formation of a limited number of 'transcription factories'. The molecular machineries involved in pre-mRNA transport, processing and storage are similarly organized as 'speckled domains' in the vicinity of such transcription factories

References

Internet Sites of Interest

'http://util.ucsf.edu/sedat/'
In vivo arrangement of chromatin and chromosomes in cell nuclei.

Research Literature

Aaronson, R.P., and Blobel, G. (1975). Isolation of nuclear pore complexes in association with a lamina. *Proc. Natl. Acad. Sci. USA* 72, 1007–1011.

Berezney, R., and Coffey, D.S. (1975). Identification of a nuclear protein matrix. *Biochem. Biophys. Res. Comm.* 60, 1410–1417.

Berezney, R., Mortillaro, M.J., Ma, H., Wei, X., and Samarabandu, J. (1995). The nuclear matrix: a structural milieu for nuclear genomic function. *Int. Rev. Cytol.* 162A, 1–54.

Bidwell, J.P., Fey, E.G., van Wijnen, A.J., Penman, S., Stein, J.L., Lian, J.B., and Stein, G.S. (1994). Nuclear matrix proteins distinguish normal diploid osteoblasts from osteosarcoma cells. *Cancer Res.* 54, 28–32.

Bode, J., and Maass, K. (1988). Chromatin domain surrounding the human interferon-beta gene as defined by scaffold-attached regions. *Biochemistry* 27, 4706–4711.

Bregman, D.B., Du, L., Zee, S., and Warren, S. (1995). Transcription-dependent redistribution of the large subunit of RNA polymerase II to discrete nuclear domains. *J. Cell Biol.* 129, 287–296.

Bridger, J.M., Kill, I.R., O'Farrell, M., and Hutchinson, C.J. (1993). Internal lamin structures within G1 nuclei of human dermal fibroblasts. *J. Cell Sci.* 104, 297–306.

Capitani, S., Girarard, P.R., Mazzei, G.J., Kuo, J.F., Berezney, R., and Manzoli, F.A. (1987). Immunochemical characterization of protein kinase C in rat liver nuclei and subnuclear fractions. *Biochem. Biophys. Res. Comm.* 142, 367–375.

Carter, C.C., Bowman, D.W., Fogarty, K., McNeil, J.A., Fay, F.S., and Lawrence, J.B. (1993). A three-dimensional view of precursor messenger RNA metabolism within mammalian nucleus. *Science* 259, 1330–1335.

Ciejek, E.M., Tsai, M.-J., and O'Malley, B.W. (1983). Actively transcribed genes are associated with the nuclear matrix. *Nature* 306, 607–609.

Cockerill, P.N., and Garrard, W.T. (1986). Chromosomal loop anchorage of the kappa immunoglobulin gene occurs next to the enhancer in a region containing topoisomerase II sites. *Cell* 44, 273–282.

Cockerill, P.N., and Yuen, M.H. (1987). The enhancer of the immunoglobulin heavy chain locus is flanked by presumptive chromosomal loop anchorage elements. *J. Biol. Chem.* 262, 5394–5397.

Cook, P.R., and Brazell, I.A. (1975). Supercoils in human DNA. *J. Cell Sci.* 19, 261–279.

Cook, P. R. (1988). The nucleoskeleton: artifact, passive framework or active site? *J. Cell Sci.* 90, 1–6.

Cook, P.R. (1994). RNA polymerase: structural determinant of the chromatin loop and the chromosome. *Bioessays* 16, 425–430.

Cremer, T., Kurz, A., Zirbel, R., Dietzel, S., Rinke, B., Schrock, E., Speicher, M.R., Mathieu, U., Jauch, A., Emmerich, P.H., Ried, T., Cremer, C., and Lichter, P. (1993). Role of chromosome territories in the functional compartmentalization of cell nucleus. *Cold Spring Harbor Symp. Quant. Biol.* 53, 777–792.

Fawcett, D. W. (1966). An Atlas of Fine Structure: The Cell, its Organelles and Inclusions. (Philadelphia: Saunders).

Ferreira, J., Paoella, G., Ramos, C., and Lamond, A.I. (1997). Spatial organization of large-scale chromatin domains in the nucleus: a magnified view of single chromosome territories. *J. Cell Biol.* 135, 1597–1610.

Fey, E.G., Krochmalnic, G., and Penman, S. (1986). The non-chromatin substructures of the nucleus: the ribonucleoprotein (RNP)-containing and RNP-depleted matrices analyzed by sequential fractionation and resinless electron microscopy. *J. Cell Biol.* 102, 1654–1665.

Fey, E., and Penman, S. (1988). Nuclear matrix proteins reflect cell type of origin in cultured human cells. *Proc. Natl. Acad. Sci. USA* 85, 121–125.

Gasser, S.M., and Laemmli, U.K. (1986). Cohabitation of scaffold binding regions with upstream/enhancer elements of three developmentally regulated genes of *D. melanogaster*. *Cell* 46, 521–530.

Gasser, S.M., and Laemmli, U.K. (1987). A glimpse at chromosomal order. *Trends Genet.* 3, 16–22.

Gerace, L., and Blobel, G. (1980). The nuclear envelope lamina is reversibly depolymerized during mitosis. *Cell* 19, 277–287.

Getzenberg, R.H., Pienta, K.J., Huang, E.Y., and Coffey, D.S. (1991). Identification of nuclear matrix proteins in cancer and normal rat prostrate. *Cancer Res.* 51, 6514–6520.

Greenleaf, A.L. (1993). Positive patches and negative noodles: linking RNA processing to transcription? *Trends Biochem. Sci.* 18, 117–122.

Hadlaczky, G., Went, M., and Ringertz, N.R. (1986). Direct evidence for the non-random localization of mammalian chromosomes in the interphase nucleus. *Exp. Cell Res.* 167, 1–15.

Harpold, M.M., Wilson, M.C., and Darnell, J.E., Jr. (1981). Chinese hamster polyadenylated messenger ribonucleic acid: relationship to non-polyadenylated sequences and relative conservation during messenger ribonucleic acid processing. *Mol. Cell. Biol.* 1, 188–198.

He, D., Nickerson, J.A., and Penman, S. (1990). The core filaments of the nuclear matrix. *J. Cell Biol.* 110, 569–580.

Herman, R.C., Weymouth, L., and Penman, S. (1978). Heterogeneous nuclear RNA-protein fibers in chromatin-depleted nuclei. *J. Cell Biol.* 78, 663–674.

Huang, S., and Spector, D.L. (1991). Nascent pre-mRNA transcripts are associated with nuclear reagions enriched in splicing factors. *Genes Dev.* 5, 2288–2302.

Ingber, D.E., Dike, L., Hansen, L., Karp, S., Liley, H., Maniotis, A., McNamee, H., Mooney, D., Plopper, G., Sims, J., and Wang, N. (1994). Cellular tensegrity: exploring how mechanical changes in the cytoskeleton regulate cell growth, migration and tissue pattern during morphogenesis. *Int. Rev. Cytol.* 150, 173–224.

Jacks, R.S., and Eggert, H. (1992). The elusive nuclear matrix. *Eur. J. Biochem.* 209, 503–509.

Jackson, D.A., and Cook, P.R. (1985). Transcription occurs at a nucleoskeleton. *EMBO J.* 4, 919–925.

Jackson, D.A., and Cook, P.R. (1986). Replication occurs at a nucleoskeleton. *EMBO J.* 5, 1403–1410.

Jackson, D.A., and Cook, P.R. (1988). Visualization of a filamentous nucleoskeleton with a 23 nm axial repeat. *EMBO J.* 7, 3667–3678.

Jackson, D.A., Yuan, J., and Cook, P.R. (1988). A gentle method for preparing cyto- and nucleo-skeletons and associated chromatin. *J. Cell Sci.* 90, 365–378.

Jackson, D.A., Dickinson, P., and Cook, P.R. (1990). The size of chromatin loops in HeLa cells. *EMBO J.* 9, 567–571.

Jackson, D.A., Hassan, A.B., Errington, R.J., and Cook, P.R. (1993). Visualization of focal sites of transcription within human nuclei. *EMBO J.* 12, 1059–1065.

Jarman, A.P., and Higgs, D.R. (1988). Nuclear scaffold attachment sites in the human globin gene complex. *EMBO J.* 7, 3337–3344.

Jiminez-Garcia, L.F., and Spector, D.L. (1993). *In vivo* evidence that transcription and splicing are coordinated by a recruiting mechanism. *Cell* 73, 47–50.

Johnson, L.H., and Holland, J.J. (1965). Ribonucleic acid and protein synthesis in mitotic HeLa cell. *J. Cell Biol.* 27, 565–574.

Kas, E., and Chasin, L.A. (1987). Anchorage of the Chinese hamster dihydrofolate reductase gene to the nuclear scaffold occurs in an intragenic region. *J. Mol. Biol.* 198, 677–692.

Keesee, S.K., Meneghini, M.D., Szaro, R.P., and Wu, Y.J. (1994). Nuclear matrix proteins in human colon cancer. *Proc. Natl. Acad. Sci. USA* 91, 1913–1916.

Keppel, F. (1986). Transcribed human ribosomal RNA genes are attached to the nuclear matrix. *J. Mol. Biol.* 187, 15–21.

Khanuja, P.S., Lehr, J.E., Soule, H.D., Gehani, S.K., Noto, A.C., Choudhury, S., Chen, R., and Pienta, K.J. (1993). Nuclear matrix proteins in normal and breast cancer cells. *Cancer Res.* 53, 3394–3398.

Kirov, N., Djondjurov, L., and Tsanov, R. (1984). Nuclear matrix and transcriptional activity of the mouse alpha-globin gene. *J. Mol. Biol.* 180, 601–614.

Klehr, D., Maass, K., and Bode, J. (1991). Scaffold-attached regions from the human interferon-beta domain can be used to enhance the stable expression of genes under the control of various promoters. *Biochemistry* 30, 1264–1270.

Kurz, A., Lampel, S., Nickolenko, J.E. Bradl, J., Benner, A., Zirbel, R.M., Cremer, T., and Lichter, P. (1996). Active and inactive genes localize preferentially in the periphery of chromosome territories. *J. Cell Biol.* 135, 1195–1205.

Lamond, A.I., and Earnshaw, W.C. (1998). Structure and function in the nucleus. *Science* 280, 547–553.

Lawrence, J.B., Singer, R.H., and Marselle, L.M. (1989). Highly localized tracks of specific transcripts in interphase nuclei visualized by in situ hybridization. *Cell* 57, 493–502.

Levy-Wilson, B., and Fortier, C. (1989). The limits of DNAase I-sensitive domain of the human apolipoprotein-b gene coincide with the locations of chromosomal anchorage loops and define the 5'-boundary and 3'-boundary of the gene. *J. Biol. Chem.* 264, 21196–21204.

Manuelides, L. (1985). Individual interphase chromosome domains revealed by *in situ* hybridization. *Hum. Genet.* 71, 288–293.

Manuelides, L. (1990). Individual interphase chromosomes. *Science* 250, 1533–1540.

Marshall, W.F., Dernberg, A.F., Harmon, B., Agard, D. and Sedat, J.W. (1996). Specific interactions of chromatin with the nuclear envelope: positional determination within the nucleus in *Drosophila melanogaster*. *Mol. Biol. Cell* 7, 825–842.

Mayer, D. T., and Gulick, A. (1942). The nature of the proteins of cellular nuclei. *J. Biol. Chem.* 46, 433–440.

McCracken, S., Fong, N., Yankulov, K., Ballantyne, S., Pan, G., Greenblatt, J., Patterson, S.D., Wickens, M., and Bentley, D.L. (1997). The C-terminal domain of RNA polymerase II couples mRNA processing to transcription. *Nature* 385, 357–360.

McKeon, F.D., Kirschner, M.W., and Caput, D. (1986). Homologies in both primary and secondary structure between nuclear envelope and intermediate filament proteins. *Nature* 319, 463–468.

Michelotti, E.F., Sanford, S., and Levens, D. (1997). Marking of active genes on mitotic chromosomes. *Nature* 388, 895–899.

Mirkovitch, J., Mirault, M.-E., and Laemmli, U.K. (1984). Organization of the higher-order chromatin loop: specific DNA attachment sites on nuclear scaffold. *Cell* 39, 223–232.

Nickerson, J.A., Krochmalnic, G., Wan, K.M., and Penman, S. (1989). Chromatin architecture and nuclear RNA. *Proc. Natl. Acad. Sci. USA* 86, 177–181.

Pardoll, D.M., and Vogelstein, B. (1980). Sequence analysis of nuclear matrix associated DNA from rat liver. *Exp. Cell Res.* 128, 466–470.

Partin, A.W., Getzenberg, R.H., Carmichael, M.J., Vindivich, D., Yoo, J., Epstein, J. I., and Coffey, D.S. (1983). Nuclear matrix protein patterns in human benign prostrate hyperplasia and prostrate cancer. *Cancer Res.* 53, 744–746.

Phi-Van, L., and Stratling, W.H. (1988). The matrix attachment regions of the chicken lysozyme gene co-map with the boundaries of the chromatin domain. *EMBO J.* 7, 655–664.

Phi-Van, L., von Kries, J.P., Ostertag, W., and Stratling, W.H. (1990). The chicken lysozyme 5' matrix attachment region increases transcription from a heterologous promoter in heterologous cells and dampens positional effects on the expression of transfected genes. *Mol. Cell. Biol.* 10, 2302–2307.

Pienta, K.J., and Lehr, J.E. (1993). A common set of nuclear matrix proteins in prostrate cancer cells. *Prostrate* 23, 61–67.

Raap, A.K., van de Rijke, F.M., Dirks, R.W., Sol, D.J., Boom, R., and van der Ploeg, M. (1991). Bicolor fluorescence in situ hybridization to intron and exon mRNA sequences. *Exp. Cell Res.* 197, 319–322.

Razin, S.V., Yarovaia, O.V., and Georgiev, G.P. (1985). Low ionic strength extraction of nuclease-treated nuclei destroys the attachment of transcriptionally active DNA to the nuclear skeleton. *Nucleic Acids Res.* 13, 7427–7444.

Razin, S.V., Hancock, R., Iarovaia, O., Westergaard, O., Gromova, I., and Georgiev, G.P. (1993). Structural-functional organization of chromosomal DNA domains. *Cold Spring Harbor Symp. Quant. Biol.* 58, 25–25.

Robinson, S.I., Nelkin, B.D., and Vogelstein, B. (1982). The ovalbumin gene is associated with the nuclear matrix of chicken oviduct cells. *Cell* 28, 99–106.

Salditt-Georgieff, M., Harpold, M.M., Wilson, M.C., and Darnell, J.E., Jr. (1981). Large heterogeneous nuclear ribonucleic acid has three times as many 5' caps as polyadenylic acid segments and most caps do not enter polyribosomes. *Mol. Cell. Biol.* 1, 170–187.

Shermoen, A.W., and O'Farrell, P.H. (1991). Progression of the cell cycle through mitosis leads to abortion of nascent transcripts. *Cell* 67, 303–310.

Stief, A., Winter, D.M., Stratling, W.H., and Sippel, A.E. (1989). A nuclear attachment element mediates elevated and position-independent gene activity. *Nature* 341, 343–345.

Stuurman, N., van Driel, R., de Jong, A.M.L., and van Renswoude, J. (1989). The protein composition of the nuclear matrix of murine P19 embryonal carcinoma cells is differentiation-state dependent. *Exp. Cell Res.* 180, 460–466.

Stuurman, N., Meijne, A.M.L., van der Pol, A.J., de Jong, L., van Driel, R., and van Renswoude, J. (1990). The nuclear matrix from cells of different origin: evidence for a common set of matrix proteins. *J. Biol. Chem.* 265, 5460–5465.

Sui, H., and Spector, D.L. (1991). Nascent pre-messenger RNA transcripts are associated with nuclear regions enriched in splicing factors. *Genes Dev.* 5, 2288–2302.

van Wijnen, A.J., Bidwell, J.P., Fey, E.G., Penman, S., Lian, J.B., Stein, J.L., and Stein, G.S. (1993). Nuclear matrix association of multiple sequence-specific DNA binding activities related to SP-1, ATF, CCAAT, C/EBP, Oct-1 and AP-1. *Biochemistry* 32, 8397–8402.

Vogelstein, B., Pardoll, D.M., and Coffey, D.S. (1980). Supercoiled loops and eukaryotic DNA replication. *Cell* 22, 79–85.

Wansink, D.G., Schul, W., Kraan, V.D., van Steensel, B., van Driel, R., and de Long, L. (1993). Fluorescent labeling of nascent RNA reveals transcription by RNA polymerase II in domains scattered throughout the nucleus. *J. Cell Biol.* 122, 283–293.

Wansink, D.G., Sibon, O.C., Cremers, F.F., van Driel, R., and de Jong, L. (1996). Ultrastructural localization of active genes in nuclei of A431 cells. *J. Cell. Biochem.* 62, 10–18.

Wei, X., Mortillaro, M.J., Kim, S., Frego, L., Bucholtz, L., Nakayasu, H., and Berezney, R. (1995). p250 is a novel nuclear matrix protein that co-localizes with splicing factors. *J. Cell. Biochem. Suppl.* 21B, 142.

Xing, Y., and Lawrence, J.B. (1991). Preservation of specific RNA distribution within the chromatin-depleted nuclear substructure demonstrated by *in situ* hybridization coupled with biochemical fractionation. *J. Cell Biol.* 112, 1055–1063.

Xing, Y., Johnson, C.V., Dobner, P.R., and Lawrence, J.B. (1993). Higher level organization of individual gene transcription and RNA splicing. *Science* 259, 1326–1330.

Zeng, C., He, D., and Brinkley, B.R. (1994). Localization of NuMA protein isoforms in the nuclear matrix of mammalian cells. *Cell Motil. Cytoskel.* 29, 167–176.

Zlatanova, J.S., and van Holde, K.E. (1992). Chromatin loops and transcriptional regulation. *CRC Crit. Rev. Eukaryotic Gene Express.* 2, 211–224.

Chapter 9

Gene Expression Dynamics and Global Genome Transcription Patterns

The previous chapters focused on the structural and functional aspects of the various components of the transcriptional machineries. But, just as focusing only on the hardware part of a computer will not reveal how a particular program works, gene expression patterns need to be investigated to understand the transcriptional machinery in action. Many of the tools for investigating transcription patterns on a large ('global') scale have only become generally available during the last five years. Here we will explore the technical background to certain key methods, and then consider some of the results obtained from the initial applications of large-scale transcription pattern studies. While much of the experimental work in this area is currently still carried out on simple model organisms to test and improve the methodology, it is already possible to see that this approach will soon lead to dramatic advances in our understanding of genetic regulatory circuits even in more complex organisms. The resulting insights into the genetic 'software' that controls the flow of information from the largely invariant genomes will almost certainly have profound implications for our understanding of core aspects of cell- and developmental biology. It is also becoming increasingly apparent that the capacity of the pharmaceutical industry to develop new drugs and therapeutic treatments will be greatly enchanced through the availability of such novel methods.

9.1. The Molecular Quest for Global Genome Transcription Patterns

The Transition from 'Structural' to 'Functional' Genomics

Ever since the discovery showing that each cell contains the complete genetic information capable of encoding the entire organism, biologists have tried to uncover the mechanisms that are used to unravel this tightly packed information in a controlled manner. Much of modern biological research is currently focused on various genome projects that aim at a comprehensive molecular description of the genetic material of a number of different model organisms. Complete genome sequences from many small pro- and eukaryotic organisms are already available and have yielded fascinating insights into the macromolecular diversity of living organisms. Similarly extensive sequence information from higher eukaryotic genomes (including *D. melanogaster*, *A. thaliana, M. musculus* and *H. sapiens*) will be forthcoming in the forseeable future.

One of the challenges arising from these projects is the need to make mechanistic sense of this flood of sequencing data in order to achieve a detailed understanding of the biological information stored in the genomes. This transition from 'structural' genomics (i.e. genome mapping and large scale sequencing) to 'functional' genomics is already well underway and will pick up in speed and significance over the coming decade (see e.g. Hieter and Boguski, 1997). Structural genomics has clearly defined and achievable endpoints that culminate in the acquisition of the complete DNA sequence of a given organism. This information is, however, only the starting point for functional genomics, which aims to discover the biological functions of gene diversity, arrangements, expression patterns and phenotypic effects in a systematic and comprehensive manner. In many ways, functional genomics started a long time ago with the application of 'classical' and molecular genetical techniques to study individual biological problems. The important difference between such conventional approaches and functional genomics is, however, the scale on which such studies are carried out.

The Technological Development of 'Massively Parallel' Detection Systems

Traditionally, molecular biologists have studied the functions of individual genes or, at most, small groups of genes. We therefore know in great detail the temporal and spatial expression patterns for several thousand transcription units through the application of conventional techniques, such as Northern blotting and *in situ* hybridization, but this still represents only a small percentage of the total number of genes. The human genome is thought to encode approximately 80,000 different transcripts (Fields *et al.*, 1994), and new technology is now emerging that enables us to look at changes in the transcription patterns of *thousands of genes simultanously* during important key events, such as cellular differentiation and carcinogenesis. The reasoning behind such experiments is simple: if we want to understand the regulatory interactions between different genes, we need to study the expression patterns of the *majority* of genes (rather than a small subset). Such data would give us new insights into the higher-order genetic programs that emerge from the activities and regulatory interactions of various genetic control circuits, and those that may not be apparent from studies of gene expression measurements based on only a few genes.

Below we will look at several key techniques that have been successfully used for studying the steady-state and changing expression patterns of large groups of genes. Particular methods, such as 'differential display' and SAGE, are useful for discovering individual transcripts among the general mRNA population of a cell that undergo drastic 'up' or 'down' regulation during events influencing the overall cell state. Other techniques, such as DNA microarrays (often referred to as 'DNA microchips') and high resolution 2D-protein electrophoresis, are especially powerful for displaying simultanously the expression levels of thousands of genes and allow a detailed characterization of global changes in gene expression occurring within relatively short time frames (minutes to hours).

Together these techniques form a solid basis for determining the 'molecular phenotype' (i.e. the complete description of the RNA and protein expression pattern) of genomes in various expression states, and will help to bring the static information from structural genome projects to life.

9.2. 'Proteome' Projects

High Resolution 2D-Protein Electrophoresis Systems

Some of the earliest successful attempts to identify the molecular basis of tissue-specific gene expression were based on the analysis of the precise protein composition of different tissues. Due to the high degree of the complexity of different proteins expressed simultanously in living cells (>10,000), high resolution electrophoresis techniques capable of separating large number of proteins had to be specifically developed. Among the various two-dimensional electrophoresis strategies the O'Farrell system has emerged as the most successful one (O'Farrell, 1975 and 1977). Protein mixtures are first separated by charge differences and subsequently according to molecular weight. After silverstaining (or autoradiography of ^{35}S-labelled protein samples), this procedure yields up to a thousand discrete spots representing the protein expression pattern of the sample. Since its initial conception, comparative 2D-gel electrophoresis technology has been substantially improved due to the introduction of immobilized pH gradients, which allow a reproducible separation of proteins in the first dimension (Bjellqvist *et al.*, 1982; Blomberg *et al.*, 1995). In combination with technical advances in 'spot identification' (based on amino acid composition analysis, peptide microsequencing or MALDI/tandem mass spectroscopy), detailed cataloging of tissue-specific protein expression patterns and comparisons of proteins expressed during cell differentiation and growth has become feasible (e.g. Dunn, 1993; Shevchenko *et al.*, 1996; Wilkins *et al.*, 1996).

Proteome Projects Complement Nucleic Acid-based Approaches

Figure 9.1 shows a typical example of the protein 2D-electrophoresis pattern of human liver cells with many of the major spots uniquely identified. Such 2D-protein electrophoresis gels can potentially yield confirmatory and complementary information to the nucleic acid-based methods (as discussed below) for studying the diversity of gene expression patterns in various tissues. Results from systematic 2D-electrophoretic analyses of samples derived from various subcellular fractions make it is possible, in principle, to determine the organellar distribution of specific proteins. Furthermore, the post-translational modification patterns (such as phosphorylation and glycosylation) of individual

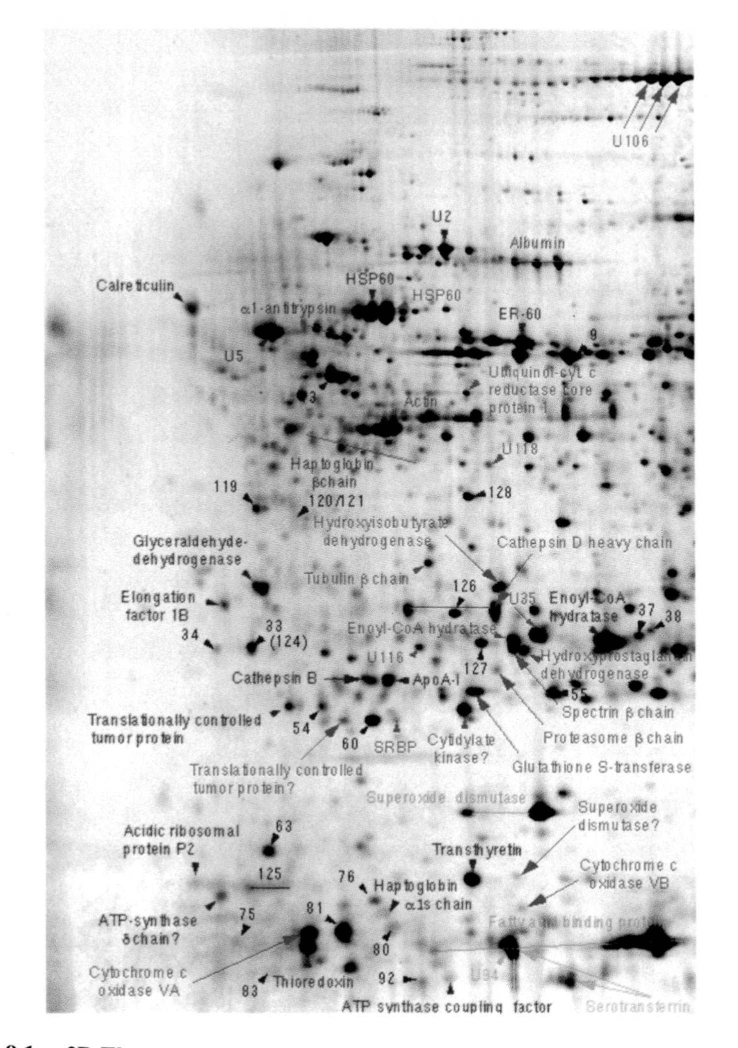

Figure 9.1. 2D Electrophoresis Gel of Human Liver Cells.
Advances in peptide sequencing technology and mass spectroscopy have allowed the unambiogous identification of many of the more abundant polypeptides on 2D electrophoresis gels. Note that several proteins (such as albumin, α1-antitrypsin and superoxide dismutase) are present in multiple spots due to differential post-translational modifications. This gel and several other examples can be inspected in detail on an internet site (http://www.expasy.ch/ch2d/).

proteins can be characterized, and the relative quantities of specific isoforms precisely quantitated, because they migrate as separate spots with slightly altered mobilities. Phosphorylation changes the overall charge of the modified protein, resulting in 'trains' of spots along the isoelectric focusing axis, and glycosylation often negatively affects the migration speed of proteins along the second dimension. Proteome projects can therefore offer us a series of high-resolution 'snap-shots' of the diversity of the protein expression pattern, which goes beyond simply stating which proteins are present. Many of the PCR-based methods for characterizing complex mRNA populations are intrinsically more sensitive for the detection of the expression of rare gene products in cells as compared to 2D-protein electrophoresis. They are, however, incapable of yielding information about the amount and ultimate destiny of proteins synthesized from such transcripts.

In the forseeable future, detailed databases will become available that will allow direct comparisons of the quantities, modification status and subcellular localization of many distinct proteins in many different types of normal and diseased human tissues. Such information will not only have a substantial impact on our biological understanding, but also provide an invaluable resource for medical research.

9.3. Molecular Characterization of Complex mRNA Populations

Subtractive Hybridization

Mainly due to technical reasons, the development of techniques for studying the mRNA populations during cell growth and differention has until recently lagged substantially behind the opportunities offered by 2D-protein gel electrophoresis to study the molecular diversity of genome expression patterns on a global scale. One of the first methods, specifically designed to identify mRNAs that change their expression quantitatively during particular cellular events, was the 'subtractive hybridization' technique (e.g. Lee *et al.*, 1991). This approach (implemented in numerous technical variations) results in the production of cDNA libraries, that are substantially enriched for the mRNAs specifically expressed in a particular cell type or at a defined developmental stage.

Differential Display ('DD') of mRNA Populations

With the more widespread application of the polymerase chain reaction (PCR), new ways of studying the diversity and complexity of mRNA patterns have become more commonly applicable. The 'differential display' ('DD') technique combines PCR using 'random' primers with high resolution polyacrylamide gel electrophoresis to produce specific ladders of PCR products (Liang and Pardee, 1992). This technique has the theoretical capacity to sample ('display') the full complexity of the transcription pattern of a eukaryotic cell, and can be implemented routinely in most molecular biology laboratories equipped with standard molecular biology instrumentation.

DD is a method for statistically amplifying a small group of individual mRNA molecules (via cDNA copies) with the help of two specific primers. One primer (the '3' anchoring primer') consists of an oligo dT stretch followed by a specific 2 nucleotide sequence. There are 12 possible primers of that structure (Figure 9.2), but only one particular primer is usually chosen in a particular experiment to prime cDNA synthesis on purified mRNA with reverse transcriptase (RT). Theoretically we can expect $^{1}/_{16}$ of the mRNA population to contain a poly(A) junction complementary to the chosen primer. These mRNAs are converted into cDNAs, leading to a reduction of the overall complexity of the mRNA population by a factor of 12. Next, PCR is carried

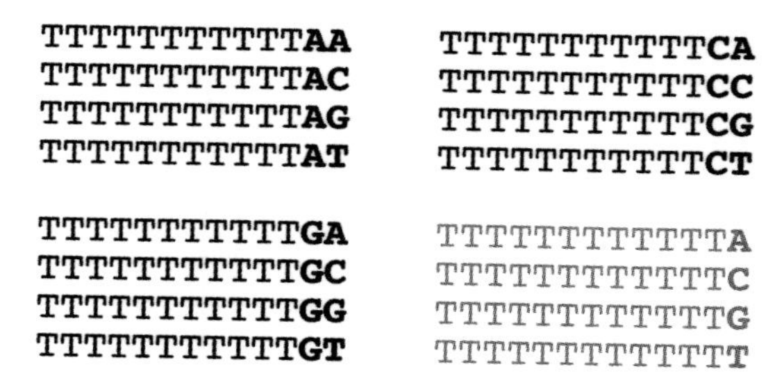

Figure 9.2. 'Anchoring' Primers to Produce cDNA Pools for Differential Display. Eleven 'T' residues are followed by a specific dinucleotide sequence. These oligonucleotides are designed to hybridize to the junctions between the poly(A) tail and coding region of eukaryotic mRNAs. Note that in practice only 12 of the 16 possible primers can be used in DD because the presence of 'T' in the first position of the dinucleotide sequence in *D* merely extends the oligo-'T' sequence and reduces the selectivity of these primers from 1/16 to 1/4.

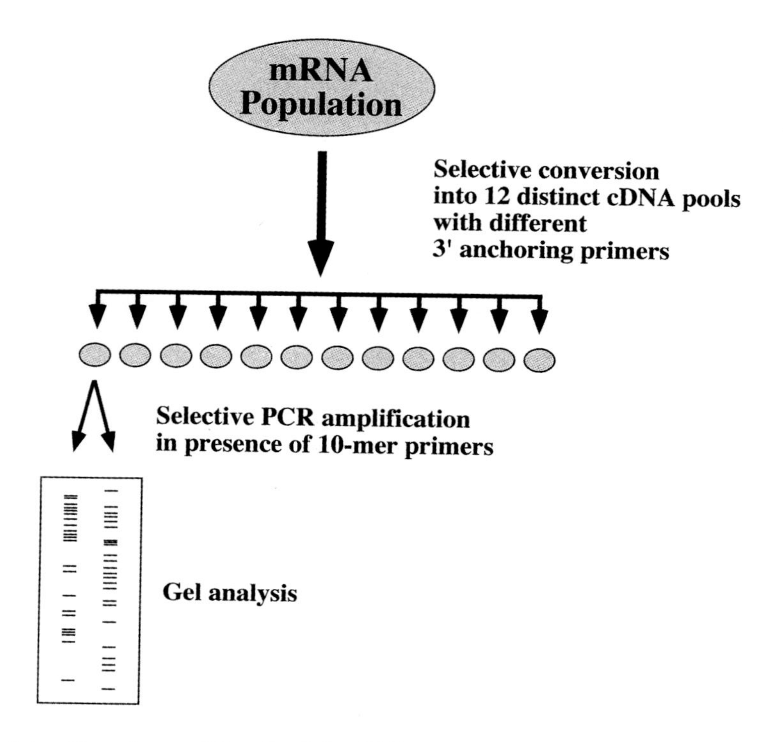

Figure 9.3. Experimental Details of the 'Differential Display' Technique.
Purified mRNA is reverse transcribed in the presence of up to 12 different 3'anchoring primers (see Figure 9.2). Specific subsets of these cDNAs are amplified in the presence of 10-mer primers and radiolabelled nucleotides. The resulting PCR products are then 'displayed' as specific bands on denaturing polyacrylamide gels.

out on these cDNAs in presence of an additional oligonucleotide that hybridizes within some of the coding regions. This is achieved by using a 'random' 10-mer primer of arbitrary sequence that is capable of specifically hybridizing to a subset of cDNA molecules. PCR using the 10-mer primer, in connection with the previously used 3' anchoring primer, leads on average to the amplification of approximately 100 cDNAs. The PCR products are separated on high resoution polyacrylamide gels (similar to gels used for separating DNA sequencing reaction products) and yield a 'fingerprint' derived from a defined subset of mRNA molecules present in the original mRNA population. If this procedure is repeated sufficiently often in the presence of different arbitrary 10-mer primers, it is possible to have a statistically significant chance to detect every different mRNA molecule that was represented in the original subset of cDNAs (Figure 9.3). Furthermore, selection of different anchoring primers with different dinucleotide combinations allows the characterization of all 12 different subsets of cDNAs. Comparison of the differential display patterns obtained from cells growing under different conditions pinpoints individual transcripts that are up- or downregulated (Figure 9.4). In principle, DD allows the precise dissection and subdivision of natural mRNA populations down to

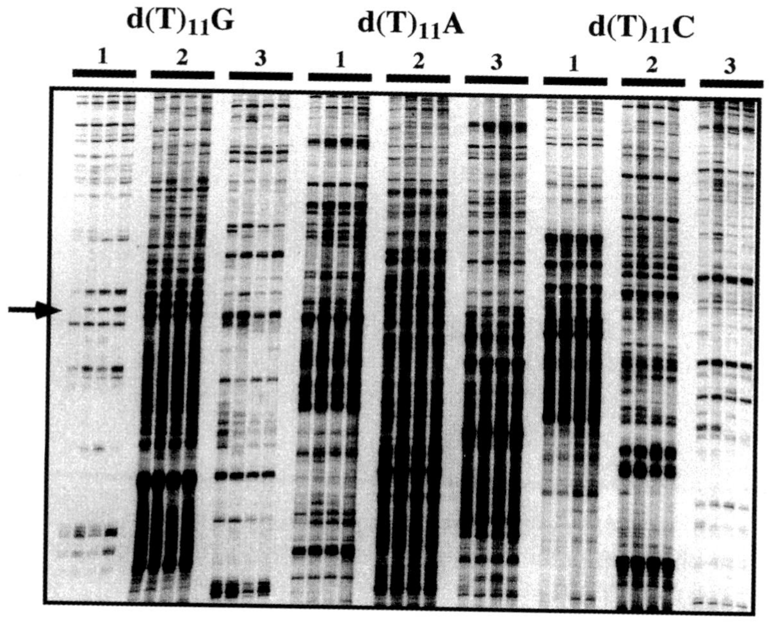

Figure 9.4. Differential Display Gel.
mRNAs samples extracted from one normal and three tumorigenic rodent fibroblast cell lines were analyzed by differential display with all combinations between three different anchoring primers 9d$[T]_{11}$G; d$[T]_{11}$A; d$[T]_{11}$C) and three arbitrary primers ('1', '2' and '3'). The arrow highlights a transcript that is present in all three tumor cell lines, but is absent in normal cells. Differential Display using the 'RNAimage' kit from GenHunter Corporation (courtesy of GenHunter Corporation, Nashville, TN, USA).

the level of single mRNA species. If, on average, 100 different mRNA species can be 'fingerprinted' in each gel lane, it is possible to calculate that DD reactions using approximately 200 different primer combinations are probably sufficient to display most (>95%) of mRNAs present in an complex eukaryotic cell.

A similar strategy has been developed for prokaryotic cells lacking polyadenylated transcripts. RNA arbitrarily primed PCR ('RAP') involves PCR with short arbitrarily selected oligonucleotide primers on reverse transcribed RNA populations to yield distinct banding patterns (Ralph *et al.*, 1993; Wong and McClelland, 1994; reviewed in McClelland *et al.*, 1995).

Serial Analysis of Gene Expression ('SAGE')

Some of the earliest studies giving an indication of a genome-wide mRNA expression profile came from from large-scale sequencing studies of cDNA clones, that were picked in an unbiased manner from cDNA libraries (Adams *et al.*, 1991; Okubo *et al.*, 1992; Adams *et al.*, 1993). This approach was primarily designed to identify new genes, rather than as method for studying tissue-differential transcription patterns. One major drawback of the random sequencing approach was that it quickly leads to 'information overkill'. To identify a cDNA clone, only a very small part of the sequence needs to be obtained for unambigous identification, rather than the more extensive sequence of several hundred nucleotides determined by automatic DNA sequencers. The 'Serial Analysis of Gene Expression' ('SAGE') method is a logical extension of the 'identification by sequencing' approach, designed to overcome this problem by characterizing only a short nucleotide sequence tag of around 9–10 basepairs originating from within a defined position within a cDNA clone. Through a series of cunning DNA manipulations, a molecularly 'punctuated' string of sequence tags (each tag derived from a random cDNA clone) is constructed, sequenced and analyzed (Figure 9.5). This approach identifies in principle up to 20–30 different tags in a single sequencing reaction, and thus can be used to identify at least some of the more abundant mRNAs present within a tissue within a reasonable amount of time (Velculescu *et al.*, 1995). Detection and comparison of mRNAs present at relatively high levels (0.5%) can be achieved within a single day, and even documentation of the expression

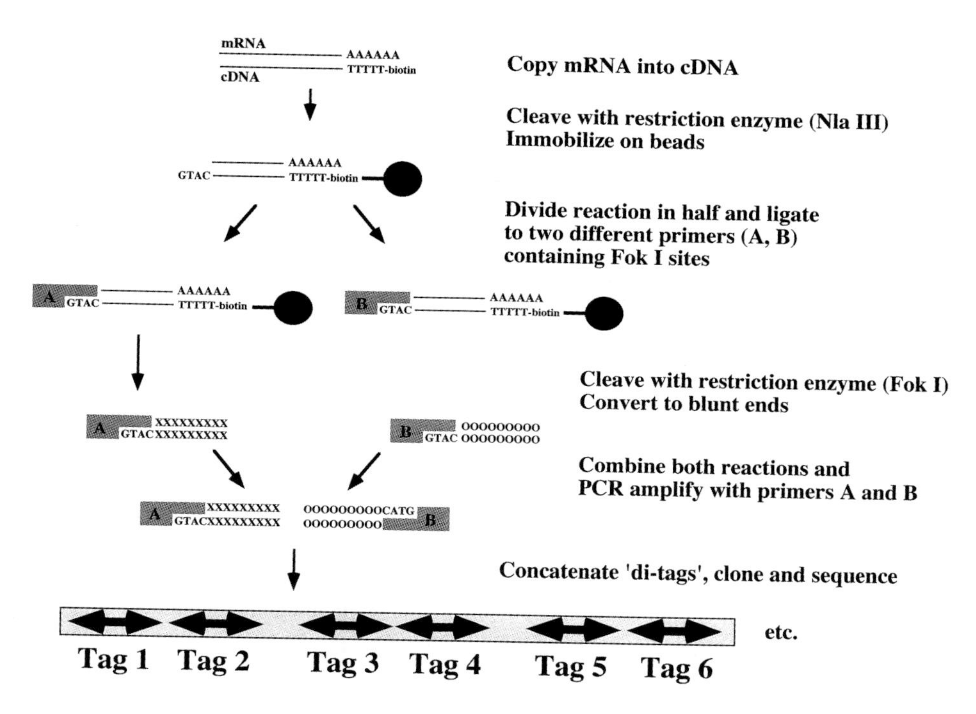

Figure 9.5. Experimental Details of the 'SAGE' Technique.
mRNAs are reverse transcribed into double-stranded cDNA and cleaved with the restriction enzyme Nla III. After coupling of the cleaved 3' ends of the various cDNAs to streptavidin beads via a biotin residue present on the oligo d(T) primer, the reaction is then divided into two separate aliquots and each is ligated in the presence of two different oligonucleotides ('A' and 'B') containing Fok I sites. The Fok I restriction endonuclease has the useful property that it cleaves 9 nucleotides away from its recognition site. Cleavage with Fok I and filling in the ends of the resulting fragment with DNA polymerase will thus result in the attachment of a specific portion of the cDNA adjacent to oligonucleotides A and B. The two aliquots are at this stage combined and PCR amplified in the presence of primers A and B to generate 'di-tags', which are ligated together, cloned and sequenced. The sequence of SAGE clones containing di-tag concatenates is then split up by computational means to count the frequency of specific cDNA sequence fragments.

levels of relatively rare transcripts (100 copies/cell or 0.025% of total mRNA) can be carried out reliably within a few months. Similar to the DD method, accurate quantitative data strongly depends on obtaining a sufficiently high sample number before statistically reliable conclusions can be drawn.

One of the most interesting applications of SAGE has been the first systematic comparison of the genes expressed in normal and tumor cells (Zhang *et al.*, 1997). This study was based on identifying more than 300,000 individual transcripts from cDNA libraries that had been prepared from normal and malignant human colorectal cells. The huge collection of tags sequenced during that project reflected an underlying diversity of approximately 49,000 different mRNA transcripts, ranging in abundance from 1 to 5,300 copies per cell. By comparing the levels of different types of transcripts between normal and cancerous cells it became clear that the expression levels of the vast majority of genes was either unchanged, or varied less than tenfold (Figure 9.6). In contrast, 181 mRNAs were present at substantially reduced levels, and 108 mRNAs were much more abundant in colon cancer cells than in normal cells. Many of these genes (but not all!) that were differentially expressed in primary colorectal cancer cells were also similarly affected in their transcription pattern in colorectal cancer cell lines grown in tissue culture. This important observation suggests that tumor cell lines grown *in vitro* in tissue culture provide reasonably good, but definitely not perfect, model systems for studying the biochemical properties of primary cancers growing *in vivo*. Another important conclusion that could be drawn from the data was that the decreased or increased expression levels of certain genes in many different types of primary cancers were quite similar. Genes that were consistently overexpressed often encoded components of the translational machinery and the glycolytic metabolic pathway (Zhang *et al.*, 1997). These findings are consistent with the fast growth rate of cancer cells, that puts a high demand on the cellular machineries responsible for protein synthesis and the generation of metabolic energy from glucose under anaerobic conditions.

DNA Microarrays Allow Simultanous Quantitation of Thousands of Different mRNAs

Although DD and SAGE are very powerful techniques for analyzing highly complex mRNA populations, one of the main problems with these methods is

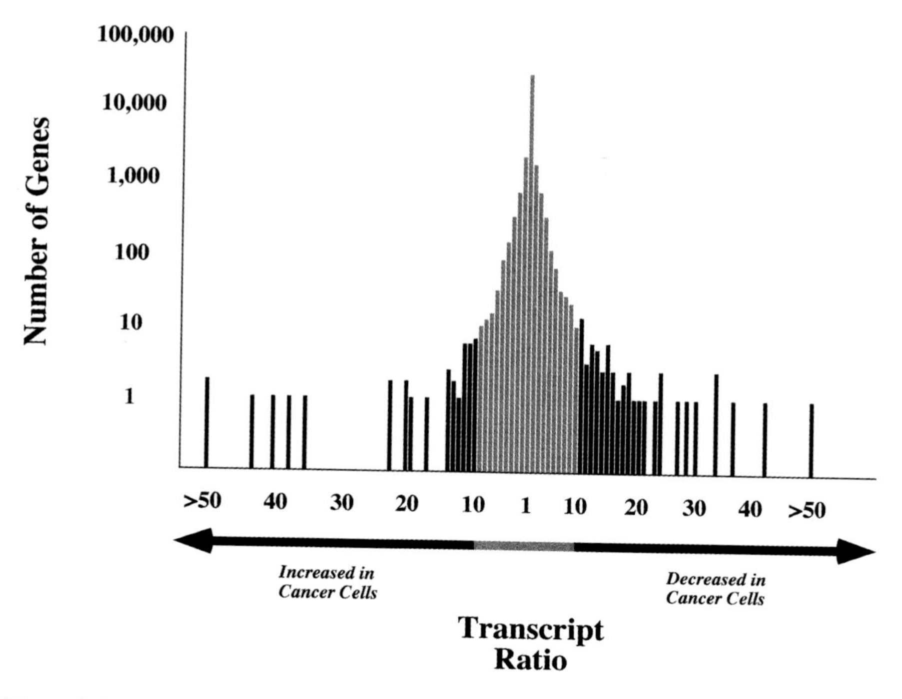

Figure 9.6. Comparison of Global Transcription Patterns in Normal and Cancer Cells. The ratio of the number of specific transcripts (shown along the x-axis) in normal cells and cancer cells is is plotted against the number of genes displaying particular ratios (semi-logarithmic scale on y-axis). Note that the expression ratio for most genes is between 1 and 10, i.e. similar between the two cell states (grey histogram bars). The transcription of 51 genes *decreased* by more than 10 fold and transcription of 32 genes *increased* by more than 10 fold (black histogram bars) in cancer cells. After Zhang *et al.* (1997).

that identification and characterization of differentially expressed transcripts typically takes weeks, if not months. With the advent of large amounts of genome sequence data, an alternative approach has become feasible that allows the simultanous quantitation of thousands of transcripts within a few hours (Figure 9.7). DNA microarrays (sometimes also referred to as 'DNA chips') contain thousands of specific DNA fragments that are arranged in a predefined pattern on a solid support surface. The manufacture of DNA microarrays containing thousands of specific DNA probes has become possible through

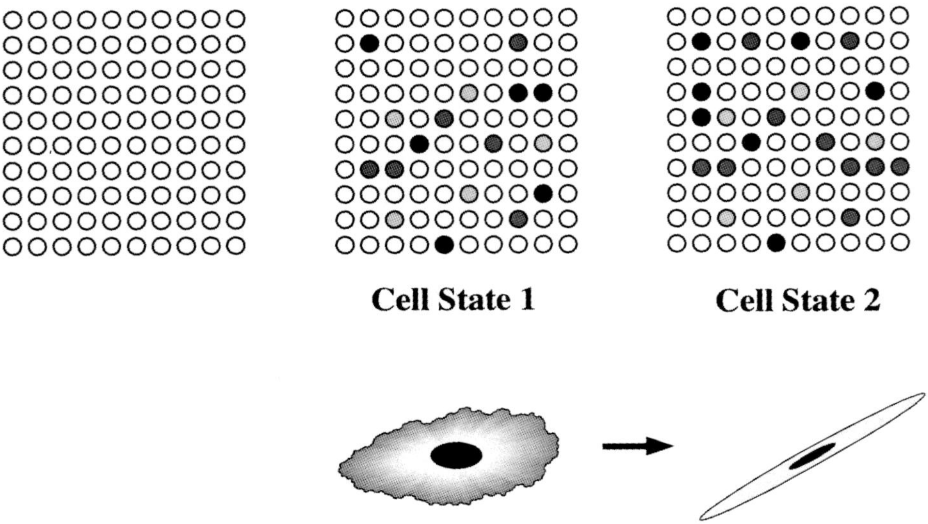

Figure 9.7. Analysis of Cell-Specific Transcription Patterns with DNA Chips.
DNA chips contain hundreds (or thousands; see Figure 9.8.) of individual gene-specific DNA probes of known identity arranged in a regular pattern on a solid support (usually glass). In this figure each circle symbolizes an individual gene-specific probe. Hybridization of DNA chips with fluorescently-labelled cDNA reverse-transcribed from the mRNA population of cells in 'cell state 1' will result in the appearance of a distinct pattern of labelled DNA dots (middle panel). The intensity of labelling of each dot reflects (within limits) the amount of mRNA from a particular gene present in the total mRNA population. When cells undergo changes in the gene expression program (towards 'cell state 2'; e.g. as a consequence of differentiation, nutrient availability or exogenous signals) the pattern of fluorescently labelled DNA dots reflects the changing concentrations of the mRNAs transcribed from different genes (compare the middle and right hand panels).

the development of two key alternative technologies. In one case, a high-density array of oligonucleotides with distinct and known sequences is chemically synthesized on a support surface, using techniques similar to those used during the manufacture of high density electronic microcircuits. Light-directed blocking of the sites allows the specific *in situ* synthesis of hundreds or thousands of olignonucleotides in parallel (photolithography, see Fodor *et al.*, 1991; Chee *et al.*, 1996; Lockhart *et al.*, 1996). An alternative approach is based on the use of robots to spot different DNA samples as small dots on specially treated glass surfaces (Schena *et al.*, 1995; 1996; De Risi *et al.*, 1997). The tip used for spotting bears a close resemblance to the tip of a fountain pen and fulfills a similar function: continuous delivery of small amounts of DNA from the gap within the tip to the glass surface. The tip is moved to a defined position with a robotic system that allows very precise positioning of the different DNA samples. A typical example is the spotting of more than 6,000 different PCR-generated DNA samples on a glass slide measuring 18 mm by 18 mm (DeRisi *et al.*, 1997). The DNA is initially electrostatically retained by a thin coat of of positively-charged poly L-lysine on the glass surface and then permanently immobilized by drying and UV-crosslinking.

Each of the DNA spots on a DNA microarray normally represents the coding region of one particular gene. Hybridization of such an array with fluorescently labelled cDNA (generated by transcribing an mRNA population with reverse transcriptase in the presence of a fluorescently-labelled nucleotide precursor) allows the cDNA molecules to hybridize to the arrayed DNA fragments they are complementary to. The amount of DNA immobilized on each spot is sufficiently large, so that the extent of the hybridizing cDNA depends solely on the amount of specific cDNA present in the probe mixture. After washing away excess probe, the DNA array is scanned with a fluorescent imaging system to reveal the amount of fluorescent cDNA specifically bound to each spot. Since the identities of all immobilized DNA spots are known, it is possible to correlate the amount of fluorescent signal obtained after hybridization with the quantity of specific transcripts present in the original mRNA population from which the fluorescent cDNAs were derived. DNA microarrays are thus ideally suited for measuring the quantity of individual transcripts within complex mRNA mixtures for a large number of different genes.

The DNA microarray technique allowed De Risi *et al.* (1997) to study for the first time the effects of environmental stimuli on the overall gene expression program of a complete eukaryotic genome. They generated more than 6,000 different DNA probes by PCR, representing every gene from the completely sequenced genome of *S. cerevisiae* (yeast) and immobilized them in defined patterns on a small glass slide. Hybridization of this gene array with fluorescently labelled cDNA pools allowed a precise quantitation of the expression level of each gene under various growth conditions, both in wildtype and mutant cells. Furthermore, as a consequence of technical improvements, De Risi *et al.* were able to carry out a simultanous comparison of two different mRNA populations on the same microarray. The cDNA produced from the first mRNA pool was labelled to produce a green fluorescent probe ('probe 1'). After hybridization of the DNA microarray with this probe, it was hybridized again with another cDNA pool labelled with a red fluorochrome ('probe 2'). DNA spots that only hybridized to cDNA present in 'probe 1' thus emitted a green signal, those that were complementary to cDNA exclusively present in 'probe 2' gave a red signal, and the DNA spots that hybridized to both probes resulted in a yellow signal due to the simultanous fluorescence of the green and red fluorochromes. Spectral analysis, measuring the relative amounts of of the red and green fluorescence emerging from each DNA array element, allowed a simultanous comparison of specific mRNA levels in two different cDNA pools complementary to a single DNA probe (Figure 9.8). In the following two sections we will investigate some of the fascinating results obtained in this study in more detail.

Changes in *S. cerevisiae* Genomic Transcription Patterns in Response to Different Growth Conditions

Most microorganisms spend their life in recurring cycles of excess food followed swiftly by starvation. These rapidly changing environmental factors require continuous adjustments in the metabolic machineries to allow the cells to survive and maximize the available resources. Yeast cells, such as *S. cerevisiae*, are typical in their responses for adapting to changing circumstances. If a small population of cells is placed into a medium that is rich in sugars, the cells will initiate an exponential growth phase and the metabolic machinery catabolizes the available glucose to produce ethanol as a 'waste' product. This process occurs under anaerobic ('oxygen-free') conditions and provides quickly the energy and anabolic components required for maintaining a high rate of cell division. Once the glucose is used up, a radical switch to a very different metabolic mode occurs. The previous 'waste' product, ethanol, suddenly becomes an attractive energy source that requires the presence of oxygen for further metabolism. This change from the anaerobic to aerobic metabolic state

Figure 9.8. Yeast Genome DNA Chip Hybridized with Fluorescent cDNAs Prepared from Cells at Two Different Growth Conditions.

This DNA chip contains probes representing all the genes present in the yeast genome. Each dot corresponds to DNA derived from an individual open reading frame. See text for more explanation. This image (and other pictures of hybridized DNA chips), together with quantitative data can be inspected in detail on an internet site (http://cmgm.stanford.edu/pbrown/).

is often referred to as 'diauxic shift,' and is fully reversible depending on the environmental conditions (Figure 9.9). During the diauxic shift major changes in carbon metabolism, protein synthesis and carbohydrate storage occur as a consequence of widespread changes in the underlying gene expression pattern. Many of these processes, especially those concerning the behaviour of cells under different forms of environmental stress, have substantial consequences for the industrial use of yeasts (recently reviewed in Attfield, 1997).

As a demonstration of the feasibility of exploring genome-wide changes of gene expression patterns, the induction and repression of all known genes from *S. cerevisiae* in cells undergoing such a diauxic change were studied (DeRisi *et al.*, 1997). DNA microarrays containing approximately 6400 different DNA fragments, representing the coding regions of all known and predicted yeast proteins (Figure 9.8; Goffeau *et al.*, 1996; 1997), were hybridized with fluorescently-labelled cDNAs derived from growth stage-specific mRNA populations. During the anaerobic growth stage with glucose as the main carbon source, overall gene expression was shown to settle into a largely invariant and stable pattern with less than 0.3% of mRNAs differing more than two-fold in their abundance over a two hour sampling interval. As the rapid growth rate continued and glucose levels diminished, significant changes in the stable anearobic pattern become clearly detectable: mRNA levels increased for 710 genes, and decreased for 1030 genes by at least two-fold (and in many cases up to four-fold). From these figures it was possible to conclude that the diauxic shift affected the transcription of 27% of the genes present in the *S. cerevisae* genome. The precise functions of approximately half of these genes are currently still unknown, and it is therefore not possible at this stage to assess their precise role in adapting the cellular machinery to growth on different nutrients.

Many intriguing conclusions could, however, be drawn from the other 50% of genes with known functions. The major changes included a substantially increased rate of transcription of the mRNAs encoding the enzymes required for reversing the flow of metabolites through the glycolytic pathway (gluconeogenesis), and for converting the newly generated glucose into storage carbohydrates (trehalose and glycogen; Figure 9.9). Similarly, in preparation for energy production under aerobic conditions, genes for enzymes participating in the TCA cycle and cytochrome biosynthesis were strongly induced

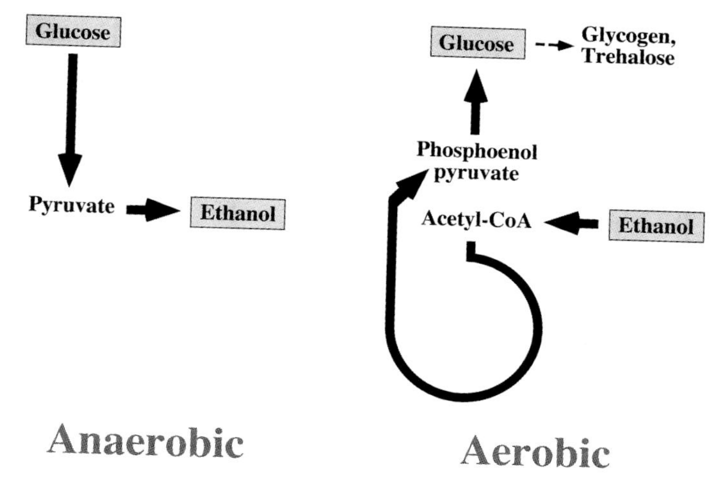

Figure 9.9. Metabolic Carbon-Flow in Yeast Under Different Growth Conditions. Yeast cells can grow in the presence ('aerobic' growth) and absence ('anaerobic' growth) of oxygen. In the absence of oxygen glucose is quickly fermented to the 'waste' product ethanol via the glycolytic pathway. At a later stage (in the presence of oxygen) ethanol can serve as a nutrient to produce glucose and the storage carbohydrates glycogen and trehalose by the gluconeogenic pathway. The change from anaerobic to aerobic growth is often referred to as diauxic shift and drastically changes the direction of carbon-flow.

(Figure 9.10). In contrast, the expression of genes encoding components of the translational machinery (ribosomal proteins, tRNAs, translation initiation and elongation factors) was found to be substantially down-regulated. Apart from the overall quantitative changes in mRNA expression levels, many groups of genes displayed distinct temporal expression patterns (below and Figure 9.10). Such differences in the expression levels of many genes were also observed between cells growing in minimal or rich growth media (Wodicka *et al.*, 1997).

9.4. Detection of Genetic Regulatory Circuits and Identification of Target Genes

The complex changes in the gene expression patterns in cells undergoing changes in growth conditions highlights the key role of transcriptional

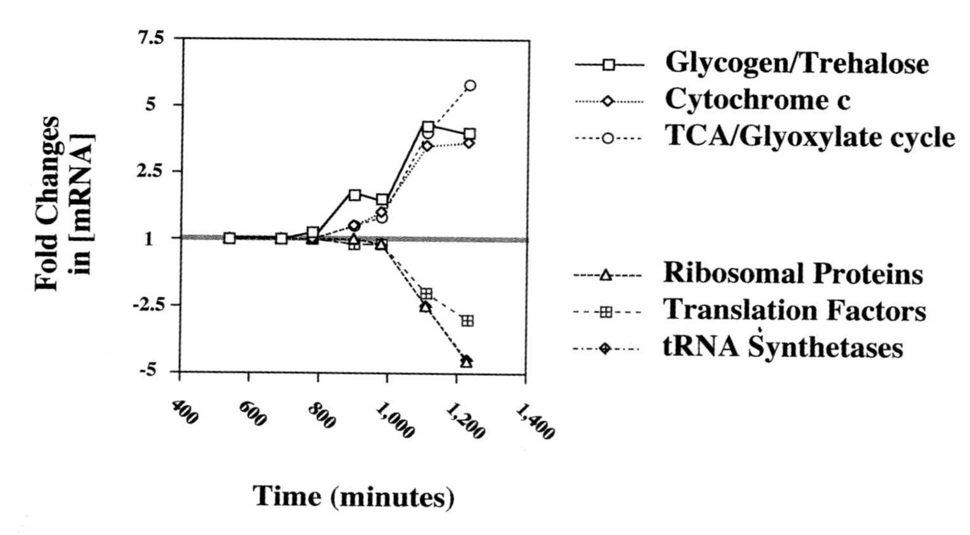

Figure 9.10. Changes in Specific mRNA Levels in Cells Undergoing Diauxic Shift. During the shift from anaerobic to aerobic metabolism the gene expression patterns within cells are adjusted. The quantity of enzymes involved in the production of the storage carbohydrates glycogen and trehalose, components of the TCA/glyoxylate cycle and proteins participating in the respiratory chain increases substantially. Due to a general slow-down in protein synthesis the level of the various components of the translational machinery decreases. From DeRisi *et al.* (1997). An internet site (http://cmgm.stanford.edu/pbrown/) contains a complete data set for expression levels of all yeast genes in cells undergoing a diauxic shift.

regulatory proteins that orchestrate such transitions in gene-specific transcription patterns. Although we are now beginning to understand the fundamental structural features of such transcription factors and the way they interact with components of the basal transcriptional machinery (Chapters 3 and 4), it is still difficult to give any firm estimates how many target genes are regulated by such proteins. Since most gene-specific transcription factors have the ability to bind to specific DNA sequences, it would in principle be possible to simply isolate and map DNA fragments containing suitable binding sites and postulate that these are the genes regulated by the factor. One of the main problems with this approach is that, at least under *in vitro* conditions, the specificity of the binding of many transcription factors is relatively low. In most cases it appears that high affinity sites can be found on average every

4–6 kilobases of DNA sequences. This is partially due to the fact that many gene-specific transcription factors are not as selective as other proteins that display absolute sequence-specificity (e.g. restriction nucleases), but recognize a variety of motifs with only few highly conserved residues (see Chapter 3). Another important consideration is that *in vivo* many gene-specific transcription factors interact cooperatively with other proteins on the promoters and/or bind to tissue specific-components of the basal machinery (such as the TRF-complex, see Chapter 4). Finally, most potential target sites on DNA are almost certainly not available for interaction with gene-specific transcription factors under *in vivo* conditions, because they are packed into an inaccessible chromatin configurations within the nucleus. These circumstances have made it a very difficult and dubious business to experimentally identify functionally important target sites *in vitro*, that are actually used *in vivo* for regulating gene expression. Other ways of identifying DNA motifs that are specific targets for certain transcription factors, including immunopurification of native transcription factor/chromatin complexes, have been successfully used in isolated instances, but are technically very challenging and difficult to apply on a routine basis (Gould *et al.*, 1990; Gould and White 1992; White *et al.*, 1992).

The Application of DNA Microarray Technology to Identify Transcription Factor Regulatory Networks

DNA microarray technology offers a new way of identifying candidate genes that are regulated by particular transcriptional activators and repressors. De Risi *et al.* (1997) observed that several groups of genes displayed distinct temporal transcription profiles that correlated well with the presence of binding motifs for certain transcription factors in the promoter regions of these genes. In one exemplary case it was found that all of the seven genes that were maximally expressed during the onset of the late stage of the diauxic shift contained several repeats of the so-called stress response elements ('STRE') upstream of their transcription start site. In another case, six out of seven genes, whose transcription is strongly repressed by glucose, contained a reconizable carbon source response element ('CSRE'). Interestingly, there are four other copies of CSRE elements in the yeast genome, but they are situated near genes that do not appear to be subject to repression by glucose. This observation supports a point made earlier, namely that the presence of a consensus sequence for a transcription factor in a stretch of DNA sequence does not automatically imply that this is a functional control element for regulating the transcriptional activity of nearby genes. With our current knowledge it is therefore not possible to examine sequences emerging from the genome projects and deduce accurate and reliable regulatory information from the arrangement of transcription factor binding sites.

Apart from detecting genes with similar expression profiles, DNA microarray technology can also be used to investigate the changes of transcription patterns in cells containing mutant transcription factors. De Risi *et al.* (1997) tested this strategy by analyzing the genome-wide changes in the transcription of genes in yeast cells mutant for the transcription factor TUP1. TUP1 plays a major role in a general transcription repression pathway that is controlled by the presence of glucose and other environmental factors. By comparing the transcription of genes in wildtype or TUP1 mutant cells grown under identical conditions, it was possible to detect genes that failed to be repressed in the TUP1 mutant and which can therefore be classified as direct (or possibly indirect) targets of the TUP1 transcriptional repressor. Conversely, overexpression of a yeast transcription factor YAP1, a factor involved in protecting cells against the damaging effects of heavy metals, osmotic shock and hydrogen peroxide resulted in the increased transcription (between 3 and 13 fold) of 17 different genes. This set of genes included a remarkably high percentage of members of a small family of functionally related proteins (oxidoreductases) demonstrating the specificity of the transcriptional genomic response to high levels of YAP1.

It is important to note, however, that at least some of the changes in the transcription patterns observed in mutant cells could be the consequence of indirect regulatory effects. We know from numerous genetic studies in different experimental systems that many genetic control circuits are arranged as cascades, i.e. where expression of a few transcription factors can radically alter the expression of many 'downstream' target genes. These altered levels of downstream products could themselves have a major effect on the metabolism and differentiation state of a cell and may alter the overall cellular transcription pattern. In other words, it becomes very difficult to distinguish between 'cause' and 'effect' if key transcriptional regulators are expressed at abnormal levels.

The examples described above nevertheless demonstrate the power of identifying and dissecting the regulatory function of individual gene-specific transcription factors on the global genome expression pattern. DNA microarrays are relatively new tools that will almost certainly become more widespread as premade micoarrays and fluorescent imaging scanners for reading the results become more widely commercially available in the future.

9.5. Defining Genome Transcription Patterns of Single Cells

Laser Capture Microdissection (LCM)

Many of the studies described in this chapter aimed at comparing genome-wide transcription patterns have been carried out either on microorganisms (e.g. diauxic shift in *S. cerevisiae*; Figure 9.10) or on clonal cell lines grown in tissue culture (e.g. SAGE comparison of normal and cancer cells, Figure 9.6). A single glance at any histology textbook will, however, immediately reveal that tissues of higher organisms are not made up of homogeneous cell types. Brain tissue, for example, contains a large variety of neuronal and glial cell types with very different morphologies and physiological properties. Comparisons between a 'brain' transcription pattern and a 'liver' transcription pattern therefore become a bit meaningless, because the tissue-specific mRNAs themselves are derived from rather heterogenous cell populations within each tissue.

For a while it looked as if such a comparison of such 'average' expression patterns was the best we could hope for in our quest to understand tissue-specific gene expression programs on the biochemical level. A lot of detailed and interesting information is lost this way. Fortunately, new techniques for dissecting complex histological samples down to the single cells level promise a solution to this dilemma. Laser capture microdissection ('LCM') can be used to procure small, histologically homogeneous populations (or even single cells) from tissue sections in a form suitable for analysis by molecular biological techniques (Emmert-Buck *et al.*, 1996; reviewed in Bonner *et al.*, 1997). The LCM strategy involves placing a thermoplastic polymer film containing infrared-absorbing dyes directly over the area on the tissue section of interest.

A brief burst of highly localized irradiation from a gallium-arsenide laser fuses the polymer film to the underlying cells (without damaging DNA and mRNA molecules contained in them) and the attached cells can be precisely removed without disturbing surrounding tissue (Figures 9.11 and 9.12). The current state of the art allows spots with diameters as small as 3 μm to be reproducibly dissected from tissue sections (i.e. at single-cell resolution; most cells range in diameter from 5–10 μm). Since the amount of recovered material is often minute this approach is best suited to PCR-based techniques for analyzing transcription patterns, such as differential display and SAGE.

The ability to study mRNA populations from single cells within complex tissues will require new ways of interpreting the gene expression information

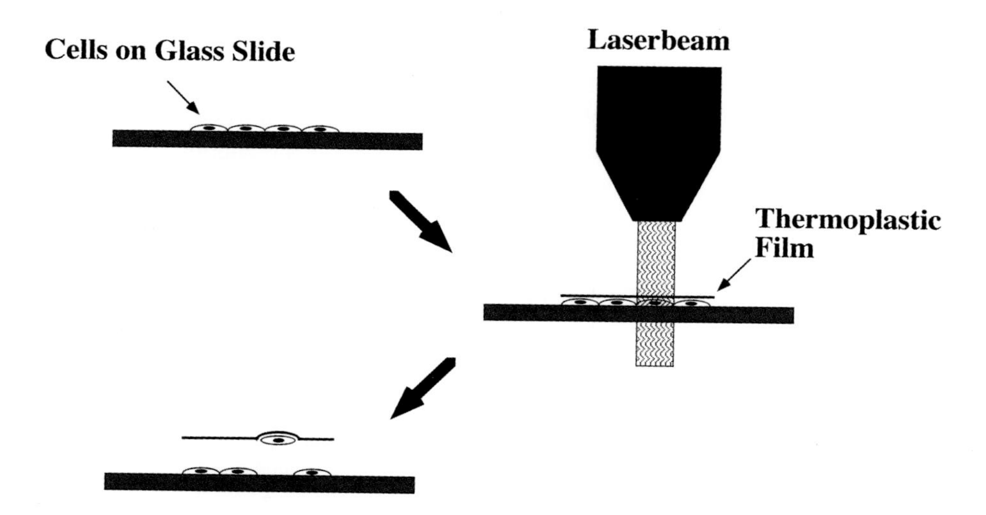

Figure 9.11. Experimental Detail of Laser Capture Microdissection.
Top Left: Schematic cross-section of a histological tissue-section containing cells for microdissection on a glass slide. *Middle Right*: The tissue section is covered with a thermoplastic film and the area of interest is briefly illuminated with an infrared laserbeam. The local increase in temperature fixes the cells to the plastic film. *Lower Left*: The cells specifically attached to the thermoplastic film are lifted away from the rest of the tissue section and are ready for further processing (such as mRNA extraction for investigating cell type-specific gene expression patterns).

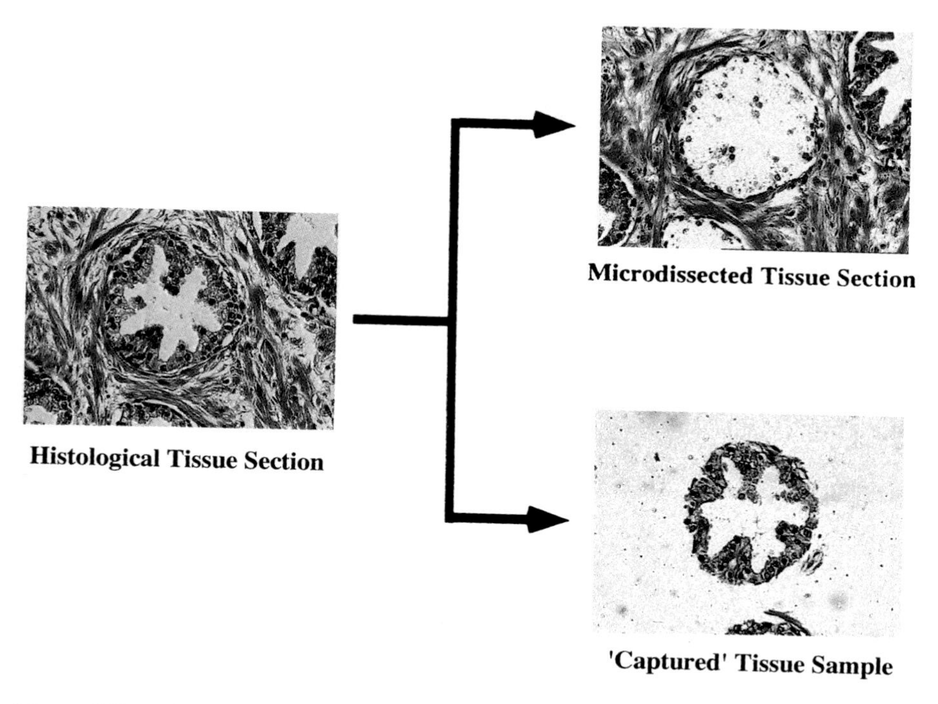

Figure 9.12. Recovery of Specific Cell Types by Laser Capture Microdissection. This and further examples can be found on the NIH internet site (http://dir.nichd.nih.gov/lcm/lcm.htm).

obtained from such experiments. Normally, in a large homogeneous population of cells any cell cycle-specific differences in the transcription patterns between individual cells are averaged out and become esentially undetectable. When individual cells are analyzed, such differences become a major source of variability that needs to be considered and taken into account, especially in situations where differences in the transcription patterns between distinct cell types are sought. Ironically perhaps, any conclusions about the composition of single cell mRNA populations will in the end need to be based on results obtained from multiple samplings of several cells displaying the same histological phenotype to avoid such single cell-specific variations.

9.6. The Ultimate Aim: Monitoring Gene Expression in Real Time

The techniqes described above are designed to measure the gene expression pattern of cells in a static manner, i.e. mRNA has to be extracted and analyzed for the expression of various distinct transcripts. As can be seen from the application of DNA microarray technology to investigate the metabolic switching of gene transcription patterns under different growth conditions, it is possible to extract an approximate temporal sequence out of the data based on the analysis of staged mRNA populations. In many cases it is important, however, to measure the expression of a gene continuosly within a living cell, or even an intact organism. Recently developed techniques for imaging particular proteins in living cells have made such studies a distinct possibility.

Monitoring the expression on individual genes under *in vivo* conditions depends on the ability to easily visualize, in a non-destructive manner, the production of a particular gene product (usually a protein) from outside the cell. This is most easily achievable by optical means and through the use of reporter genes encoding optically detectable products. A popular choice is 'green fluorescent protein' (GFP; derived from jellyfish) that fluoresces in a vivid shade of green in the presence of UV light (reviewed in Cubitt *et al.* 1995). A recent interesting application of this technique was the demonstration that a gene, *period*, has a cyclically fluctuating expression pattern in the various parts of the fruitfly *D. melanogaster*, indicating the presence of independent molecular clocks in certain cells types (Plautz *et al.*, 1997).

An alternative approach, possibly more suited to monitoring gene expression in tissue culture cells, has been reported by Zlokamik *et al.* (1998). The reporter gene product is the enzyme β-lactamase which is used to specifically cleave a membrane-permeable fluroescent dye (Figure 9.13). Cleavage of the dye molecule induced a wavelength shift of the emitted fluorescent light from green to blue which is easily detectable by eye (Figure 9.14).

Both approaches have drawbacks in their current form, that limit the interpretation of 'live' gene expression dynamics. GFP assays are relatively insensitive, and 10^5 to 10^6 molecules of GFP would have to produced in a single cell before the signal becomes clearly detectable above background fluorescence levels (Zlokarnik *et al.*, 1998). In comparison, the enzymatic assays, where each enzyme catalyzes the conversion of multiple substrate molecules, are far more sensitive and 50 molecules of β-lactamase per cell catalyze sufficient substrate conversion over 16 hours to be easily detected (Zlokarnik *et al.*, 1998). GFP assays therefore require strong constitutive promoters that are not necessarily representative of many cellular promoters. Both approaches suffer from an additional disadvantage caused by the stability of reporter gene product. β-lactamase has a half-life of approximately 3 hours (similar to another reproter enzyme, luciferase; Thompson *et al.*, 1991), which

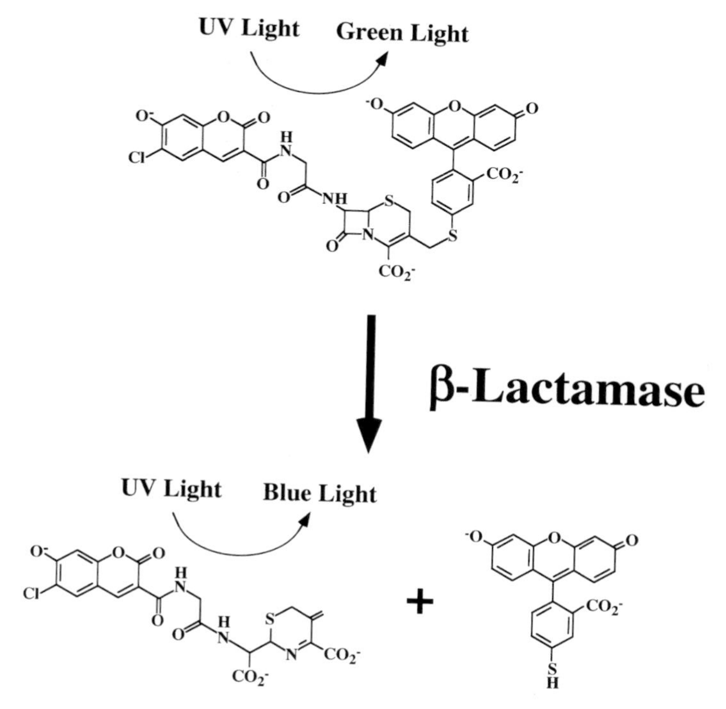

Figure 9.13. Cleavage of Fluorogenic Substrate CCF2 by β-Lactamase.
CCF2 fluoresces green (520 nm) when illuminated with UV light (409 nm) through fluorescence resonance energy transfer from the left half of the molecule ('donor') to the right half ('acceptor'). After cleavage of the lactam ring of CCF2 by β-lactamase the separated donor fluoresces blue (447 nm). Cleavage of CCF2 thus results in a distinct change of the emitted light from green to blue.

Figure 9.14. Monitoring Gene Expression in Living Cells.
From Zlokarnik *et al.* (1998)

distorts the correlation between quantity of enzyme present and the rate of reporter gene transcription, especially if the transcription rate of the reporter gene slows down. It is possible, however, that these disadvantages can be minimized in the future through protein engineering approaches designed to *lower* the stability of the reporter proteins. A reporter with a high catalytic substrate turnover-rate and short half-life would indeed make it possible to visualize gene expression from a reporter construct almost in real time.

9.7. Conclusions

The technical advances achieved during the last few years have allowed the detailed molecular characterization of mRNA populations to proceed at an awesome rate. New methods for studying the expression levels of thousands of genes *simultanously* are beginning to reveal new organizational principles of genetic control circuits that are difficult to detect with conventional approaches. Similar to the fact that it is currently impossible to predict the structure and function of a protein from its primary structure, it is likely that the results from structural genomics will need to be supplemented with studies of the gene expression dynamics. Many of the key technologies required for such a task are already in place and have performed successfully in 'simple' model systems. Other novel techniques, such laser capture microdissection, will allow exciting new types of experiments to be carried out within the forseeable future at single cell resolution.

The understanding gene expression dynamics will almost certainly have major implications for the detection and treatments of cancer and and other disease states. It is likely that at some stage it will become possible to use gene expression pattern as precise diagnostic tools, and to detect the onset of diseases or infections at early stages. This will allow medical treatment to commence before the appearance of distinct (and possibly irreversible) symptoms and increase the chances of successful therapeutic intervention.

References

Internet Sites of Interest

'http://www.ncbi.nlm.nih.gov/ncicgap/'
The Cancer Genome Anatomy Web site. Technological approaches for investigating molecular characteristics of normal, pre-cancerous and malignant cells. Laser microdisection from histological sections on 'http://dir.nichd.nih.gov/lcm/lcm.htm'.

'http://www.protogene.com'
Information about DNA microarray production.

'http://www.affymetrix.com'
On-chip oligonucleotide synthesis directed by photolithography.

'http://cmgm.stanford.edu/pbrown/'
DNA microarray production; yeast genome chips; complete numeric data for changes in transcription patterns for every gene in *S. cerevisiae* cells undergoing a diauxic shift.

'http://www.genomix.com'
Differential display methods and examples.

'http://www.lsbc.com/'
Website of a company specializing in the development of techniques for proteomics and 2D-protein gel electrophoresis.

'http://www.genome.ad.jp/kegg/'
A useful catalog of molecules, genes, metabolic pathways and genome maps.

Selected Patents

U.S. Patent No. 5,474,796 'Method and apparatus for conducting an array of chemical reactions on a support surface'

Original Research Papers:

Adams, M.D., Kelley, J.M., Gocayne, J.D., Dubnick, M., Polymeropoulos, M.H., Xiao, H., *et al.* (1991). Complementary DNA sequencing: expressed sequence tags and human genome project. *Science* 252, 1651–1656.

Adams, M.D., Kervalage, A.R., Fields, C., and Venter, J.C. (1993). 3,400 new expressed sequence tags identify diversity of transcripts in human brain. *Nature Genet.* 4, 256–267.

Anderson, L., Seilhamer, J. (1997). A comparison of selected mRNA and protein abundances in human liver. *Electrophoresis* 18, 533–537.

Attfield, P.V. (1997). Stress tolerance: the key to effective strains of industrial baker's yeast. *Nature Biotechnology* 15, 1351–1357.

Blomberg, A., Blomberg, L., Fey, S.J., Mose Larsen, P., Roepstorff, P., Degand, P., Boutry, M., Posch, A., Gorg, A. (1995). Interlaboratory reproducibility of yeast protein patterns analyzed by immobilized pH gradient two-dimensional gel electrophoresis. *Electrophoresis* 16, 1935–1945.

Bonner, R.F., Emmert-Buck, M., Cole, K, Pohida, T., Chaqui, R., Goldstein, S., and Liotta, L.A. (1997). Laser capture microdissection: molecular analysis of tissue. *Science* 278, 1481–1483.

Chee, M., Yang, R., Hubbell, E., Berno, A., Huang, X.C., Stern, D., Winkler, J., Lockhart, D.J., Morris, M.S., and Fodor, S.P.A. (1996). Accessing genetic information with high-density DNA arrays. *Science* 274, 610–614.

Cubitt, A.B., Heim, R., Adams, S.R., Boyd, A.E., Gross, L.A. and Tsien, R.Y. (1995). Understanding, improving and using green fluorescent proteins. *Trends Biochem. Sci.* 20, 448–455.

DeRisi, J.L., Iyer, V.R., and Brown, P.O. (1997). Exploring the metabolic and genetic control of gene expression on a genomic scale. *Science* 278. 680–686.

Dunn, M.J. (1993). Gel electrophoresis of proteins. BIOS Scientific Publishers Ltd., Oxford.

Fields, C., Adams, D., White, O. , and Venter, J.C. (1994). How many genes in the human genome? *Nature Genet.* 7, 345–346.

Fodor, S.P., Rava, R.P., Huang, X.C., Pease, A.C. Holmes, C.P. and Adams, C.L. (1993). Multiplexed biochemical assays with biological chips. *Nature* 364, 555–556.

Goffeau, A., Barrell, B.G., Bussey, H., Davies, R.W., Dujon, B., Feldmann, H., *et al.* (1996). Life with 6000 genes. *Science* 274, 546–567.

Goffeau, A., *et al.* (1997). The yeast genome directory. *Nature* 387 (Suppl.1), 1–105.

Gould, A.P., Brookman, J.J., Strutt, D.I., and White, R.A. (1990). Targets of homeotic gene control in *Drosophila. Nature* 348, 308–312.

Gould, A.P., and White, R.A. (1992). Connectin, a target of homeotic gene control in *Drosophila. Development* 116, 1163–1174.

Heller, R.A., Schena, M., Chai, A., Shalon, D., Bedilion, T., Gilmore, J., Woolley, D.E., and Davies, R.W. (1997). Discovery and analysis of inflammatory disease-related genes using cDNA microarrays. *Proc. Natl. Acad. Sci. USA* 94, 2150–2155.

Hieter, P., and Boguski, M. (1997). Functional genomics: it's all how you read it. *Science* 278, 601–602.

Lee, S.W., Tomasetto, C., and Sager, R. (1991). Positive selection of candidate tumor-suppressor genes by subtractive hybridization. *Proc. Natl. Acad. Sci. USA* 88, 2825–2829.

Liang, P. and Pardee, A.B. (1992). Differential display of eukaryotic messenger RNA by means of the polymerase chain reaction. *Science* 257, 967–971.

Lockhart, D.J., Dong, H., Byrne, M.C., Follettie, M.T., Gallo, M.V., Chee, M.S., Mittmann, M., Wang, C., Kobayashi, M., Horton, H. and Brown, E.L. (1996). Expression monitoring by hybridization to high-density oligonucleotide arrays. *Nature Biotechnology* 14, 1675–1680.

McClelland, M, Mathieu-Daude, F., and Welsh, J. (1995). RNA fingerprinting and differential display using arbitrarily primed PCR. *Trends Genet.* 11, 242–246.

O'Farrell, P.H. (1975). High resolution two-dimensional electrophoresis of proteins. *J. Biol. Chem.* 250, 4007–4021.

O'Farrell, P.Z., Goodman, H.M., and O'Farrell, P.H. (1977). High resolution two-dimensional electrophoresis of basic as well as acidic proteins. *Cell* 12, 1133–1142.

Okubo, K., Hori, N., Matoba, R., Niiyama, T., Fukushima, A., Kojima, Y., and Matsubara, K. (1992). Large scale cDNA sequencing for analysis of quantitative and qualitative aspects of gene expression. *Nature Genet.* 2, 173–179.

Plautz, J.D., Kaneko, M., Hall, J.C., and Kay, S.A. (1997). Independent photoreactive circadian clocks throughout *Drosophila. Science* 278, 1632–1635.

Ralph, D., McClelland, M., and Welsh, J. (1993). RNA fingerprinting using arbitrarily primed PCR identifies differentially regulated RNAs in mink lung (My1Lu) cells growth arrested by transforming growth factor β1. *Proc. Natl. Acad. Sci. USA* 90, 10710–10714.

Schena, M., Shalon, D., Davis, R.W., and Brown, P.O. (1995). Quantitative monitoring of gene expression patterns with a complementary DNA microarray. *Science* 270, 467–470.

Schena, M., Shalon, D., Heller, R., Chai, A., Brown, P.O., and Davis, R.W. (1996). Parallel human genome analysis: microarray-based expression monitoring of 1000 genes. *Proc. Natl. Acad. Sci. USA* 93, 10614–10619.

Shevchenko, A., Jensen, O.N., Podtelejnikov, A.V., Sagliocco, F., Wilm, M., Vorm, O., *et al.* (1996). Linking genome and proteome by mass spectrometry: large-scale identification of yeast proteins from two-dimensional gels. *Proc. Natl. Acad. Sci. USA* 93, 14440–14445.

Thompson, J.F., Hayes, D.B., and Lloyd, D.B. (1991). Modulation of firefly luciferase stability and impact on studies of gene regulation. *Gene* 103, 171–000.

Varmus, H. (1989). In 'Oncogenes and the Molecular Origin of Cancer'. R.A. Weinberg (Ed.). Cold Spring Harbor Laboratory Press, New York.

Velculescu, V.E., Zhang, L., Vogelstein, B., and Kinzler, K.W. (1995). Serial analysis of gene expression. *Science* 270, 484–487.

Velculescu, V.E., Zhang, L., Zhou, W., Vogelstein, J., Basrai, M.A., Bassett, D.E. Jr., Hieter, P., Vogelstein, B., and Kinzler, K.W. (1997). Characterization of the yeast transcriptome. *Cell* 88, 243–251.

White, R.A., Brookman, J.J., Gould, A.P, Meadows, L.A., Shashidhara, L.S., Strutt, D.I., and Weaver, T.A. (1992). Targets of homeotic gene control in *Drosophila. J. Cell Sci. Suppl.* 16, 53–60.

Wilkins, M.R. et al. (1996). From proteins to proteomes: large scale protein identification by two-dimensional electrophoresis and amino acid analysis. *Nature Biotechnology* 14, 61–65.

Wodicka, L., Dong, H., Mittmann, M., Ho, M.-H., and Lockhart, D.J. (1997). Genome-wide expression monitoring in *Saccharomyces cerevisiae. Nature Biotechnology* 15, 1359–1367.

Wong, K.K, and McClelland (1994). Stress-inducible gene of *Salmonella typhimurium* identified by arbitrarily primed PCR of RNA. *Proc. Natl. Acad. Sci. USA* 91, 639–643.

Zhang, L., Zhou, W., Velculescu, V.E., Kern, S.E., Hruban, R.H., Hamilton, S.R., Vogelstein, B., and Kinzler, K.W. (1997). Gene expression profiles in normal and cancer cells. *Science* 276, 1268–1272.

Zlokarnik, G., Negulescu, P.A., Knapp, T.E., Mere, L., Burres, N., Feng, L., Whitney, M., Roemer, K., and Tsien, R.Y. (1998). Quantitation of transcription and clonal selection of single living cells with β-lactamase as reporter. *Science* 279, 84–88.

Chapter 10

Appearing on the Horizon: Medical Applications Focusing on Transcriptional Control Mechanisms

Ultimately, one of the main reasons for researching transcription mechanisms must be the goal to understand the control of gene expression in sufficient detail, so that the resulting knowledge can be used for alleviating and curing medical conditions. This chapter reviews some of the most successful steps towards modifying the expression of specific genes through drugs that influence directly the function of the transcriptional machinery.

10.1. The Need for Drugs Affecting Gene Expression Patterns

Much of the current pharmacological research effort is directed at characterizing the biochemical and physiological properties of cell surface receptors present on a variety of cell types. The logic for such an approach is straightforward: Cell surface receptors are in an exposed position, where they are ideally placed for continuously sampling the extra-cellular environment for a variety of signalling molecules. This provides the easiest entry point for drugs designed to change cellular behaviour. Moreover, cell surface receptors are linked with efficient intracellular signal processing and transmission pathways, so that relatively small effects can be swiftly amplified into a concerted and widespread response affecting a multitude of cellular activities.

In contrast, drugs that have their effect on the level of gene expression either need to penetrate cell membranes directly, or need to be specifically linked to other molecules to get them efficiently transported from the outside of the cell to their ultimate site of action, the nucleus. This transport problem does not only affect the effectiveness of a drug at the final stages of its development cycle, but causes major problems during preliminary screens used to identify chemicals that might be suitable candidates for additional modifications and fine tuning ('lead drugs', see below). It could be realistically imagined that, although drugs affecting specific transcription factors may exist in the screening libraries, they are never detected simply because they are not taken up into the cell in sufficiently high levels to exert any activity. This is an especially important consideration in drug screens based on *in vivo* assays with tissue culture cells.

It therefore does not come as a big surprise that only a limited range of the drugs affecting transcription of specific genes exists at this point in time. It is nevertheless becoming increasingly clear, mostly through the activities of several biotechnology companies in the field, that the interest in identifying therapeutic targets and carrying out large scale drug screens is now growing rapidly. Despite the difficulties described above, there is a very strong case to be made for an increased effort to identify drugs that target various components of the transcriptional machinery (Cai *et al.*, 1996; Latchmann, 1997). It is well established that wrongful expression of certain gene products, such as transcriptional regulators, cell cycle control proteins, growth factors and accidental fusions from chromosomal breakpoints (lymphomas), are a major cause for many forms of human disease (Latchman, 1996). Also, it is likely that viral infections can probably be dealt with effectively on the transcriptional level. Many viruses encode strong transcriptional activators (such as the herpes virus VP16 protein), which are prime targets for drugs interfering specifically with their interaction with basal factors and coactivators.

Below we will have a closer look at the state of the art of applying our knowledge of gene expression mechanisms to solving real medical problems.

First, high-throughput drug screening procedures suitable for identifying promoter-specific and transcription-modifying drugs will be described to illustrate the technological challenges and general direction of the field. Next, we will have a look at some specific examples where transcription-based approaches have either already been successful, or are providing promising new ways of solving medical problems.

10.2. Strategies for Identifying Drugs Interfering with Gene-specific Transcriptional Regulators

'Random' Drug Screen Strategies: Chemical Libraries, Rainforests and Coral Reefs Full of Goodies!

In an ideal world drugs would be designed rationally on the basis of extensive knowledge of the structure of target molecules, their function within the cells and in the intact organism (e.g. Blundell, 1996). In reality, most of the drugs that are nowadays in common use were discovered by identifying the active ingredients in traditional folk medicine, or by systematic tests of thousands of chemical compounds and natural extracts for relevant activities. For this strategy relatively little prior knowledge of the molecular targets is required, and 'random' molecules are tested for their effectiveness in high-throughput screening procedures. The composition of a representative industrial 'screening compound library' is shown in Figure 10.1. Natural extracts, mostly derived from plants, fungi and marine organisms, contain large varieties of ready-made complex molecules that would be too time-consuming and difficult to synthesize with conventional organic chemistry techniques. A classic example is the discovery of the anti-cancer drug 'taxol' (useful for breast cancer treatment), a molecule that would probably have been too complex and too expensive to produce in a chemical laboratory for screening purposes. Such drugs are available now for treatment only because they happened to occur in natural extracts and were 'discovered' to be effective. The large variety of unusual chemicals in natural extracts is complemented in screening compound libraries by collections of defined organic compounds. These are often synthesized by combinatorial methods, so that a large variety of 'simple' molecules can be

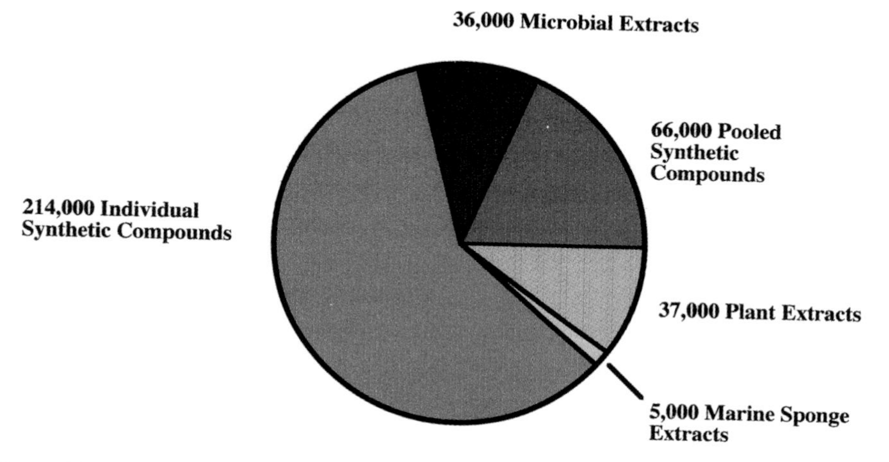

Figure 10.1. Composition of a Typical Pharmaceutical Screening Library Used for High-throughput Strategies.
A large quantity of well-defined organic synthetic compounds is complemented by collections of complex extracts derived from microbes, plants and marine sponges. Data based on information shown on http://www.tularik.com.

produced that might be particularly useful for oral applications and amenable to large scale chemical synthesis.

Ideally, extensive screens with such libraries will lead to the identification of 'lead compounds' already showing limited activity, but that can subsequently be refined further through chemical modifications to increase their potency and to produce clinically useable drugs. The large number of extracts and synthesized compounds (usually >100,000) that need to be screened to identify lead compounds demonstrates very clearly the need for efficient and sensitive screening methods. In the next section we will look at a variety of screening strategies that have been specifically developed for assaying drugs that influence gene-specific transcription.

In Vivo or *In Vitro*? Cell-based versus Protein-Based Assays

There are two main considerations affecting the design philosophy of an assay for testing the activity of 'random' drugs on the transcription of specific genes: the most important question is whether the assay should be based on an *in vivo* or *in vitro* system. This is a fundamental decision based on a number of considerations that depends greatly on the nature of the targeted disease, the availability of suitable 'model' cell lines, or the existence of an already defined *in vitro* system.

In Vitro Drug Screening Assays

For several genes involved in controlling medically-relevant processes in the human body the molecular details of their mode of regulation are beginning to become reasonably well understood. Especially in the case of many viruses we have now sufficient information regarding their infection cycle and the control features of their transcriptional programs, so that it becomes possible to develop drugs that are capable of specifically neutralizing key transcription factors. The 'Tat' protein encoded by the human HIV virus is a good example for this approach, and below assays based on disrupting the activity of Tat will be described. If it is clear that a specific transcription factor, such as Tat, plays a major functional role, it becomes possible to use this knowledge to develop technically straightforward biochemical assays that are designed to detect specific interactions between possible lead compounds and portions of the target protein. Such interactions, especially if they interfere with the function of DNA-binding motifs or activation domains, could identify drugs that might be effective in combating viral infections under *in vivo* conditions.

In more sophisticated *in vitro* assays this concept is taken one step further by using functional assays to differentiate between drugs that only display specific interactions, and those that specifically inactivate the function of a transcription factor. Such functional assays could include electrophoretic mobility shift assays (EMSA) and *in vitro* reconstituted transcription reactions. It is possible to scale up the capacity of all these methods to carry out the large number of parallel screens required for the random drug screening strategies described above.

The major advantage of *in vitro* assays is that they can be completely controlled and are usually highly reproducible. With the advent of a flood of molecular data from the various ongoing genome sequencing projects, more and more potential target molecules, including some that can be identified by sequence homologies to proteins of known function, become readily available for *in vitro* screens (reviewed in Cohen, 1997; Gelbert and Gregg, 1997). Nevertheless, *in vitro* screens are usually only used as a first step towards drug identification, because lead compounds that display apparently specific effects in such systems may fail to exert a similar function within intact cells. Such problems are often due to transport problems (especially inability to cross cell membranes), or general toxicity effects.

In Vivo Drug Screening Cell-Based Assays

Because of the potential pitfalls of *in vitro*-based approaches, the development of cell-based screening procedures has flourished during the last few years. Although they are often more complex to set up and can be erratic in their behaviour, cell-based assays offer a more realistic experimental setup by closely mimicking the environment in which the drugs are normally expected to perform. For high-throughput screening purposes, genetically modified human cells are often grown in 96-well microtiter dishes, which can be robotically handled throughout the procedure, thus minimizing variations between individual wells and reducing the chances of contamination.

One of the key requirements for developing cell-based screening assays with high-throughput capacity is a way of monitoring transcription levels from a defined promoter in a relatively straightforward manner. This is mostly achieved through the use of reporter genes that are expressed from a promoter regulated by the target transcription factor. A reporter gene can be defined as any type of gene encoding a macromolecule that can be easily assayed for by biochemical means. Traditionally, some of the most successful reporter molecules were enzymes, such as alkaline phosphatase and β-galactosidase, capable of catalyzing substrates detectable by colorimetric assays. The amount of reporter enzyme produced within cells can be monitored by fixing the cells histologically and incubating them, after permeabilization, with the colorimetric substrate. For high-throughput screens the measurement of intracellular enzyme levels is too cumbersome, and secretable forms of e.g. alkaline phosphatase have been developed that allow the more convenient measurement of reporter enzyme activity in tissue culture supernatants. Another notable improvement has been the application of the light-emitting reporter enzyme luciferase, whose activity can be measured continously and accurately in intact cells through the amount of emitted light.

Case Study: A Random Drug Screen Reveals Lead Drugs Interfering with the Transcription from an Activated Proto-Oncogene Promoter

Many of the principles described above are illustrated in a report describing the results obtained in a large-scale drug screen that aimed at identifying lead drugs interfering with the transcription of a specific promoter. This report is unusual, but especially illuminating, because it illustrates the potential of such approaches and shows that a surprisingly large number of compounds in a screening library display very specific effects on gene-specific transcription patterns (Dhundale and Goddard, 1996).

In the described screen, each individual component of the screening library was screened in triplicate in the presence of appropriate positive and negative controls. A robotic setup allowed 500,000 cell-based reporter assays to be carried out per week, resulting in the acquisition of quantitative data of the effect of nearly 100,000 different drugs on the expression of the selected reporter construct. The target promoter tested in the screen was the *bcr*-promoter. Although *bcr* expression is generally not a medical problem, there are certain instances were active transcription from this promoter directs transcription of an oncogene, resulting in forms of chronic myelogenous leukemia (the 'Philadelphia' chromosome [Ph^1]). The cytological phenotype of Ph^1 is a reciprocal chromosomal translocation where a small fragment from the tip of the long arm of chromosome 9 is exchanged for a portion of the long arm of chromosome 22, resulting in two mutant chromosomes. Interestingly, the translocation breakpoints always involve very defined regions (<50 kb) on both chromosomes, resulting in very similar translocations occuring in different patients independently. The carcinogenic effect of the Ph^1 chromosome is due to the creation of a fusion gene at the translocation breakpoint that joins the proto-oncogene *abl* (normally located on chromosome 9) to the promoter and N-terminal exons of the *bcr* gene (on chromosome 22). The resulting *bcr-abl* fusion gene produces a protein product, p210, containing a constitutively active tyrosine kinase activity that interferes with normal cell cycle control mechanisms in myeloid stem cells (Figure 10.2). Drugs that specifically down-regulate transcription from the *bcr* promoter could therefore prove clinically useful for treatments of cancers involving Philadelphia chromosomal translocations.

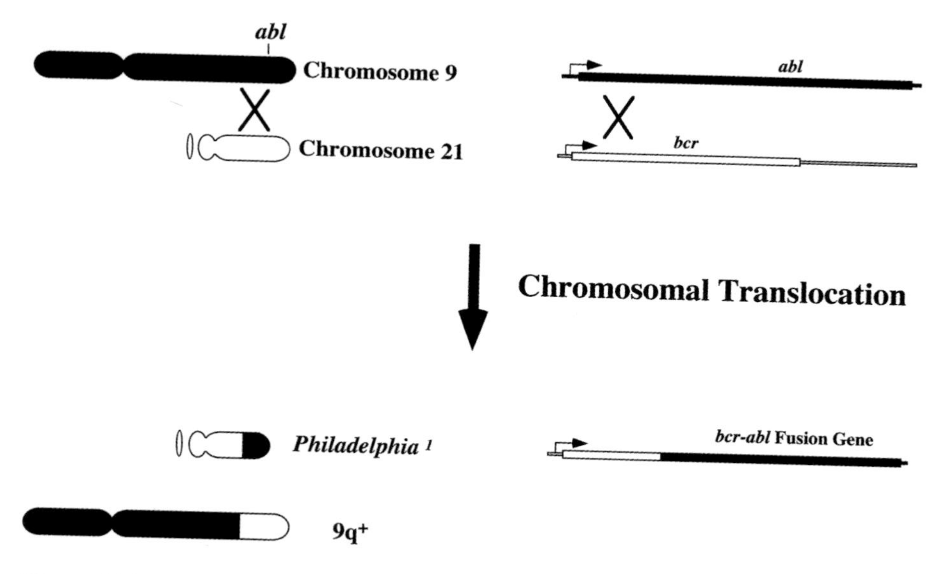

Figure 10.2. Generation of Oncogenic *bcr-abl* Fusion Gene through 'Philadelphia' Chromosome Translocation.
The proto-oncogene *abl* is located on the long arm of chromosome 9 and another gene, *bcr*, on the long arm of chromosome 22. Both genes are normally appropriately expressed from their own promoters (chromosome digrams in left half of the figure; schematic diagrams of the *abl* and *bcr* genes on the right). Reciprocal translocations involving specific break points in chromosomes 9 and 22 lead to the generation of fusion chromosomes ('Philadelphia' and $9q^+$) that express an oncogenic *bcr-abl* fusion protein from the *bcr* promoter in a misdirected manner and cause myelogenous leukemia.

In order to identify such compounds, four kilobases of the *bcr* promoter were linked up to the coding region of firefly luciferase and used in a stably transfected human leukemia cell line to screen 87,000 different compounds. The results are summarized in Figure 10.3. Although most tested compounds

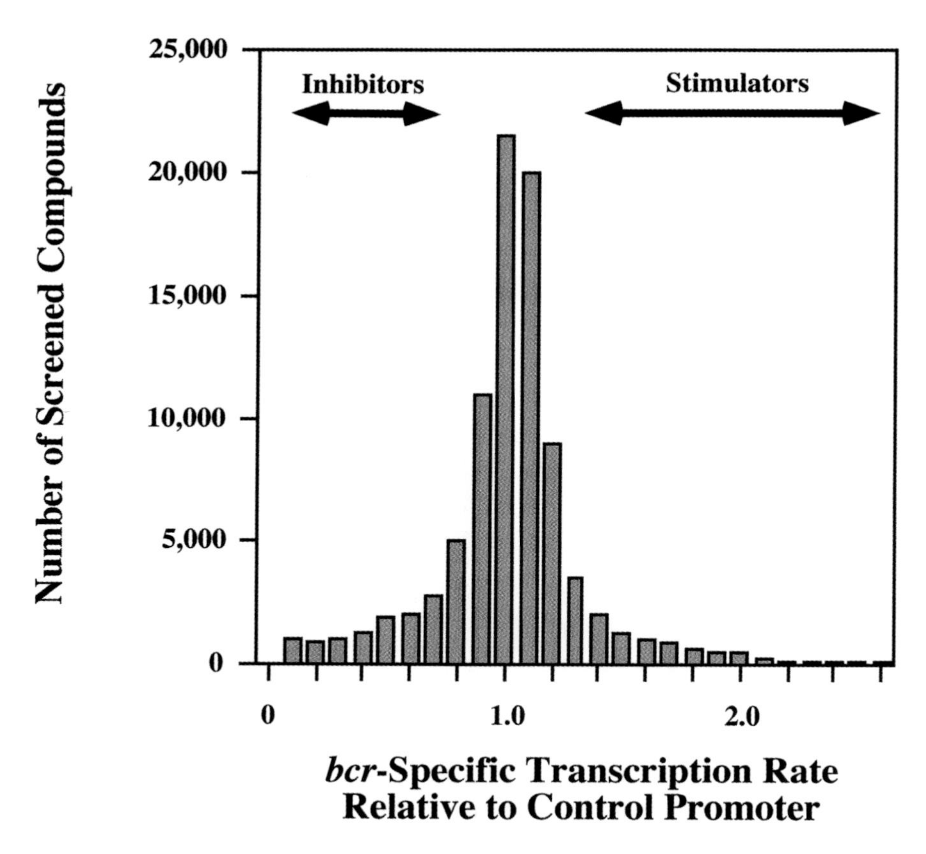

Figure 10.3. Results of a Search for Compounds that Specifically Modulate Transcription from a Target Promoter.

87,000 different compounds were screened in triplicate for their effects on the transcription from a reporter system containing the *bcr*-promoter. The results are displayed as the ratio (TIR; transcription inhibition/induction ratio) between the transcriptional activities directed by the *bcr*-promoter and a negative control promoter. A TIR of 1.0 indicates that the tested compounds had no measureable *bcr* promoter-selective activity, whereas a TIR of 0.3 means that the transcription from the *bcr* promoter in comparison to a control promoter was inhibited by 70%. Note the large numbers of identified *bcr* promoter-specific inhibitors (located in the TIR 0.1–0.3 range) on the left side of the graph. There are also several hundred components that specifically stimulate transcription from the *bcr*-promoter around two-fold or more. After Dhundale and Goddard (1996).

had little or no effect, a startling number of lead drugs were identified that either substantially inhibited or stimulated transcription from the *bcr* promoter in comparison to the control promoter. The collection of inhibitory compounds and extracts (with varying degrees of specificity) identified in this screen will form a solid basis for identifying and isolating the active ingredients which, possibly after further chemical modifications, could become available for animal and clinical trials.

The results of most screens of this type are commercially sensitive and therefore not usually publicly available. It seems clear, however, from the single example shown above, that if a suitable reporter system can be devised (containing the 'right' part of a promoter in an appropriate cell line), the high-throughput screening approach with random screening libraries can reliably lead to the identification of numerous interesting leads for transcription-selective drugs. Any high-troughput screening strategy will result in some artefactual results and/or miss active compounds due to a variety of conceivable artefacts (cytotoxicity, non-linearity of the reporting system, indirect effects). But as Dhundale and Goddard (1996) point out memorably: '*As for those hits that might not be detected with an artificial promoter-reporter assay, we believe that in a world where literally hundreds of thousands of compounds can be (and have been) accurately screened in live cell reporter assays in a matter of weeks, perhaps our attention should be more appropriately directed toward the power and speed with which we can discover, sort and prioritize those very high quality leads that we KNOW are present in the enormous libraries now available to discovery programs!*'. The preliminary results certainly seem to justify this positive view.

Transcription 'Collision' Assays for Detection of Gene-specific Inhibitors *In Vivo* and *In Vitro*

The screening study on the *bcr* promoter described above illustrates the power of transcription assays to identify lead drugs that interfere with transcription from specific promoters. A recent technological improvement of the screening strategy promises an even quicker and more sensitive way of identifying promoter-specific inhibitors through the use of transcriptional 'collision' constructs' in reporter assays (Del Rosario *et al.*, 1996). In such a collision construct, a reporter enzyme is transcribed by a constitutively active promoter (p1). An additional promoter (p2; derived from the gene of interest) is inserted downstream of the same reporter gene, which faces in the opposite direction to p1 (Figure 10.4). The RNA polymerases sent into the reporter gene coding region via the two opposing promoters will therefore proceed directly towards each other and collide somewhere in the middle of the gene, disrupting the production of complete 'sense' or 'antisense' transcripts. Under these conditions no functional reporter 'sense' mRNA will be produced until transcription from p2 is specifically inhibited, while transcription from p1 is actively maintained. Such a collision construct is therefore ideally suited for

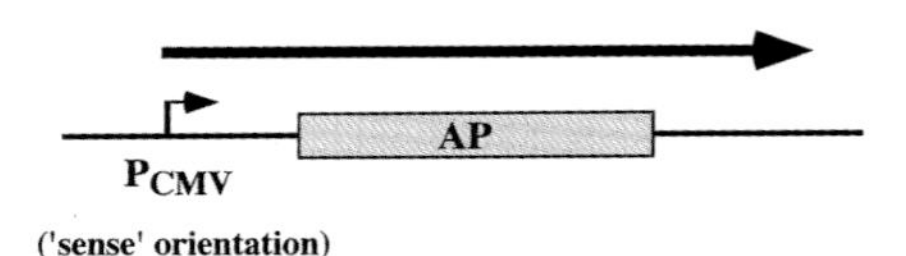

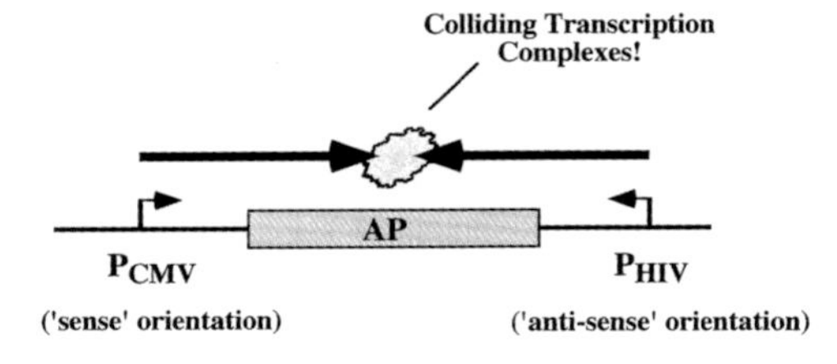

Figure 10.4. Principle of a Transcription Collision Construct.
For many transcription assays a reporter gene (such as alkaline phosphatase, AP) is transcribed from a single promoter, similar to the construct shown on top. Transcription from the P_{CMV} promoter results in the production of an mRNA encoding the complete coding region of AP. In the transcription collision construct shown at the bottom, another promoter (in this case P_{HIV}) is inserted downstream of the transcribed region and faces in the opposite direction. The elongating RNAP molecules initiated by both promoters will collide in the middle of the AP coding region (and thus fail to produce a complete AP-encoding mRNA), unless the activity of the antisense promoter is specifically inhibited. This experimental set-up can therefore be used to screen for drugs that specifically inhibit transcription from P_{HIV}. After Del Rosario *et al.* (1996).

the detection of lead drugs that prevent transcription from the test promoter (p2), but do not have a general inhibitory effect on transcription from other promoters (such as p1). Del Rosario *et al.* tested the principle of this strategy with the HIV-1 promoter, which is positively regulated by Tat (Feinberg *et al.*, 1991; Jones and Peterlin, 1994). The HIV-1 promoter was inserted downstream of an alkaline phosphatase (AP) reporter gene, whose transcription was driven by a constitutively active cytomegalovirus (CMV) promoter. After transfection of this construct into the human HeLa cell line, expression of AP was observed at high levels, due to the fact that such cells do not contain the Tat activator required to stimulate high levels of transcription from the HIV-1 antisense promoter. Co-transfection of the collision construct with increasing amounts of a Tat-expressing plasmid resulted in the expected decrease of AP production as a consequence of Tat-mediated activation of transcription from the HIV-1 promoter, causing the production of collision transcripts (Figure 10.5).

For large scale screening purposes, HeLa cells are stably transformed with plasmids containing the collision construct and expressing a suitable amount of Tat to activate the expression of the HIV-1 promoter. After exposure of these cells to various compounds from screening libraries (Figure 10.1), only the cells that were exposed to a highly-specific inhibitor affecting HIV-1 promoter- or Tat function will produce large amounts of the alkaline phosphatase reporter. From such an (up to now still hypothetical) example, the major technical advantages of collision screening become obvious. Firstly, a promoter-specific inhibitory event is converted into a positive reporter signal, which substantially increases the detection reliability in large-scale robotic screening set-ups. Secondly, each experiment has a built-in positive control designed to eliminate compounds that have a toxic effect on some of the general components of the transcriptional machinery, because transcription from the constitutive CMV promoter still has to proceed normally to produce the 'sense' transcript from the reporter gene. And finally, like in any of the other cell-based systems, potential lead drugs can be picked up in this assay without any prior detailed molecular knowledge of the gene-specific transcription factors acting on the promoter of interest. Any drug that specifically interferes with any of the numerous promoter-specific features, regardless whether it affects protein-protein or protein-nucleic acid contacts, or initiation- and elongation efficiency, can in principle be detected with such an assay.

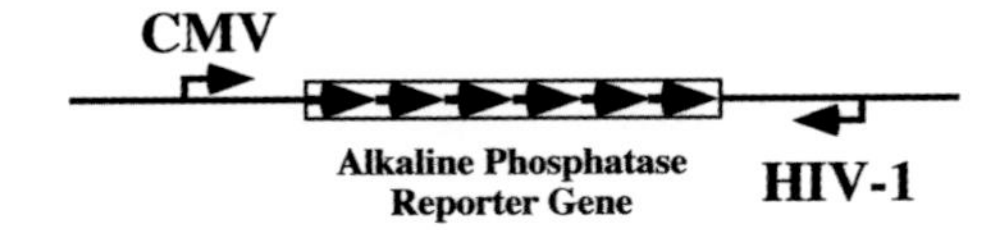

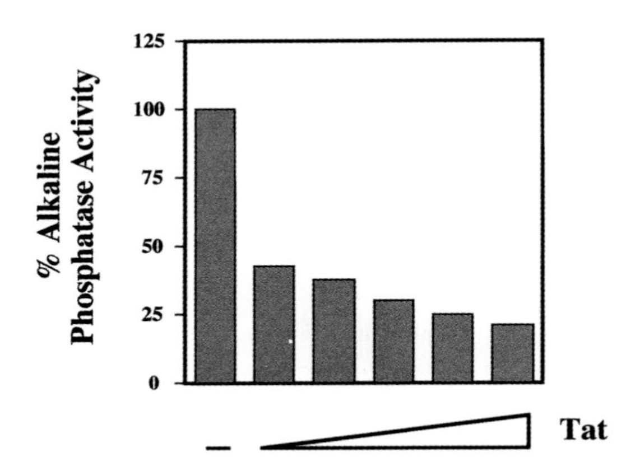

Figure 10.5. Experimental Verification of the Collision Construct Strategy. Transcription from the P_{HIV} promoter in the collision construct shown in Figure 10.4. is strongly dependent on the presence of the Tat transcriptional activator. In this experiment the collision construct was transfected into HeLa cells that do not produce Tat. In the absence of Tat the transcription from the sense CMV promoter allows the production of high levels of AP (100%). Co-transfection of the collision construct with increasing amounts of plasmid producing the Tat-protein *in trans* results in the stimulation of the antisense HIV promoter, and thus leads to a substantial reduction of AP production from the collision construct.

10.3. Reprogramming Transcription Patterns I: Drugs that Control Expression of Genes *In Vivo*

In the previous section we encountered some of the *in vitro* and *in vivo* screening strategies that are used to detect new types of drugs affecting the expression of specific target genes. These techniques will, without doubt, ultimately contribute greatly to the overall drug discovery effort. It takes, however, on average 15 years to develop a clinically applicable drug because only one in 1,000 discovered components proceeds all to way to human clinical trials, and of these only one in 20 is finally approved. It is therefore perhaps not surprising that no therapeutically useful treatments have yet emerged that are based on compounds discovered through transcription-based screening technology. There is, however, a well-studied range of drugs (developed by conventional means), that were only discovered fairly recently to influence the function of a particular family of gene-specific transcription factors: the estrogen receptors. The interaction between the estrogen receptors and estrogen-like antagonists is currently the only known paradigm of a direct drug-transcription factor interaction and therefore deserves special attention.

Nuclear Receptors as *Par Excellence* Targets for Transcription-Modifying Therapy

Nuclear receptors constitute a large protein family with numerous members that are involved in mediating changes in gene expression in response to steroid hormones, thyroid hormone, retinoids, and vitamins A and D (see Table 10.1). Upon specifically binding their ligands these receptors are transformed from an inactive conformation to an active transcription factor, capable of regulating transcription of numerous target genes containing appropriate binding sites in their promoters (reviewed in Evans, 1988; Beato, 1989; McDonnell *et al.*, 1993). One of the most interesting examples is the estrogen receptor (ER), a major target for therapeutic intervention in cancer and various endocrine disorders (e.g. King, 1991). Estrogen controls many important functions through the ER, including bone growth, development of female reproductive organs and breast cell proliferation. The inactive receptor is held in the cytoplasm, complexed with three different heat-shock proteins and immunophilins. Upon binding of estrogen the ER undergoes a conformational change, resulting in the freeing of the factor from the complex and formation of a ER homodimer with sequence-specific DNA-binding abilities (Figures 10.6 and 10.7; see review by McDonnell, 1995 for more details). ER dimers translocate to the nucleus, bind to estrogen response elements (EREs) and activate transcription through at least two different activation domains with the assistance of a series of cofacors (Halachmi *et al.*, 1994; Cavailles *et al.*, 1995; Metzger *et al.*, 1995;

Table 10.1. Overview of the Nuclear Receptor Superfamily.
Only the best studied examples are shown. More than 50 other members have been identified by sequence homology, but in most cases their specific ligands are unknown ('orphan receptors'). Based on review by McDonnell *et al.*, 1993.

Receptor Type	Location of Inactive Form	Special Features
Glucocorticoid receptor (GR)	cytoplasm	Ligand binding releases sequestered,
Mineral corticoid receptor (MR)	cytoplasm	transcriptionally inactive receptors
Progesterone receptor (PR)	cytoplasm	from large complex containing heatshock
Androgen receptor (AR)	cytoplasm	proteins and cyclophilins. Members of this
Estrogen receptor (ER)	cytoplasm	family bind to DNA as homodimers
Thyroid hormone receptor (TR)	nucleus	Transcriptional repressors in absence of
Retinoic Acid receptor (RAR)	nucleus	ligands, conformational changes in
Vitamin D receptor (VDR)	nucleus	presence of ligands alter heterodimers in
		various combinations.

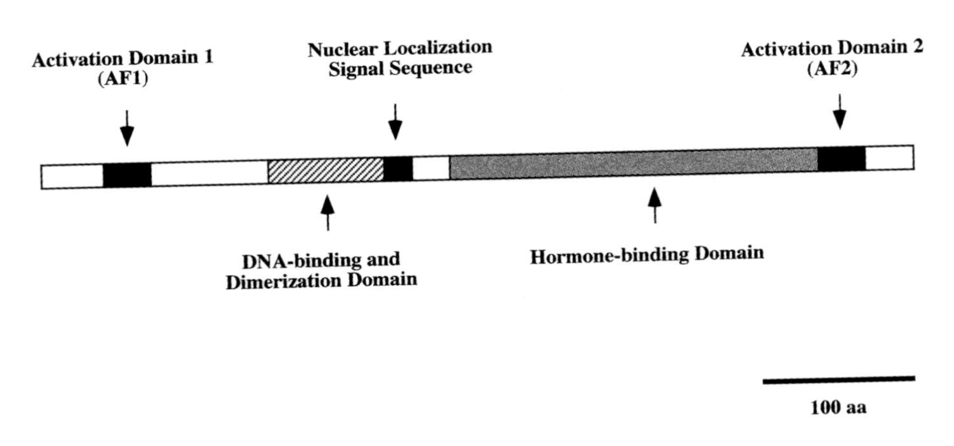

Figure 10.6. Molecular Anatomy of the Estrogen Receptor.
The various functional regions of ER have been mapped by mutagenesis experiments. Note the presence of the two activation domains (AF-1 and AF-2) near the N- and C-termini, respectively. The total length of ER is 585 amino acids.

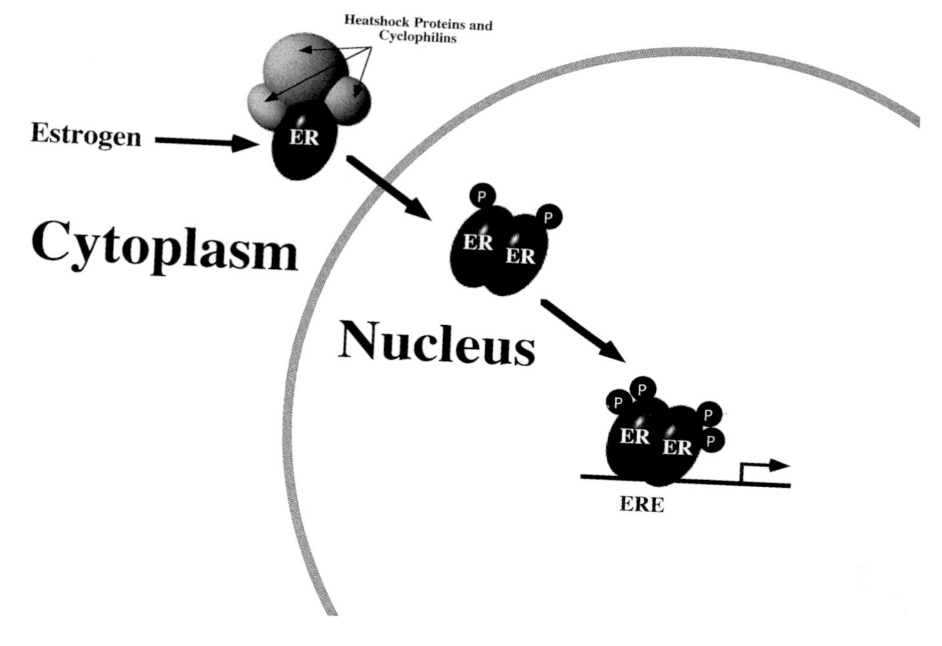

Figure 10.7. Generic Nuclear Receptor Signalling Pathway as Exemplified by the Estrogen Receptor.

The hormone estrogen diffuses freely from blood vessels across cell membranes into the cell interior and is specifically bound by the estrogen receptor (ER). ER (shown in black) is normally sequestered in a monomeric inactive form (complexed with various other proteins; shown in dark grey) predominantly in the cytoplasm. Upon binding of estrogen, ER undergoes phosphorylation, dimerization and translocates to the nucleus where it regulates the expression of genes containing estrogen response elements (EREs).

Vom Baur *et al.*, 1996; reviewed in Mangelsdorf *et al.*, 1995). Other activities of ER (in conjunction with another cellular transcription factor, AP1) are independent of EREs (Webb *et al.*, 1995; Uht *et al.*, 1997).

Although ligand-binding is clearly a key event in the mechanism of nuclear receptor function, there is amazingly little information available about the molecular details involved. Early models hypothesized that the sequence-specific binding of ER to DNA depended on ligand-induced ER dimerization, but later studies revealed that a mutation in the recombinant protein used in these experiments altered the functional properties of the recombinant protein used for these studies (Tora *et al.*, 1989). Subsequent reports of ligand-independent binding of ER (Reese and Katzenellenbogen, 1991), and ligand dose-dependency studies (Xing and Shapiro, 1993), deepened the mystery of ER-DNA interactions even further. Recently, a possible solution has emerged, based on kinetic studies concerning the interaction of ER-ligand complexes with DNA that reported measureable differences in the binding kinetics, rather than changes in affinity (Cheskis *et al.*, 1997). These studies confirmed that there were no detectable differences between the affinity of liganded or free ER for DNA. Analysis of ER-DNA interaction in real time, however, revealed a 50-fold increase in the rate of ER-DNA complex formation in the presence of ligand. This observation might account for some of the effects of estrogen-mediated changes in cellular transcription patterns.

Drugs Targeting Nuclear Receptor Function

Anti-estrogen therapeutic approaches are based on the complete or partial antagonization of ER function and fall into different three major categories (McDonnell *et al.*, 1995). Certain antagonists, such as 'ICI Compound 164,384', inhibit ER function strongly by specifically preventing dimerization of ER after ligand binding, a step that is crucial for the ability of ER to bind to and activate its target genes (Dauvois *et al.*, 1992; Fawell *et al.*, 1990). Another class, the clinically useful partial antagonists (such as 'tamoxifen'; Figure 10.8.), modify ER activity in more subtle manner. Tamoxifen is mainly used as a prophylactic against breast cancer in healthy women and changes the conformation of the ligand-binding domain of the estrogen receptor (ER). This prevents one of the two activation domains ('AF-2') of ER from functioning, stimulates the activity of another activation domain ('AF-1'), and ultimately results in a functionally-modified form of ER which has an inhibitory effect on the expression of transforming growth factor α (TGF-α) in breast cancer cells (Saeki *et al.*, 1991; Noguchi *et al.*, 1993; reviewed in Jordan, 1994). Tamoxifen also stimulates the expression of transforming growth factor $\beta3$ (TGF-$\beta3$), an important growth factor involved in bone maintainance, and thus can be used to prevent bone loss in osteoporosis patients (Yang *et al.*, 1996).

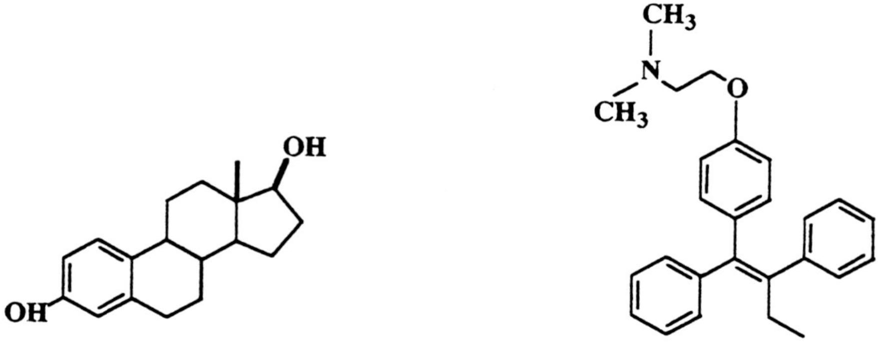

Estradiol **Tamoxifen**

Figure 10.8. Structure of the Estrogen Estradiol and the Estrogen Antagonist Tamoxifen. Although the overall structure of tamoxifen differs quite substantially in many respects from estradiol, the overall geometry of the aromatic and hydrophobic ring structures are comparable.

Development of Antiestrogen Resistance in Breast Cancer: A Transcription-based Diagnostic System

Although antiestrogens, such as tamoxifen, are valuable drugs for the treatment of hormone-responsive forms of breast cancer (50–60% of patients with with ER-positive breast cancers respond to tamoxifen treatment; Ferno *et al.*, 1996), they often become ineffective in individual patients due to the development of drug resistance (reviewed in Katzenellenbogen *et al.*, 1997). The tumor-growth inhibitory role of tamoxifen creates an environment where tumor cell clones that have become insensitive to the drug will have a high selective advantage, and in fact approximately 40% of tamoxifen-treated breast cancer patients will eventually develop tamoxifen-resistant tumors (Saez *et al.*, 1989). A subset of tamoxifen-resistant tumors are the consequence of mutations in the protein-coding portion of the ER gene itself (Miksicek, 1994), whereas others are due to alterations in the many other proteins ER interacts with during its normal mode of action (including transcriptional co-factors and kinases involved in ER post-translational modifications). The majority of the ER tamoxifen resistance mutations are, however, caused by changes in the pre-mRNA splicing pattern that result in the synthesis of constitutively active or inactive proteins. Especially the loss of exon 5 has been strongly linked with development of tamoxifen resistance during long-term drug treatment (Daffada *et al.*, 1995).

The clinically significant problem of tamoxifen resistance can, in principle, be tackled on a number of different levels. Ultimately, the development of antiestrogens with a different mode of interaction with ER will provide the best solution for the problem. In the meantime, an imaginative application of a transcription-based assay, 'functional analysis of separated alleles in yeast' or FASAY, could provide an important diagnostic tool for detecting mutant ER variants in tumor samples (Ishioka *et al.*, 1993; van Dijk *et al.*, 1997). The ER-FASAY assay is based on the fact that human ER is fully active in yeast and transactivates from an artificial ERE-containing reporter gene in an estrogen-dependent manner (Metzger *et al.*, 1988; Wrenn and Katzenellenbogen, 1993). As the first step, mRNA is extracted from breast cancer biopsies, reverse transcribed into cDNA, followed by the specific amplification of the ER coding region by the polymerase chain reaction (RT-PCR). The amplified PCR product is recombined *in vivo* into an expression vector and expressed in a yeast strain containing a ERE site upstream of a gene encoding an enzyme essential for histidine biosynthesis (HIS3). Plating of individual yeast transformants on medium lacking histidine will allow the cells to grow if human ER activates transcription from the histidine gene. In cells expressing a wildtype ER, growth on minimal medium will only be possible in presence of estrogen. Cells that contain constitutively active ER will also be able to grow in absence of estrogen and, conversely, cells with constitutively negative ER proteins will be unable to grow on minimal medium under any conditions. By studying the growth characteristics of the yeast transformants obtained in the FASAY assay it is thus possible to determine within a reasonable short time (7 days) whether tamoxifen treatment of a patient has resulted in the production of constitutively active forms of ER, indicating drug resistance of the biopsied tumor sample. Early diagnosis of the onset of tamoxifen resistance can increase the chance of survival of the patient because alternative forms of treatments, or surgical intervention, can be initiated at a substantially earlier stage. Patients that do not develop constitutively active ERs can continue in their treatment, safe in the knowledge that the tamoxifen treatment remains effective.

10.4. Reprogramming Transcription Patterns II: Drug-Mediated Activation of Alternate Members of Multigene Families

Most genes are only present in single copies in simple eukaryotes, such as yeast, but often in multiple copies as members of multigene families in higher eukaryotes. A general 'rule of thumb' suggests that a single gene in a simple eukaryote is usually present in duplicated form in lower eukaryotes (*D. melanogaster*, *C. elegans*), and in 3–5 copies in mammals. This high degree of redundancy has become very clear during the last few years because of the surprisingly low success rate of studies aimed at obtaining mice strains homozygously deficient for important genes that display substantial phenotypic effects. Mutations in most genes are efficiently compensated for by the activity of other members of the same multigene family. Obviously this is not always the case because there are many genetic diseases affecting individual members in multigene families that cause substantial medical problems in afflicted patients. Often the proteins expressed from individual genes in multigene families are, however, functionally very similar and could provide a replacement of the defective gene product, if they were transcribed at the right time and in the proper place. The case study described below illustrates current attempts to 'reactivate' expression of fetal globin genes as a replacement for defective adult globins. It is likely that similar approaches will become feasible in the near future as a possible therapeutic approach for a range of other genetic diseases.

Case Study: Reactivation of Fetal γ-Globin Expression for Treating Hemoglobinopathies

β-thalassaemias and sickle-cell syndrome constitute some of the most widespread and medically important blood diseases. Normal human hemoglobin is tetramer consisting of two α-globin and two β-globin chains (α_2/β_2). Total absence of β-globin production is lethal, but in many clinically relevant β-thalassaemia cases small amounts of β-globin are produced in patients, that allow the assembly of low levels of functional adult hemoglobin. Apart from the low overall hemoglobin levels, one of the main problems in thalassemic patients is the toxic build-up of excess α subunits due to the lack of a suitable protein interaction partner.

During the last decade much of the ongoing research has focused on the idea of developing methods for reactivating the γ-globin genes in thalassaemic adults (reviewed in Loukopoulos, 1997). γ-globins are normally expressed only during the fetal stage of human development and are replaced shortly after birth with β-globin (Figure 10.9; Wood *et al.*, 1985). By artifically reversing

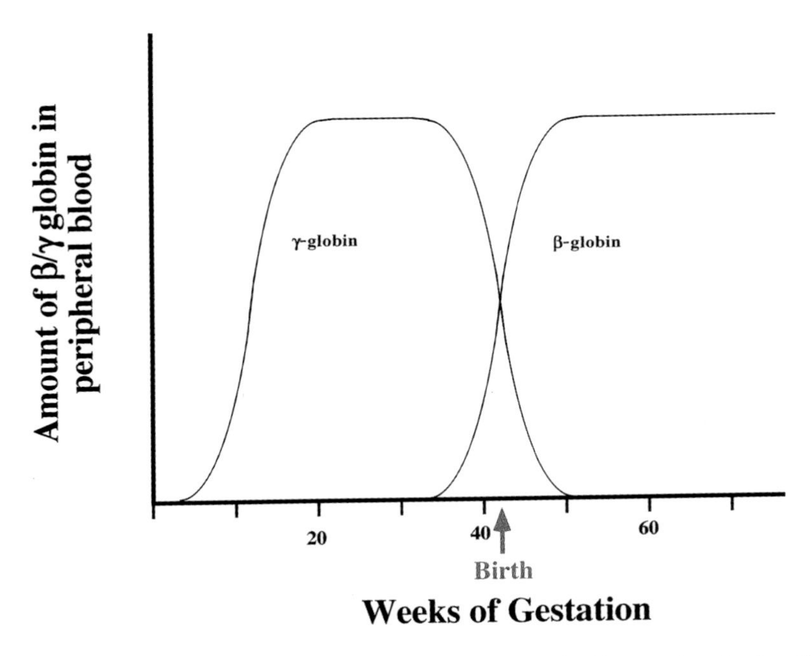

Figure 10.9. Differential Regulation of β and γ Globin Expression during Human Development.
Most of the red blood cells circulating during the fetal stage contain γ-globin. Several weeks before birth this is gradually (and normally irreversibly) substituted by the adult version, β-globin. β- and γ-globin are functionally quite similar, which stimulated the idea of artificially reactivating fetal γ-globin gene expression to improve the condition of β-thalassaemia patients.

this developmental switch it should, in principle, be possible to produce fetal hemoglobin (α_2/γ_2) in adults and thus prevent the damaging build up of excess α-globin. The strategy for activating fetal γ-globin gene expression is also of major importance for the treatment of sickle-cell anaemia. In this disease the substitution of a single amino acid near the amino terminus of the β-chain causes oxygenated hemoglobin containing the mutant protein to form long aggregates. The presence of aggregated hemoglobin leads to dramatic changes in the shape of red blood cells, which impedes their flow through narrow capillaries and eventually prevents efficient oxygenation of tissues. Activation of fetal hemoglobin synthesis looks like a particularly advantagous strategy for treatment of sickle cell anaemia due to the increased stability of erythrocytes containing a mixture of sickle-cell and 'normal' fetal hemoglobin. Presence of fetal hemoglobin prevents the aggregates from growing to a size where they affect cell morphology, so that relatively small increases in γ-globin transcription lead to substantial increases in the level of functional erythrocytes in the blood (Perrine *et al.*, 1994).

So what can be done to switch on expression of a gene in adults that has ceased to be actively transcribed since birth? Transcription from the dormant γ-globin promoters ($^{A}\gamma$ and $^{G}\gamma$) can actually be directly reactivated with butyrate-derived therapeutic agents in 50–85% of β-thalassemia patients, and has in some cases eliminated the need for continuous blood transfusions (Figure 10.10; Perrine *et al.*, 1990; Faller and Perrine, 1995). *In vivo* studies in transgenic mice have shown that the 1.3 kb upstream sequence of the human $^{A}\gamma$-globin gene promoter contains at least two 'butyrate response elements' (BREs) that might be responsible for this effect (Pace *et al.*, 1996). Although we do not currently understand the molecular basis and physiological significance of the butyrate-mediated activation of γ-globin transcription it can be hoped that the mapping and analysis of the BREs will eventually lead to the identification of the transcription factor(s) mediating the response. This will allow the development of drugs that are possibly more specific than butyrate-derivatives for controlling the switching process between different members of the β-globin-like multigene family.

Recent strategies geared towards a possible alleviation of the effects of Muscular Dystrophy (MD) are based on very similar ideas. In MD patients the expression of a large protein (dystrophin) is disrupted, causing severe structural defects in skeletal muscle structure and function. A homolog of dystrophin, utrophin, is structurally very similar and might be able to substitute adequately for defective dystrophin, if the expression of utrophin could be activated in the right place and time (e.g. Tinsley and Davies, 1993; Blake *et al.*, 1996). A systematic high-throughput screen for drugs that specifically stimulate transcription from the utrophin promoter in human muscle cells is currently underway and it remains to be seen whether the strategy will turn out to be successful.

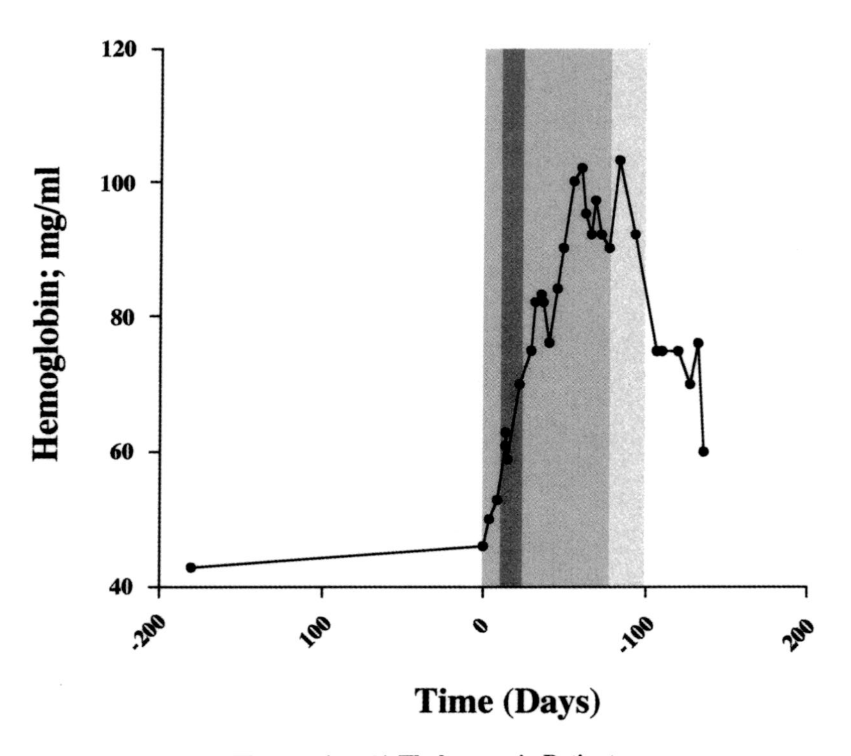

Figure 10.10. Butyrate Therapy in a β^{+}-Thalassaemia Patient.
The level of functional hemoglobin in peripheral blood cells increases substantially after the start of butyrate administration (between day '0' and day '100'; the frequency of intravenous butyrate injections is indicated by the intensity of the grey background shading) and decreases again after cessation of the treatment. After Faller and Perrine (1996).

10.5. Nucleic Acid-Based Gene-Specific Therapeutic Agents

Although many drug screening strategies aim at the identification of 'small molecule drugs' that interact specifically with the protein components of the transcriptional machinery, it is also possible to target the DNA and RNA molecules involved in the specific expression of genes directly. Nucleic acid molecules are chemically not as diverse as proteins and thus present fewer opportunities for small molecules to interact with them in a sequence-specific manner. Nevertheless, because of the ability of nucleic acids to hybridize to each other through specific base-pairing, a new generation of oligonucleotide-based 'designer' drugs to interfere with gene-specific transcription ('triple-helix' approach), or mRNA use and stability ('antisense' approach) is currently emerging.

Triple Helix-Forming Oligonucleotides Block Transcription of Targeted Genes

Although DNA is normally present in biological systems in either single- or doublestranded form, there are circumstances where three polynucleotide chains can intertwine to form a triple helix. Under these conditions an additional polypyrimidine DNA strand, consisting exclusively of thymine and 5'-methylcytosine, is incorporated into a 'normal' DNA double helix through a specific hydrogen bonding pattern (Figure 10.11). The mechanism of triple helix-mediated repression of transcription relies on the simple hypothesis that such a structure will interfere with transcription from a promoter containing it, leading to the expression of such a gene being switched off. This effect occurs with a high degree of sequence specificity and has been shown to interfere with gene transcription in several well-defined *in vitro* and *in vivo* experimental systems (Moser and Dervan, 1987; Duvalvalentin *et al.*, 1992; Maher *et al.*, 1992; Thuong and Helene, 1993; Aggarwal *et al.*, 1996; Lavrosky *et al.*, 1996). The formation of triple-helical DNA structures actually seems to repress transcription by stiffening the DNA, rather than by preventing transcriptional activators from binding to their target sites. The presence of a triple-helical structure in a promoter also represses basal transcription in the absence of gene-specific activators, and transcription factors (such as Sp1) can not counteract this effect (Maher *et al.*, 1992).

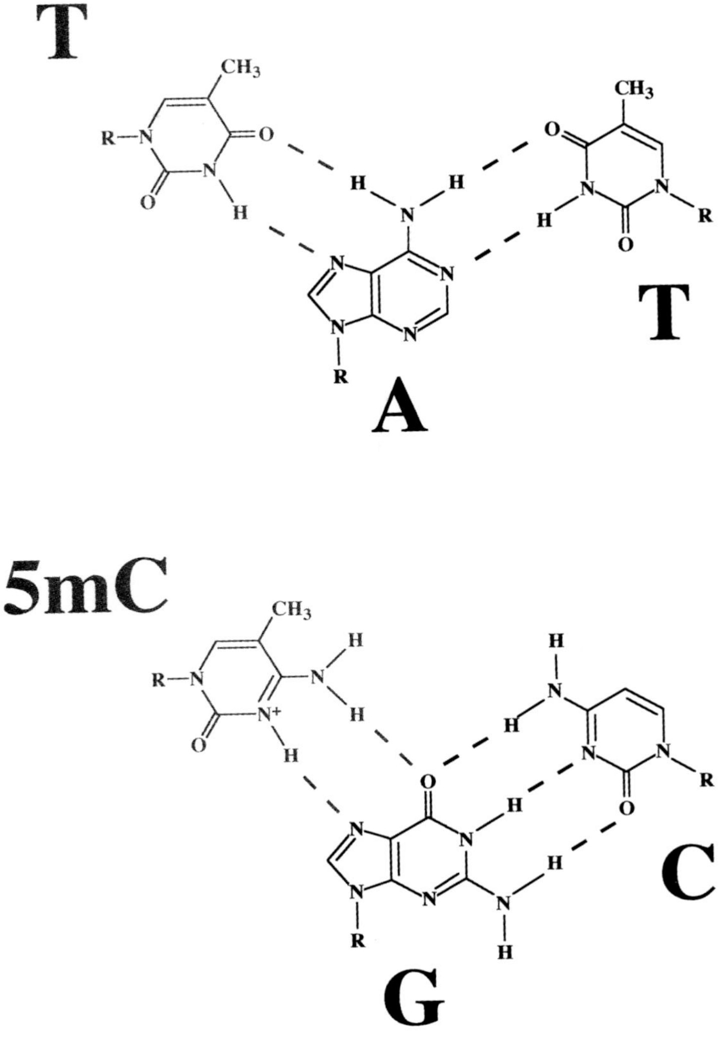

Figure 10.11. Hydrogen Base-Pairing Involved in Triple Helix Formation.
The base-pairing arrangments in normal double-stranded DNA are shown in black. It is possible to specifically hybridize an additional polypyrimidine-strand (containing only thymidine and methylcytosine) to a polypurine stretch of double-helical DNA to form a triple-helix. The additional bases and hydrogen bonding patterns involved in triple-helix formation are shown in grey.

Apart from the advantages outlined above, there are several drawbacks which limit any possible clinical applications of triple helix-based transcription-modifying drugs. One of the main disadvantages is that triple-helices can not be formed on just any piece of DNA, but require a relatively long stretch of DNA sequences consiting exclusively of purines (remember that the third strand to be added is exclusively made up of pyrimidine bases!). Because of this stringent sequence requirement it may therefore not be feasible to obtain triple helix-mediated repression of certain promoters that lack suitable polypurine motifs near the transcription start site (repression by triple-helical complexes increases if they are located in the proximity of the TATA-element). Furthermore, some experimental evidence suggests that there are some yet unknown enzymatic activities present in nuclear extracts that specifically prevent (or destabilize) the formation of triple-helical DNA. The effective clinical use of triple helix-based drugs may therefore be limited to a relatively small subset of promoters and may require additional treatments with other drugs, such as distamycin. Distamycin has been found to stabilize triple-helical DNA complexes and could therefore play a valuable assistant role in such therapeutic applications (Durang and Maurizot, 1996).

Logical Extension of Triple Helix Concept: Sequence-specific Pyrrol-Imidazole Polyamines

The theoretical and practical limitations of the triple-helix approach to inhibit the expression of specific promoters under *in vivo* conditions have recently been overcome with the help of a polyamides (Gottesfeld *et al.*, 1997). Polyamides can, in principle, be designed to bind with a high degree of sequence-specificity and affinity to any pre-determined DNA target sequence (Mrksich *et al.*, 1992; Wade *et al.*, 1992). Although the application of polyamide-based drugs has not been demonstrated within intact organisms, preliminary data from an $RNAP_{III}$ model system look extremely promising (Gottesfeld, *et al.*, 1997). An eight-ring polyamide was specifically designed to bind to the DNA target sequence 'AGTACT', which constitutes a part of the TFIIIA recognition sequence (Figure 10.12). Quantitative footprinting revealed that the affinity of this compound for this target sequence was very similar to that of transcription factors TFIIIA, and that it specifically suppressed transcription

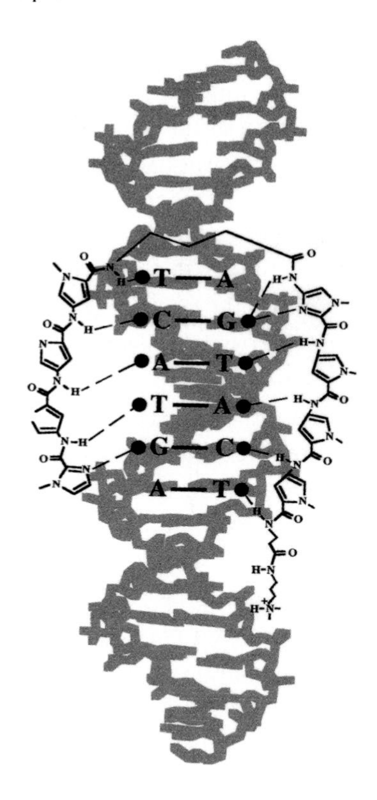

Figure 10.12. Sequence-Specific Interaction between a Polyamide and Double-Stranded DNA.

The sequence-specific contacts made between a polyamide and a stretch of double-stranded DNA are schematically shown. Note that different polyamide variants are used to interact specifically with the different bases present in the DNA, which allows the design of polyamide variants suitable for binding to precisely predetermined DNA sequences. After Gottesfeld *et al.* (1997).

of a 5S RNA template in a competitive manner (Figure 10.13). Polyamide-based probes therefore offer a number of advantages in comparison to the traditional triple-helix approach described in the previous section. At this stage it remains to be seen, however, whether it will be possible to apply this technology successfully to modify the rate of transcription from specific promoters within an intact organism.

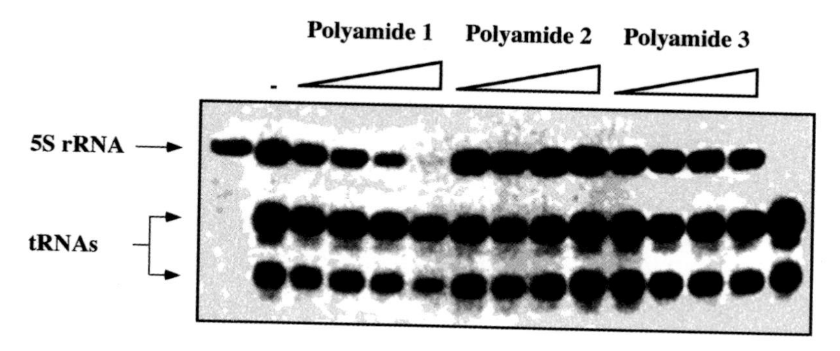

Figure 10.13. Specific Inhibition of Gene Transcription With a Polyamide. Three different polymamides were incubated with DNA templates containing the 5S rRNA and tRNA genes. Polyamide 1 specifically interacts with the DNA-motif recognized by TFIIIA, which is a gene-specific transcription factor for the 5S rRNA gene (see Chapter 6). Increasing amounts of polyamide 1 specifically interfere with the transcription of the 5S rRNA gene, whereas the tRNA genes (which do not depend on TFIIIA; Chapter 6) are not affected. Polyamides 2 and 3, which do not bind to the TFIIIA binding sites, also show no effect thus proving the specificity of the inhibition. From Gottesfeld *et al.* (1997).

The Antisense Approach: Post-transcriptional Inhibitory Effects

Up to now we have seen a cross-section of the current efforts of the pharmaceutical and biotechnology industry to identify drugs that specifically control (inhibit) transcription from target genes, either through influencing the function of gene-specific transcription factors directly, or by modifying the DNA motifs they interact with. The preliminary successes suggest that the search for small molecule drugs influencing gene expression is a valid approach, and provides a sufficient degree of specificity necessary for controlling the expression of single (or small groups of) genes. In this section we will investigate the progress achieved using an alternative approach, namely inhibition of gene expression through antisense nucleic acids at the post-transcriptional level.

What is antisense-control? The fundamental idea of the antisense-technique has been around for quite some time and was initially suggested by the existence of naturally-occurring antisense control mechanism in bacteria (reviewed in Wagner and Simons, 1994). Most importantly, antisense inhibition of gene expression is thought to act mainly by affecting mRNA processing and the efficiency of translation. It is based on the simple premise that crucial areas in mRNA molecules need to be available in single-stranded form to carry out their functions. Sequence-specific binding of a short synthetic oligonucleotide to a stretch of mRNA, e.g. near the translation initiation site could, in principle, prevent translation and thus down-regulate the amount of the encoded protein. Alternatively, binding of a oligonucleotide to a region of an mRNA molecule involved in the formation of specific loops could severely destabilize the three-dimensional mRNA conformation and shorten the half-life of the molecule. Antisense-agents thus prevent effective use of mRNA molecules that have already been synthesized and could therefore be used in conjunction with drugs interfering with gene-specific transcription.

Despite the apparent simplicity of the antisense concept, the practical realization of the idea has been fraught with difficulties (see e.g. Stein and Cheng, 1993; Gewirtz *et al.*, 1996). Initially, a major problem was stability of the antisense oligonucleotides in cells, because endogenous nucleases break down very quickly any oligonucleotides introduced into the cell. The use of oligonucleotide variants containing chemically modified internucleotide bonds (such as phosphorothioate) has overcome this shortcoming (e.g. Peyman *et al.*, 1997).

In addition to the chemical stability of antisense oligonucleotides, the need for targeting accessible sites on RNAs has recently been recognized as a major factor for the success of antisense approaches (Lima *et al.*, 1992; Ho *et al.*, 1998; Patzel and Sczakiel, 1998). RNA molecules are capable of assuming extensive secondary and tertiary structures, and it can be safely assumed that mRNAs are no exception to this general rule. In many cases oligonucleotides used for antisense inhibition experiments fail to hybridize to their target mRNA, a fact that has been convincingly demonstrated in at least two different systematic studies. Only one out of 34 oligonucleotides tested had significant *in vivo* antisense activity in lowering the amount of mRNA encoding *C-raf* kinase (Monia *et al.*, 1996), and only six out of one hundred oligonucleotides were effective in inhibiting the *in vivo* propagation of herpes simplex virus in tissue culture cells (Peyman *et al.*, 1995). These *in vivo* results correlate well with the observation that only a small number out of 2000 oligonucleotides hybridized efficiently to β-globin mRNA under *in vitro* conditions (Figure 10.14 and 10.15; Milner *et al.*, 1997).

The increased experience in antisense oligonucleotide design and successful applications in simple model system has lead to the successful application of antisense technology to interfere with the expression of specific genes *in vivo* (Dean *et al.*, 1996; Neurath *et al.*, 1996). The technology thus hold great promise for the clinically-applied regulation of gene expression in the forseeable future.

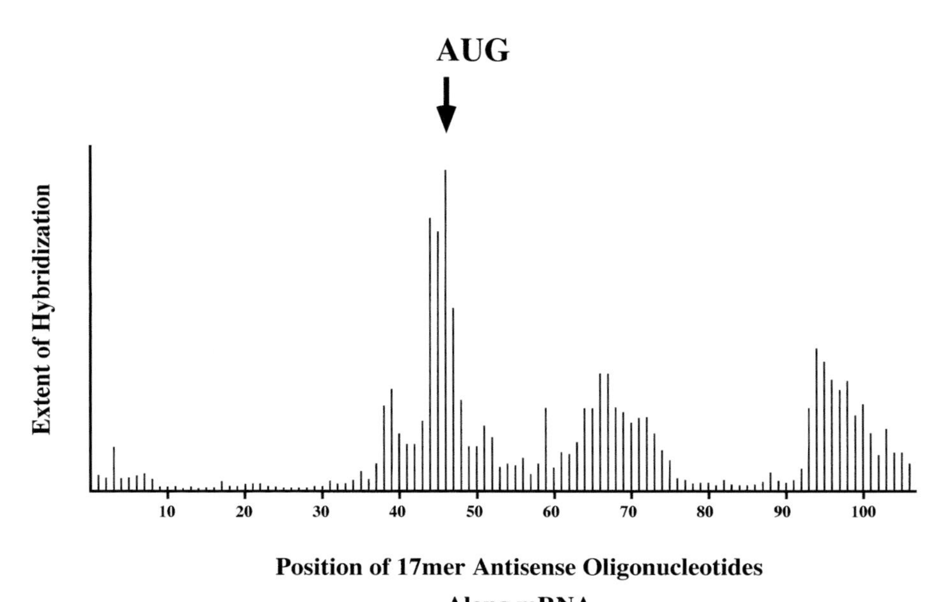

Figure 10.14. Systematic Hybridization Scan of 120 different Oligonucleotides to the 5' End of Rabbit β-Globin mRNA.

Each of the bars represents the extent of hybridization of a specific oligonucleotide covering a stretch of 17 nucleotides to the 5' end region of globin mRNA. The mRNA region spanning nucleotide positions 1 to 37 and 76 to 90, respectively, is not accessible to hybridization by oligonucleotides as indicated by the absence of clearly detectable hybridization signals. The region surrounding the AUG initation codon (indicated by arrow) is maximally available for hybridization to antisense oligonucleotides. Data from Milner *et al.* (1997).

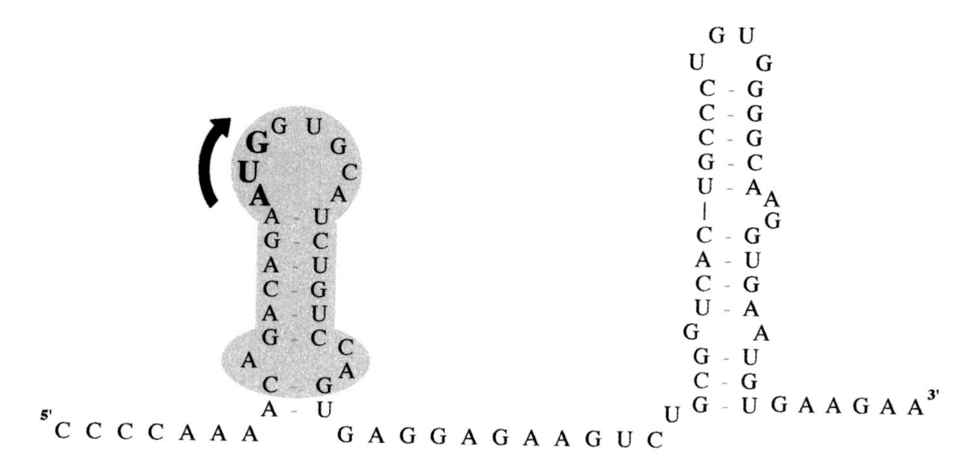

Figure 10.15. Secondary Structure Prediction of the Region Surrounding the Initiation Codon of Rabbit β-Globin mRNA.
The initiation codon is shown in bold and is located near the top end of a loop. The grey-shaded area indicates the surrounding nucleotide sequence that has been successfully used as the target region for antisense oligonucleotides. Interestingly, at least 12 nucleotides of the target area are predicted to participate in a stable, base-paired stem structure. Data from Milner *et al.* (1997).

10.6. Drugs That Control Gene Transcription by Indirect Means

Some of the most interesting discoveries that have been made during the last few years are based on the growing realization that several drugs in widespread use act on the transcriptional level, although often in a rather indirect manner (Table 10.2). Instead of modulating directly the function of gene-specific transcription factors, such drugs often target various signalling pathways that ultimately control the activity of whole transcriptional networks in the nucleus. The expression of many genes is indirectly controlled through ligand/receptor interactions at the cell surface, followed by changes in second messenger levels, and changes in post-translational modification of key regulatory proteins in the cytoplasm (especially by phosphorylation). A particularly well-understood example that illuminates such an indirect transcriptional control effect is provided by the effects of glucocorticoids and salicylate drugs on NF-κB, a key transcriptional activator involved in inflammatory responses.

Table 10.2. Drugs Affecting Gene Expression By Indirect Means.
After Cai *et al.*, 1997.

Drug	Disease
Aspirin	Inflammation
Accutane; Retin-A	Dermatology
Calcitriol	Renal disease
Cyclosporin A	Immunosuppression
Fibrates	Lipid lowering
Flutamide	Prostate cancer
Hydrocortisone	Inflammation
Lovastatin	Cholesterol regulation
Ortho-novum	Contraception
Premarin	Osteoporosis
Synthroid	Hyperthyroidism
Troglitazone	Diabetes

Glucocorticoids Prevent NFκB-mediated Transcription of Genes Involved in Immune-Response

Glucocorticoids have been used for decades to control the progression of serious inflammatory diseases, such as rheumatoid arthritis, by supressing immune reactions. Until recently the mode of action of glucocorticoids was unknown, although it was clear that they supressed transcription of many of the genes that are required for immune response functions. In 1994 and 1995 first hints started to appear, suggesting that glucocorticoids acted through stimulating the production of a inhibitory cofactor of a key transcriptional activator, NF-κB. Originally discovered as a protein binding upstream of the gene encoding the human κ immunoglobulin light chain, NF-κB is now recognized as a key regulator of many genes encoding cytokines and proteins involved in the immune response and inflammation (Sen and Baltimore, 1986; reviewed in Baldwin, 1996; Bauerle and Baltimore, 1996). In immunologically quiescent cells NF-κB is mostly complexed in an inactive form with another protein, IκB, and stored as such in the cytoplasm (Figure 10.16). Various immunostimulating agents cause the phosphorylation of IκB through a complex

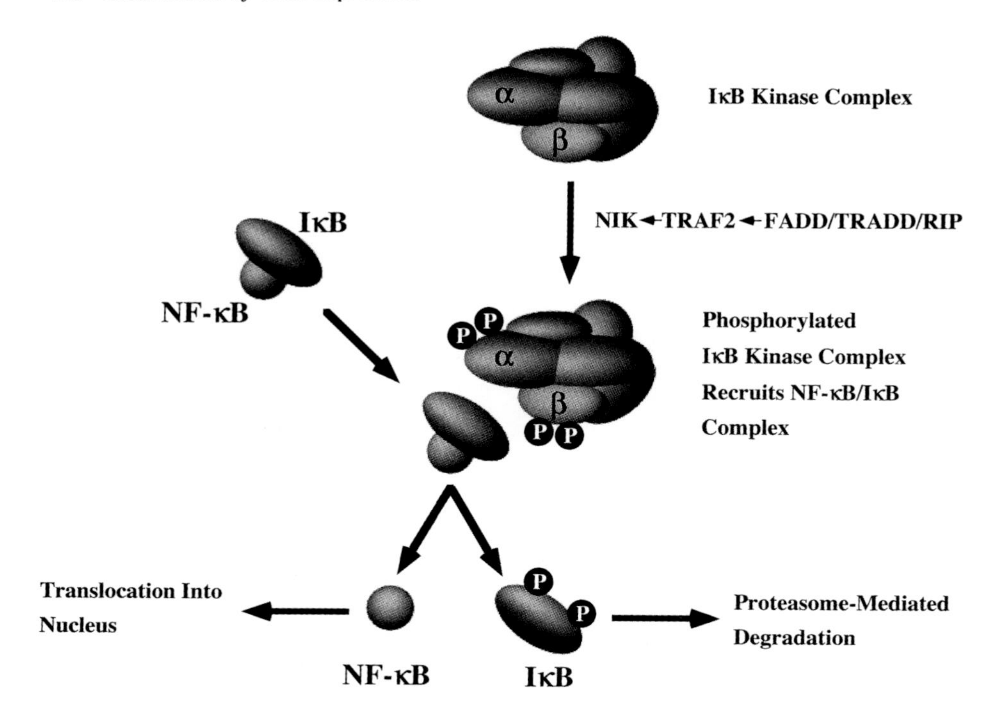

Figure 10.16. Regulation of NF-κB Availability through the IκB Phosphorylation Pathway.
In unstimulated cells the bulk of NF-κB is kept in an inactive form in the cytoplasm through association with the inhibitor IκB. IκB can be phosphorylated by a large (approximately 700 kD) multiprotein complex that contains two subunits (IKK-α and IKK-β) subject to activation by phosphorylation through a complex kinase cascade emanating from cell-surface receptors. Phosphorylation of IκB leads to its ubiquitin-directed proteolysis and allows NF-κB to translocate into the nucleus in order to activate a number of target genes.

kinase cascade, resulting in the release of NF-κB from the complex (reviewed in Stancovski and Baltimore, 1997). In the absence of IκB, NF-κB is free to translocate into the nucleus where it activates transcription from a wide range of putative target genes. Glucocorticoids stimulate the production of high levels of IκB and thus diminish the pool of free NF-κB available for migrating to the nucleus. This sequestration of NF-κB has substantial consequences for the ability of cells to participate in a full scale immune response. The discovery of IκB phosphorylation as a central control point for the indirect regulation of NF-κB now opens the way towards development of new classes of drugs with a more specific mode of action that avoid the side effects often associated with long-term glucocorticoid and hydrocortisone treatments.

Interestingly, the anti-inflammatory action of another widely-used drug, aspirin (acetylsalicylic acid), is also at least partially based on preventing the activation of NF-κB from the NF-κB/IκB complex (Kopp and Ghosh, 1994; Weber *et al.*, 1995). In addition, it is known that sodium salicylate induces the production of human heat shock transcription factor (HSTF) in tissue cell cultures (Jurivitch *et al.*, 1992), supporting the general notion that salicylates exert their anti-inflammatory effect in a significant manner through the transcriptional level .

10.7. Conclusions

The recent development of powerful *in vitro* and *in vivo* transcription-based screening assays has set the scene for the discovery of new types of drugs with the ability to interfere specifically with transcription from defined promoters. Such drugs will provide valuable additions to the currently available pharmaceutical armoury and provide new types of antiviral and antitumor treatments. Many available drug treatments affect gene expression through indirect means. The estrogen-antagonist tamoxifen represents the only clinically-proven drug that *directly* affects a family of gene-specific transcription factors. Alternative strategies aimed at inhibiting transcription from certain target promoters have focused on various forms of triple-helix and polyamide-nucleic acid hybridization methods. The combination of small molecule drugs directed against specific transcription factors, triple-helix approaches and post-transcriptional antisense-methods will enable us in future to modulate the expression of numerous genes at will. With particular types of genetic diseases, such as the β-thalassaemias, the selective stimulation of transcription of functionally homologous gene family members may present a straightforward alternative to somatic gene therapy.

References

Internet Sites of Interest

'http://www.osip.com/'
Homepage of OSI Pharmaceuticals (formerly 'Oncogene Scientif.), one of the major biotechnology companies specializing in developing high-troughput screens for drugs affecting expression of specific genes.

'http://www.tularik.com/'
A company with a strong emphasis on developing gene-expression modifying drugs. Good description of the basic philosophy and technology behind the concept of high-throughput screens.

'http://www.datanet.hu/comgenex/'
A company specializing in providing screening libraries of natural and synthetic compounds for lead drug identification.

'http://www.emaxhost.com/'
The 'Society for Biomolecular Screening'.

'http://www.signal.com/signal.research.html'
A company investigating approaches to modulating gene expression for treatment of various diseases.

Patents

Foulkes, J.G., Liechtfried, F.E., Pieler, C., Stephenson, J.R., and Case, C. (1996). Methods for determining chemicals that modulate expression of genes associated with cardiovascular disease. *U.S. Patent* 5,580,722.

Foulkes, J.G., Leichtfried, F.E., Pieler, C., Stephenson, J.R., and Franco, R. (1997). Method of discovering chemicals capable of functioning as gene expression modulators. *U.S. Patent* No. 5,665,543.

Froehler, B., Wagner, R., Matteucci, M., Jones, R.J., Gutierrez, A.J., and Pudlo, J. (1997). Enhanced triple-helix and double-helix formation with oligomers containing modified pyrimidines. *U.S. Patent* 5,645,985.

Lamarco, K., Wilson, A., and Herr, W. (1996). Herpes simplex virus drug screen. *U.S. Patent* No. 5,585,239.

Tjian, R., Comai, L., Dynlacht, B.D., Hoey, T., Ruppert, S., Tanese, N., Wang, E., and Weinzierl, R.O.J. (1996). TATA-binding protein associated factors drug screens. *U.S. Patent* No. 5,534,410.

Research Literature

Aggarwal, B.B., Schwarz, L., Hogan, M.E., and Rando, R.F. (1996). Triple helix-forming oligodeoxyribonucleotides targeted to the human tumor necrosis factor (TNF) gene inhibit TNF production and block the TNF-dependent growth of human glioblastoma tumor cells. *Cancer Res.* 56, 5156–5164.

Baldwin, A.S. (1996). The NF-κB and IκB proteins: new discoveries and insights. *Annu. Rev. Immunol.* 14, 649–681.

Bauerle, P.A., and Baltimore, D. (1996). NF-κB: ten years after. *Cell* 87, 13–20.

Beato, M. (1989). Gene regulation by steroid hormones. *Cell* 56, 335–344.

Blake, D.J., Tinsley, J.M., and Davies, K.E. (1996). Utrophin: a structural and functional comparison to dystrophin. *Brain Pathol.* 6, 37–47.

Blundell, T.L. (1996). Structure-based drug design. *Nature* 384 (Suppl.), 23–26.

Cai W., Hu L., and Foulkes, J.G. (1996). Transcription-modulating drugs: mechanism and selectivity. *Curr. Opin. Biotechnol.* 7, 608–615.

Cavailles, V., Dauvois, S., L'Horset, F., Lopez, G., Abbondanza, C., and Brown, M. (1994). Nuclear factor R1P140 modulates transcriptional activation by the estrogen receptor. *EMBO J.* 14, 3741–3751.

Cavalieri, F. (1996). Drugs that target gene expression: an overview. *Crit. Rev. Eukaryot. Gene Expr.* 6, 75–85.

Cheskis, B.J., Karathanasis, S. and Lyttle, C.R. (1997). Estrogen receptor ligands modulate its interactions with DNA. *J. Biol. Chem.* 272, 11384–11391.

Cohen, J. (1997). The genomics gamble. *Science* 275, 767–772.

Daffada, A.A.I., Johston, S.R.D., Smith, I.E., Detre, S., King, N., and Dowsett, M. (1995). Exon 5 deletion variant estrogen receptor messenger RNA expression in relation to tamoxifen resistance and progesterone receptor/pS2 status in human breast cancer. *Cancer Res.* 55, 288–293.

Dauvois, S., Danielian, P.S., White, R., and Parker, M.G. (1992). Antiestrogen ICI 164,384 reduces cellular estrogen receptor content by increasing its turnover. *Proc. Natl. Acad. Sci. USA* 89, 4037–4041.

Dean, N., McKay, R., Miraglia, L., Howard, R., Cooper, S., Giddings, J. *et al.* (1996). Inhibition of growth of human tumor cell lines in nude mice by an antisense oligonucleotide inhibitor of protein kinase C-α expression. *Cancer Res.* 56, 3499–3507.

Del Rosario, M., Stephans, J.C., Zakel, J., Escobedo, J., and Giese, K. (1996). Positive selection system to screen for inhibitors of human immunodeficiency virus-1 transcription. *Nature Biotechn.* 14, 1592–1596.

Dhundale, A., and Goddard, C. (1996). Reporter assays in the high throughput screening laboratory: a rapid and robust first look. *J. Biomol. Screening* 1, 115–118.

Durang, M., and Maurizot, J.C. (1996). Dystamycin A complexation with a nucleic acid triple helix. *Biochemistry* 35, 9133–9139.

Duvalvalentin, G., Thuong, N. T., and Helene, C. (1992). Specific inhibition of transcription by triple-helix forming oligonucleotides. *Proc. Natl. Acad. Sci.USA* 89, 504–508.

Evans, R. M. (1988). The steroid and thyroid receptor superfamily. *Science* 240, 889–895.

Faller, D.V., and Perrine, S.P. (1995). Butyrate in the treatment of sickle cell disease and beta-thalassemia. *Curr. Opin. Hematol.* 2, 109–117.

Fawell, S.E., White, R., Houre, S., Sydenham, M., Page, M., and Parker, M.G. (1990). Inhibition of estrogen receptor-DNA binding by 'pure' antiestrogen ICI 164,384 appears to be mediated by impaired dimerization. *Proc. Natl. Acad. Sci. USA* 87, 6883–6887.

Feinberg, M.B., Baltimore, D., and Frankel, A.D. (1991). The role of Tat in the human immunodeficiency virus life cycle indicates a primary effect on transcriptional elongation. *Proc. Natl. Acad. Sci. USA* 88, 4045–4049.

Ferno, M., Andersson, C., Fallenius, G., and Idvall, I. (1996). Oestrogen receptor analysis of paraffin sections and cytosol samples of primary breast cancer in relation to outcome after adjuvant tamoxifen treatment. *Acta Oncol.* 35, 12–22.

Gelbert, L.M., and Gregg, R.E. (1997). Will genetics really revolutionize the drug discovery process? *Curr. Opin. Biotechnol.* 8, 669–674.

Gewirtz, A.M., Stein, C.A., and Glazer, P.M. (1996). Facilitating oligonucleotide delivery: Helping antisense deliver on its promise. *Proc. Natl. Acad. Sci. USA* 93, 3161–3163.

Gottesfeld, J. M., Neely, L., Trauger, J. W., Baird, E. E., and Dervan, P. B. (1997). Regulation of gene expression by small molecules. *Nature* 387, 202–205.

Gronemeyer, H. (1991). Transcription activation by estrogen and progesterone receptors. *Annu. Rev. Genet.* 25, 89–123.

Halachmi, S., Marden, E., Martin, G., Mackay, H., Abondanza, C., and Brown, M. (1994). Oestrogen receptor-associated proteins: possible mediators of hormone-induced transcription. *Science* 264, 1455–1458.

Heguy, A. (1997). Inhibition of the HIV REV transactivator: A new target for therapeutic intervention. *Frontiers in Bioscience* 2, 283–297.

Ho, S.P., Bao, Y., Lesher, T., Malhotra, R., Ma, Fluharty, S.J., and Sakai, R.R. (1998). Mapping of RNA accessible sites for antisense experiments with oligonucleotide libraries. *Nature Biotechnology* 16, 59–63.

Ishioka, C., Frebourg, T., Yan, Y., Vidal, M., Friend, S., Schmidt, S. and Iggo, R. (1993). Screening patients for heterozygous p53 mutations using a functional assay in yeast. *Nat. Genet.* 5, 124–129.

Jones, K.A., and Peterlin, B.M. (1994). Control of RNA initiation and elongation at the HIV-1 promoter. *Annu. Rev. Biochem.* 63, 717–743.

Jordan, V.C. (1994). Molecular mechanisms of antiestrogen action in breast cancer. *Breast Cancer Res. Treat.* 31, 41–52.

Jurivitch, D.A., Sistonen, L., Kroes, R.A., and Morimoto, R.I. (1992). Effect of sodium salicylate on the human heat shock response. *Science* 255, 1243–1245.

Katzenellenbogen, B.S., Montano, M.M., Ekena, K., Herman, M.E., andMcInerney, E.M. (1997). Antiestrogens: mechanisms of action and resistance in breast cancer. *Breast Cancer Res. Treat.* 44, 23–38.

King, R.J.B. (1991). A discussion of the roles of estrogen and progestin in human mammary carcinogenesis. *J. Steroid Biochem. Mol. Biol.* 39, 811–818.

Kopp, E., and Ghosh, S. (1994). Inhibition of NF-Kappa B by sodium salicylate and aspirin. *Science* 256, 956–959.

Latchman, D.S. (1996). Transcription factor mutations and human disease. *N. Engl. J. Med.* 334, 28–33.

Latchman, D.S. (1997). How can we use our growing understanding of gene transcription to discover effective new medicines? *Curr. Opin. Biotechnol.* 8, 713–717.

Lavrosky, Y., Stoltz, R.A., Vlassov, V.V., and Abraham, N.G. (1996). *c-fos* protooncogene transcription can be modulated by oligonucleotide-mediated formation of triplex structures *in vitro*. *Eur. J. Biochem.* 238, 582–590.

Lima, W.F., Monia, B.P., Ecker, D.J., and Freier, S.M. (1992). Implications of RNA structure on antisense oligonucleotide hybridization kinetics. *Biochemistry* 31, 12055–12061.

Loukopoulos, D. (1997). New therapies for the haemoglobinopathies. *J. Intern. Med. Suppl.* 740, 43–48.

Love, D.R., Byth, B.C., Tinsley, J.M., Blake, D.J., and Davies, K.E. (1993). Dystrophin and dystrophin-related proteins: a review of protein and RNA studies. *Neuromuscul. Disord.* 3, 5–21.

Maher, J. L., Dervan, P. B., and Wold, B. (1992). Analysis of promoter-specific repression by triple-helical DNA complexes in a eukaryotic cell-free transcription system. *Biochemistry* 31, 70–81.

Mangelsdorf, D.J., Trummel, C., Beato, M., Herrlich, P., Schutz, G., Umesono, K., Blumber, B., Kastner, P., Mark, M., Chambon, P., and Evans, R.M. (1995). The nuclear receptor superfamily: the second decade. *Cell* 83, 835–839.

McDonnell, D.P., Vegeto, E., Gleeson, M.A.G. (1993). Nuclear hormone receptors as targets for new drug discovery. *Bio/Technology* 11, 1256–1261.

McDonnell, D.P. (1995). Unraveling the human progesterone receptor signal transduction pathway. *TEM* 6, 133–138.

McDonnell, D.O., Clemm, D.L., Hermann, T., Goldman, M.E., and Pike, J.W. (1995). Analysis of estrogen recptor function *in vitro* reveals three distinct classes of antiestrogens. *Mol. Endocrinol.* 9, 659–669.

Metzger, D., White, J.H., and Chambon, P. (1988). The human estrogen receptor functions in yeast. *Nature* 334, 31–36.

Metzger D., Ali, S., Bornert, J.-M., and Chambon, P. (1995). Characterization of the amino-terminal transcriptional activation function of the human estrogen receptor in animal and yeast cells. *J. Biol. Chem.* 270, 9535–9542.

Miksicek, R.J. (1994). Steroid receptor variants and their potential role in cancer. *Semin. Cancer Biol.* 5, 369–379.

Milner, N., Mir, K., and Southern, E.M. (1997). Selecting effective antisense reagents on combinatorial oligonucleotide arrays. *Nature Biotechnology* 15, 537–541.

Monia, B.P., Johnston, J.F., Geiger, T., Muller, M., and Fabbro, D. (1996). Antitumor activity of a phosphorothioate antisense oligodeoxynucleotide targeted against C-raf kinase. *Nature Medicine* 2, 668–675.

Moser, H. E., and Dervan, P. B. (1987). Sequence-specific cleavage of double helical DNA by triple helix formation. *Science* 238, 645–650.

Mrksich, M., Wade, W.S., Dwyer, T.J., Geierstanger, B.H., Wemmer, D.E., and Dervan, P.B. (1992). Antiparallel side-by-side dimeric motif for sequence-specific recognition in the minor groove of DNA by the designed peptide 1-methylimidazole-2-carboxamidenetropsin. *Proc. Natl. Acad. Sci. USA* 89, 7586–7590.

Neurath, M.F., Pettersson, S., Zumbuschenfelde, K.H.M., and Strober, W. (1996). Local administration of antisense phosphorothioate oligonucleotides to the p65 subunit of NF-kappa B abrogates established experimental colitis in mice. *Nature Medicine* 9, 998–1004.

Noguchi, S., Motomura, K., Inaji, H., Imaoka, S., and Koyama, H. (1994). Downregulation of transforming growth factor-α by tamoxifen in human breast cancer. *Cancer* 72, 131–136.

Pace, B.S., Li, Q., and Stamatoyannopoulos, G. (1996). *In vivo* search for butyrate responsive sequences using transgenic mice carrying $^{A}\gamma$ gene promoter mutants. *Blood* 88, 1079–1083.

Patzel, V. and Sczakiel, G. (1998). Theoretical design of antisense RNA structures substantially improves annealing kinetics and efficacy in human cells. *Nature Biotechnology* 16, 64–68.

Perrine, S.P., Faller, D.V., Swerdlow, P., Miller, B.A., Bank, A., Sytkowski, A.J., Reczek, J., Rudolph, A.M., and Kan, Y.W. (1990). Stopping the biological clock for globin gene switching. *Ann. N. Y. Acad. Sci.* 612, 134–140.

Perrine, S.P., Olivieri, N.F., Faller, D.V., Vichinsky, E.P., Dover, G.J., and Ginder, G.D. (1994). Butyrate derivatives. New agents for stimulating fetal globin production in the beta-globin disorders. *Am. J. Pediatr. Hematol. Oncol.* 16, 67–71.

Peyman, A., Helsberg, M., Kretzschmar, G., Mag, M., Grabley, S. and Uhlmann, E. (1995). Inhibition of viral growth by antisense oligonucleotides directed against the IE110 and the UL30 mRNA of herpes simplex virus type-1. *Biol. Chem. Hoppe-Seyler* 376, 195–198.

Peyman, A., Helsberg, M., Kretzschmar, G., Mag., M., Ryte, A. and Uhlmann, E. (1997). Nuclease stability as dominant factor in the antiviral activity of oligonucleotides directed against HSV-1 IE110. *Antiviral Res.* 33, 135–139.

Rabbits, T.H. (1994). Chromosomal translocations in human cancer. *Nature* 372, 143–149.

Reese, J.C., and Katzenellenbogen, B. (1991). Differential DNA-binding abilities of estrogen receptor occupied with two classes of antiestrogens: studies using human estrogen receptor overexpressed in mammalian cells. *Nucleic Acids Res.* 19, 6595–6602.

Rochette, J., Craig, J.E., and Thein, S.L. (1994). Fetal hemoglobin levels in adults. *Blood Rev.* 8, 213–224.

Saeki, T., Cristiano, A., Lynch, M.J., Brattain, M., Kim, N., Normanno, N., Kenney, N., Ciardello, F and Salomon, D.S. (1991). Regulation by estrogen through the 5'-flanking region of the transforming growth factor α gene. *Mol. Endocrinol.* 12, 1955–1963.

Saez, R.A., McGuire, W.L., and Clark, G.M. (1989). Prognostic factors in breast cancer. *Semin. Surg. Oncol.* 5, 102–110.

Sen, R., and Baltimore, D. (1986). Multiple nuclear factors interact with the immunoglobulin enhancer sequences. *Cell* 47, 921–928.

Stancovski, I. and Baltimore, D. (1997). NF-κB activation: The IκB kinase revealed? *Cell* 91, 299–302.

Stein, C.A., and Cheng, Y.C. (1993). Antisense oligonucleotides as therapeutic agents — is the bullet really magical? *Science* 261, 1004–1012.

Thuong, N. T., and Helene, C. (1993). Sequence-specific recognition and modification of double-helical DNA by oligonucleotides. *Angew. Chem. Int. Ed. Engl.* 32, 666–690.

Tinsley, J.M., and Davies, K.E. (1993). Utrophin: a potential replacement for dystrophin. *Neuromuscul. Disord.* 3, 537–539.

Tora, L., Mullick, A., Metzger, D., Ponglikitmongkol, M., Park, I., and Chambon, P. (1989). The cloned human oestrogen receptor contains a mutation which alters its hormone binding properties. *EMBO J.* 8, 1981–1986.

Uht, R.M., Anderson, C.M., Webb, P. and Kushner, P.J. (1997). Transcriptional activities of estrogen and glucocorticoid receptors are functionally integrated at the AP-1 response element. *Endocrinology* 138, 2900–2908.

van Dijk, M.A.J., Floore, A.N., Kloppenborg, K.I.M., and Van't Veer, L. (1997). A functional assay in yeast for the human estrogen receptor displays wild-type and variant estrogen receptor messenger RNAs present in breast carcinoma. *Cancer Res.* 57, 3478–3485.

Vom Baur, E., Zechel, C., Heery, D., Heine, M.J.S., Garnier, J.M., Vivat, V., Le Douarin, B., Gronemeyer, H., Chambon, P., and Lasson, R. (1996). Differential ligand-dependent interactions bewteen the AF-2 activating domain of nuclear receptors and the putative transcriptional intermediary factors mSUG1 and TIF1. *EMBO J.* 15, 110–124.

Wade, W. S., Mrksich, M., and Dervan, P. B. (1992). Design of peptides that bind in the minor groove of DNA at 5'(A, T)5 G(A, T)C(A, T) sequences by a dimeric side-by-side motif. *J. Am. Chem. Soc.* 114, 8784–8794.

Wagner, E.G.H., and Simons, R.W. (1994). Antisense RNA control in bacteria, phages and plasmids. *Annu. Rev. Microbiol.* 48, 713–742.

Webb, P., Lopes, G.N., Uht, R.M., and Kushner, P.J. (1995). Tamoxifen activation of the estrogen receptor/AP1 pathway: potential origin for the cell-specific estrogen-like effects of antiestrogens. *Mol. Endocrinol.* 9, 443–456.

Weber, C., Erl, W., Pietsch, A., and Weber, P.C. (1995). Aspirin inhibits nuclear factor kappa B mobilization and monocyte adhesion in stimulated human endothelial cells. *Circulation* 91, 1914–1917.

Wood, W.G., Bunch, C., Kelly, S., Gunn, Y., and Breckon, G. (1985). Control of haemoglobin switching by a developmental clock. *Nature* 313, 320–322.

Wrenn, C.K., and Katzenellenbogen, B.S. (1993). Structure-function analysis of the human estrogen receptor by region-specific mutagenesis and phenotypic screening in yeast. *J. Biol. Chem.* 268, 24089–24098.

Xing, H., and Shapiro, D.J. (1993). An estrogen receptor mutant exhibiting hormone-independent transactivation and enhanced affinity for the estrogen response element. *J. Biol. Chem.* 268, 23227–23233.

Yang, N.N., Bryant, H.U., Hardikar, S., Sato, M., Galvin, J.S., Glasebrook, A.L., and Termine, J.D. (1996). Estrogen and raloxifene stimulate transforming growth factor-b-3 gene expression in rat bone: a potential mechanism for estrogen- or raloxifene-mediated bone maintainance. *Endocrinology* 137, 2075–2084.

Index